Assessing
Forest Ecosystem Health
in the Inland West

Assessing Forest Ecosystem Health in the Inland West

R. Neil Sampson
David L. Adams
Editors

Maia J. Enzer
Assistant Editor

CRC Press
Taylor & Francis Group
Boca Raton London New York

CRC Press is an imprint of the
Taylor & Francis Group, an **informa** business

Published by

Food Products Press, 10 Alice Street, Binghamton, NY 13904-1580 USA

Food Products Press is an Imprint of The Haworth Press, Inc., 10 Alice Street, Binghamton, NY 13904-1580 USA.

Assessing Forest Ecosystem Health in the Inland West has also been published as *Journal of Sustainable Forestry*, Volume 2, Numbers 1/2/3/4, 1994.

The development, preparation, and publication of this work has been undertaken with great care. However, the publisher, employees, editors, and agents of The Haworth Press and all imprints of The Haworth Press, Inc., including The Haworth Medical Press and Pharmaceutical Products Press, are not responsible for any errors contained herein or for consequences that may ensue from use of materials or information contained in this work. Opinions expressed by the author(s) are not necessarily those of The Haworth Press, Inc.

Library of Congress Cataloging-in-Publication Data

Assessing forest ecosystem health in the Inland West / R. Neil Sampson, David L. Adams, editors.
 p. cm.
 Proceedings of the American Forests Workshop, November 15-19, 1993, Sun Valley, Idaho.
 Includes bibliographical references and index.
 ISBN 1-56022-052-X (alk. paper)
 1. Forest health–Inland Empire–Congresses. 2. Forest ecology–Inland Empire–Congress. 3. Forest management–Inland Empire–Congresses. I. Sampson, R. Neil. II. Adams, David L. (David Lewis), 1937- . III. American Forests Workshop (1993 : Sun Valley, Idaho)
SB763.I7A88 1994
574.5'2642'0979–dc20

                  94-32728
                   CIP

INDEXING & ABSTRACTING

Contributions to this publication are selectively indexed or abstracted in print, electronic, online, or CD-ROM version(s) of the reference tools and information services listed below. This list is current as of the copyright date of this publication. See the end of this section for additional notes.

- *Abstract Bulletin of the Institute of Paper Science and Technology,* Institute of Paper Science and Technology, Inc., 575 14th Street, N.W., Atlanta, GA 30318

- *Abstracts in Anthropology,* Baywood Publishing Company, 26 Austin Avenue, P.O. Box 337, Amityville, NY 11701

- *Abstracts on Rural Development in the Tropics (RURAL),* Royal Tropical Institute, 63 Mauritskade, 1092 AD Amsterdam, The Netherlands

- *Abstracts on Sustainable Agriculture,* Deutsches Zentrum fur Entwicklungstechnologien, Neuenlander Weg 23, 2725 Hemslingen, Germany

- *AGRICOLA Database,* National Agricultural Library, 10301 Baltimore Boulevard, Room 002, Beltsville, MD 20705

- *Biostatistica,* Executive Sciences Institute, 1005 Mississippi Avenue, Davenport, IA 52803

- *Environment Abstracts,* Congressional Information Service, Inc., 4520 East-West Highway, Bethesda, MD 20814-3389

(continued)

- *Forestry Abstracts; Forest Products Abstracts (CAB Abstracts),* CAB International, Wallingford, Oxon OX10 8DE, England

- *Environmental Periodicals Bibliography (EPB),* International Academy at Santa Barbara, 800 Garden Street, Suite D, Santa Barbara, CA 93101

- *GEO Abstracts (GEO Abstracts/GEOBASE),* Elsevier/GEO Abstracts, Regency House, 34 Duke Street, Norwich NR3 3AP, England

- *Human Resources Abstracts,* Sage Publications, Inc., 2455 Teller Road, Newbury Park, CA 91320

- *Referativnyi Zhurnal (Abstracts Journal of the Institute of Scientific Information of the Republic of Russia),* The Institute of Scientific Information, Baltijskaja ul., 14, Moscow A-219, Republic of Russia

- *Sage Public Administration Abstracts,* Sage Publications, Inc., 2455 Teller Road, Newbury Park, CA 91320

- *Sage Urban Studies Abstracts,* Sage Publications, Inc., 2455 Teller Road, Newbury Park, CA 91320

- *Wildlife Review/Fisheries Review,* U.S. Fish and Wildlife Service, 1201 Oak Ridge Drive, Suite 200, Fort Collins, CO 80525-5589

SPECIAL BIBLIOGRAPHIC NOTES

related to special journal issues (separates)
and indexing/abstracting

☐ indexing/abstracting services in this list will also cover material in the "separate" that is co-published simultaneously with Haworth's special thematic journal issue or DocuSerial. Indexing/abstracting usually covers material at the article/chapter level.

☐ monographic co-editions are intended for either non-subscribers or libraries which intend to purchase a second copy for their circulating collections.

☐ monographic co-editions are reported to all jobbers/wholesalers/approval plans. The source journal is listed as the "series" to assist the prevention of duplicate purchasing in the same manner utilized for books-in-series.

☐ to facilitate user/access services all indexing/abstracting services are encouraged to utilize the co-indexing entry note indicated at the bottom of the first page of each article/chapter/contribution.

☐ this is intended to assist a library user of any reference tool (whether print, electronic, online, or CD-ROM) to locate the monographic version if the library has purchased this version but not a subscription to the source journal.

☐ individual articles/chapters in any Haworth publication are also available through the Haworth Document Delivery Services (HDDS).

Papers
from the American Forests Workshop

November 14-20, 1993
Sun Valley, Idaho

DISCLAIMER

The opinions, findings, and conclusions or recommendations expressed in this publication are those of the authors and do not necessarily reflect the views or policies of the government agencies, institutions, or organizations with which these authors are affiliated.

Assessing
Forest Ecosystem Health
in the Inland West

CONTENTS

SECTION III

ECOLOGICAL AND HISTORICAL PERSPECTIVES

PROCESSES, MODELS, AND TOOLS

ABOUT THE EDITORS

R. Neil Sampson is Executive Vice President of AMERICAN FO-RESTS. Previously, he was Executive Vice President of the National Association of Conservation Districts, and spent 16 years with USDA's Soil Conservation Service. He has been very active in analyzing the implications for agriculture and forestry in the event of climate changes caused by rising levels of atmospheric "greenhouse gasses." This has led to development of an international campaign entitled *Global ReLeaf*, sponsored by AMERICAN FORESTS, that encourages individuals to begin addressing environmental problems by restoring trees and forests.

The author of 2 books on soil conservation, 20 book chapters on various aspects of resource conservation, and dozens of articles in scientific and popular magazines and journals. Mr. Sampson's research and writing have largely concentrated on land and water management policy, ranging from pollution control through forestry and resource economics.

Dave L. Adams currently teaches silviculture at the University of Idaho with emphasis on forest health, adaptive forestry applications, and agro-forestry. He worked in timber management with the US Forest Service in Wyoming, and held teaching and research positions at Colorado State University, Humboldt State University, and the University of Idaho where he also served for 15 years as Chair of the Forest Resources Department. Dr. Adams is a Fellow in the Society of American Foresters and has held several local, regional, and national offices.

Foreword

This publication emerged from a scientific workshop convened in Sun Valley, Idaho, November 14-20, 1993, by a partnership involving American Forests, Boise Cascade Corporation, Idaho Department of Lands, The Boise National Forest and the Intermountain Forest and Range Research Station of the USDA Forest Service, and the University of Idaho College of Forestry, Range, and Wildlife Sciences. The workshop brought together a diverse group of scientists and forest managers to assess the current state of scientific knowledge about the health of the forests of the Inland West. The goal was to produce a current, accurate, credible synthesis of information about forest health in the Inland West.

One overview paper and five workgroup papers were developed by the participants during the workshop to focus scientific consideration on basic questions framed by the sponsors. The workgroup papers represent the combined judgment of the authors, developed at the workshop out of information they had previously prepared, or which emerged from the synergy of working together at the task. The authors were instructed to not attempt to limit content to new knowledge, but rather to focus on the combined knowledge of people who, in general, had never consulted each other on this topic before.

The workshop focused on forests of the inland west because of a sense that these forests, while very diverse, share similar conditions and situations. Those conditions, their genesis, and the likely results of their continuation differ significantly from forests in regions such as the western slope of the Cascades. National attention to the situation in that region has translated into public opinions and public policies on ecosystem management and forest health that do not fit the forests of the Inland West.

The sponsors are deeply indebted to the participants, who took on the challenge with vigor and tenacity, often working far into the night. By week's end, the basic papers had all been written, and were in the process of final review. Over half of the scientists present also brought individual submissions to the workshop, where each paper was reviewed by three peers. Thus, the following material presents scientific insight into the Inland West forests that reflects the multiple judgments and talents of a

xix

remarkable, hard-working, dedicated group of experts. Our goal is to help people understand these great Inland West forests so that policies can reflect a constructive and realistic framework in which they can be managed for sustained forest health.

R. Neil Sampson

Preface

It is hard today to find an ecologist who does not believe that change is the natural condition in forests. The idea of a "steady state" or a "natural balance" has long since passed from the lexicon of forest ecology. But when it comes to policy, we find a totally different situation. Policy is still, for the most part, based upon outdated and false premises about how forests function.

There are two important reasons for this, in my view. First, we live with a long tradition of mythology about the "balance of nature." There is a 4,000-year history that believes that nature, undisturbed, achieves a permanency of form, substance, and proportion. That is our cultural heritage. That was reinforced at the beginning of the scientific revolution 300 years ago, and again in the 19th Century with the rise of mechanics and engineering. The mechanical revolution, which coincided with the beginnings of environmental sciences and ecology, gave rise to the concept of nature as a machine. The machine became so deep-seated in our psyche that we began using it as an explanation for how things–even natural things–function.

The second problem is that we lack good, quantitative, well-formulated terms, concepts, definitions and measures to replace the old idea of constancy. This is our challenge today–to begin to explain what we know in terms that convey a sense of scientific validity, to the general public and to policymakers.

If you say to a person on Capitol Hill "natural fire regimes aren't constant–they vary a lot," then you proceed to tell a long anecdote, the natural reaction will be "That's all very nice, but its impossible to formulate a policy on that." The result is that they then revert to the more precise-sounding "steady-state" ideas that help them to write a more definitive policy statement.

Foresters are not alone in this quandary. All of the fisheries in the world, for example, are managed on the basis of the old familiar notion of the balance of nature and the logistic growth equation. This assumes a population, when disturbed, tends to return to its former size along a logistic growth curve–that there is no important environmental change, and no other chaotic or outside influences. When there is a disturbance, the

xxi

recovery is mainly related to the type and magnitude of the disturbance. Using this theory of population dynamics as the basis for management, all the major fisheries in the world have crashed. The problem, I propose, is that we have simply not provided a good replacement for the terminology and concepts that have been proven to fail.

The two basic assumptions in the classic theory are: (1) that the system has a natural equilibrium, and (2) that when disturbed, the system will always seek to return to that equilibrium. I have searched through the ecological literature for years, seeking examples of systems that meet this criteria. I found none, except for a few laboratory experiments using microbes and insects. In nature, these assumptions simply do not hold true.

Instead, what we find is that conditions are always changing in response to climate and other environmental effects. There is no steady state, no true equilibrium. That is a problem to many scientists, because if there is no equilibrium, and no tendency to return to a known state, it is difficult to use classic methodology to make predictions about the system's future behavior. But this simply calls for a different approach to prediction. Instead of seeking to define a balance, we must use quantitative terms for system conditions that include a great deal of random and chaotic behavior. Can we do so? I believe we can.

The first concept that helps in this regard is that of persistence. Persistence suggests that a system remains within certain bounds, but tends to vary within these bounds. The variation may be deterministic or stochastic, periodic or aperiodic. This term can be expressed mathematically. What it requires, that the old idea of "balance" did not require, is knowledge. You have to know something about the history of the system, and you ought to have some way of making reasonable projections about the future conditions that will impact the system. Instead of looking for an equilibrium, you look for the trajectory over which the system is most likely to travel in the future.

In looking at these trajectories, and predicting what will happen if you intervene with management, it is important to understand at which point on the trajectory the management or disturbance will occur. A system disturbed at one point on its trajectory will react with an adjusted trajectory that may be vastly different than would result if it were disturbed at another point in time. In one case, the system may stay within normal bounds; it will just proceed on a different time course. The same disturbance at a different point may throw the system far outside the normal bounds.

Using this concept as a basis for forest management leads to a shift in the focus of concern. Instead of attempting to keep the forest within a single state, we begin to ask whether or not an action (or a no-action) will result in the forest staying within the set of bounds that seem either normal or preferred.

Many forests today–for reasons that vary greatly–are in conditions that are no longer within the normal bounds. This is a neutral statement; there is no value judgment here. Maybe the abnormal system is what is desired; perhaps the normal one was not what people wanted. This is not a matter of values, it is a statement of neutral terminology.

If we want to conserve wilderness, then we must decide how to keep an area within the normal ranges for that condition. Although one can not talk about an equilibrium state for an area like this, it is possible to explain the normal trajectory of change. Then, it is possible to ask whether or not the introduction of fire, or recreation, or other uses, will help the system remain within these normal bounds, or tend to drive it outside. Once you've made that determination, you can leave it up to the people to decide (in a democracy) which values will be given highest priority.

The second concept, recurrence, is even more useful if we accept the idea of risk, uncertainty, and randomness in forest systems. Recurrence is defined as a state that recurs or repeats. It argues that if a system exists in a particular state, it will tend to return to that state. This can occur over a finite time period, or over an infinite period of time. And this concept can be related to sustainability. A state to which a system cannot return after a disturbance is not a sustainable state, obviously.

The advantage of using this concept in forest management is that you can begin to ask questions that really matter. For instance, suppose you want the open ponderosa pine forests that were here before settlement. You can now talk about the average time it may take for the desired state of the system to recur. You can talk about the average return time if nothing is done, or if some management is undertaken. This requires projection methods, based on real historical knowledge. If you have that, however, you can make projections such as the minimum time that would be needed to return to that desirable state.

The other thing you can look at is the average time that a system tends to stay in a desirable condition. Then, your planning can look at the question of how management can lengthen that time. This leads to a set of action options that can be considered.

You know, for example, that burning a forest every year can prevent successional processes from occurring, but you also know that excluding fire all together also interferes with succession. That means that there is some burning cycle between those two extremes that will help maintain the forest in a more normal range of conditions. Just knowing that fact leads to the intelligent discussion of how often a burn should occur, under what circumstances, and for what reasons. It gets us away from the old debate about burning being either "bad" or "good." Clearly, that debate does not lead us to an intelligent answer for this forest situation.

Using terms and concepts such as persistence and recurrence can give us precise mathematical measures upon which to base predictions and plans. More importantly, they can help free people's minds from the old concept of the balance of nature and help them understand what is really going on with these forest systems. Then we can begin to have changes at many levels, leading, hopefully, to policies that are based on how forests really work.

Daniel B. Botkin

Acknowledgments

In addition to the sponsors, who made the whole endeavor possible, a planning committee worked hard to make the workshop a success. Participating in that effort over several months were Steve Mealey and Lyn Morelan, Boise National Forest; Dave Van De Graaff and Herb Malany, Boise Cascade; Stan Hamilton, Idaho Department of Public Lands; Dean Knighton and Bob Steele, Intermountain Research Station; Dave Adams and Leon Neuenschwander, University of Idaho; and Neil Sampson and Lance Clark, AMERICAN FORESTS.

During the workshop, Lance Clark and Maia Enzer of AMERICAN FORESTS and Judy Simmons of the Idaho Department of Public Lands kept things functioning smoothly.

EDITORS

R. Neil Sampson, Executive Vice-President, American Forests, Washington, DC

David Adams, Professor of Forest Resources, University of Idaho, College of Forestry, Wildlife, and Range Sciences, Moscow, Idaho

ASSISTANT EDITOR

Maia J. Enzer, Forest Policy Center, Washington DC

PROJECT COORDINATOR

Lance R. Clark, Forest Policy Center, Washington DC

SPONSORING COMMITTEE

David L. Adams, Professor of Forest Resources, University of Idaho, Moscow, Idaho

Stanley S. Hamilton, State Forester and Director, Idaho Department of Lands, Boise, ID

John Hendee, Dean, College of Forestry, Wildlife, and Range Sciences, Moscow, Idaho

Dean Knighton, Assistant Station Director, Intermountain Research Station, USDA Forest Service, Boise, Idaho

Steve Mealey, Forest Supervisor, Boise National Forest, Boise, ID

R. Neil Sampson, Executive Vice-President, American Forests, Washington, DC

Joe Ulliman, Head, Forest Resources Department, University of Idaho, Moscow, ID

Dave Van De Graaff, Regional Timberlands Manager, Boise Cascade Corporation, Emmett, ID

INVITED SCIENTISTS

Aplet, Gregory H., Forest Ecologist,The Bolle Center for Ecosystem Management, The Wilderness Society, Washington DC

Auclair, Allan N.D., Senior Scientist, Science and Policy Associates Inc. Washington DC

Blatner, Keith A. Associate Professor, Natural Resource Sciences, Washington State University

Botkin, Daniel B., Director Institute for Global Change, George Mason University, Fairfax, VA; President, Center for the Study of the Environment, University of California at Santa Barbara

Covington, W. Wallace, Professor of Forest Ecology, School of Forestry, Northern Arizona University

Daer, Tom A., Silviculturist, Garnet Resource Area, Burea of Land Management, Missoula, MT.

Erickson, John R., Wildlife Biologist, Boise National Forest

Everett, Richard L., Science Team Leader, Ecosystem Processes, Wenatchee Forestry Sciences Laboratory, USDA Forest Service, Pacific Resource Forestry Station

Ferguson, Dennis E., Research Silviculturist, USDA Intermountain Research Station

Graham, Russell T., Research Silviculturist, USDA Intermountain Research Station

Harvey, Alan E., Project Leader, Root diseases and soil microbiology, USDA Forest Service

Haufler, Jonathan B., Manager, Wildlife and Ecology, Timberland Resources, Boise Cascade Corporation, Boise, ID

Humphries, Hope C., Ecological Modeler, The Nature Conservancy, Boulder, CO

Irwin, Larry L., Wildlife Program Manager, National Council for Air and Stream Improvement, Corvallis, OR

Livingston, R. Ladd, Forest Entomologist, Idaho Department of Lands, Coeur d'Alene, ID

Malany, Herbert, Forester, Boise Cascade Corporation, Emmett, ID

Mandzak, John M., Senior Forester, Land and Water Consulting, Missoula, MT

McKetta, Charley W., Associate Professor of Forest Resources, College of Forestry, Wildlife, and Range Sciences, University of Idaho, Moscow, ID

Moore, Margaret M., Associate Professor of Forest Ecology, School of Forestry, Northern Arizona University

Morelan, Lynnette, Forest Health Coordinator, Boise National Forest, USDA Forest Service, Boise, ID

Morgan, Penelope, Associate Professor of Forest Resources, College of Forestry, Wildlife, and Range Sciences, University of Idaho, Moscow, ID

Mutch, Robert W., Research Applications Leader, Missoula Fire Sciences Laboratory, USDA Forest Service, Intermountain Research Station, Missoula, MT.

O'Laughlin, Jay, Director, Policy Analysis Group, College of Forestry, Wildlife, and Range Sciences, University of Idaho

Oliver, Chadwick D., Professor of Silviculture and Forest Ecology, College of Forest Resources, University of Washington, Seattle, WA

Steele, Robert, Research Forester, Boise Forestry Sciences Laboratory, USDA Forest Service, Intermountain Research Station, Boise, ID

Thier, Ralph, Forest Entomologist, Boise National Forest, USDA Forest Service, Boise, ID

Thornton, John, Forest Hydrologist, Boise National Forest, USDA Forest Service, Boise, ID

Toweill, Dale E., Wildlife Program Coordinator, Natural Resources Policy Bureau, Idaho Department of Fish and Game, Boise, ID

Wilson, W. Dale, Forest Soil Scientist, Clearwater National Forest, USDA Forest Service, Orofino, ID

SECTION I

This overview paper was developed by the members of the sponsoring committee as a reflection of the thoughts and discussions shared during the Sun Valley workshop. This paper is not meant to be a workshop consensus document, but rather an overview of the issues discussed by the participants and a response by the sponsoring committee to those discussions.

OVERVIEW

Assessing Forest Ecosystem Health in the Inland West

R. Neil Sampson
David L. Adams
Stanley S. Hamilton
Stephen P. Mealey
Robert Steele
Dave Van De Graaff

R. Neil Sampson is Executive Vice President, American Forests, Washington, DC.

David L. Adams is Professor of Forest Resources, College of Forestry, Wildlife, and Range Sciences, University of Idaho, Moscow, ID.

Stanley S. Hamilton is Director and State Forester, Idaho Department of Lands, Boise, ID.

Stephen P. Mealey is Forest Supervisor, Boise National Forest, USDA Forest Service, Boise, ID.

Robert Steele is Forest Ecologist, Intermountain Research Station, USDA Forest Service, Boise, ID.

Dave Van De Graaff is Regional Timberlands Manager, Boise Cascade, Emmett, ID.

The authors acknowledge the participation and contributions of all the participants in the workshop, particularly Lance Clark, American Forests; Robin Hartmann, Office of Congressman Larry LaRocco of Idaho; Herb Malany, Boise Cascade; Lyn Morelan, USDA Forest Service, Boise National Forest; and Leon Neuenschwander, U. of Idaho, in the discussions that produced this paper.

[Haworth co-indexing entry note]: "Assessing Forest Ecosystem Health in the Inland West." Sampson, R. Neil et al. Co-published simultaneously in the *Journal of Sustainable Forestry* (The Haworth Press, Inc.) Vol. 2, No. 1/2, 1994, pp. 3-10; and: *Assessing Forest Ecosystem Health in the Inland West* (eds: R. Neil Sampson and David L. Adams) The Haworth Press, Inc., 1994, pp. 3-10. Multiple copies of this article/chapter may be purchased from The Haworth Document Delivery Center [1-800-3-HAWORTH; 9:00 a.m. - 5:00 p.m. (EST)].

ABSTRACT. This paper presents an overview of the conclusions developed by 35 participating scientists and land managers at a scientific workshop held in Sun Valley, Idaho, November 14-20, 1993. The conclusions presented here are those of the authors, but reflect discussions of the entire group, and are based upon conclusions reached by those participants in working groups.

The forests of the Inland West are, over wide regions, not healthy. Remedial, restorative, and preventative treatment and management– particularly on the federal lands–is urgently needed. A brief window of opportunity, perhaps 15-30 years in length, exists. Without timely management intervention, the region is threatened by major ecological setbacks–pest epidemics and uncontrollable wildfires–that will damage resource values and convert large areas into new even-aged forest systems that set the stage for a repeat of the current problems far into the 21st Century. The scientific tools to understand these problems and mitigate them exist today, but are not being applied on the federal forests rapidly enough to meet the urgency of the situation. The current legal and procedural requirements on federal land management agencies impose time delays which, combined with public opposition to timber harvesting, prevent timely management, doom major forest areas to needless loss and damage, and impose large (and, perhaps, preventable) costs on both local and national economies.

INTRODUCTION

Forest health is a condition of forest ecosystems that sustains their complexity while providing for human needs. (O'Laughlin et al. 1994)

In many forested areas of the inland western United States, trees across large landscapes are dying faster than they are growing or being replaced (O'Laughlin 1994). In other areas, conditions exist that virtually guarantee an onset of serious forest health problems which may lead to large wildfires, reburning, erosion, and loss of habitat and property. In those areas, it is not just trees and their values that are at risk. Where terrestrial ecosystems are adversely impacted, the entire range of aquatic resources, wildlife, and other values are affected as well. The current conditions, many of which are unprecedented in recent times, demand urgent response. Forest managers on both public and private lands, even though they manage for a different combination of objectives, face a common forest health challenge when the forests in their care are deteriorating.

Within the region there are also forest areas that, either through management efforts or natural controls, represent excellent examples of healthy, thriving forest ecosystems. In places, healthy forest examples exist adjacent to lands with serious health problems. In all forests, managers are challenged to design forest health strategies that focus on: (a) the *prevention* of socially undesirable forest conditions by protecting the forest from insects, diseases, and fire within an ecological framework and (b) the *restoration* of socially desired forest conditions where needed (O'Laughlin et al. 1994). This management needs to reflect a strong commitment to maintaining healthy conditions for many values, as well as to the study and understanding of both the historical range of variability and the variability occurring currently in forest ecosystems (Morgan et al. 1994).

Forest health is defined in this context as a condition of forest ecosystems that sustains their complexity while providing for human needs, and it is clear that many of the forests in the Inland West fail the test. In areas where insects, disease, and wildfire are causing total or near-total tree mortality, the evidence of forest health problems is visual and stark. In other areas, the visual evidence and widespread mortality may be lacking, but the onset of major ecosystem setbacks are assured by the existence of conditions that inevitably lead to large, stand-replacing wildfires (Covington et al. 1994). Managers are challenged to take rapid preventative action to restore these forests to conditions more similar to their historical range of variability, or, where that is judged not possible or desirable, to strive for another sustainable condition.

Without the application of needed silvicultural treatments and other tools consistent with ecosystem management within a fairly short time (15-30 years), there is great danger that over the next century this region's forest legacy will be a series of large, uniform landscapes recovering from wildfires and other ecosystem setbacks on a scale unprecedented in recent evolutionary time (Covington et al. 1994). These landscapes will present future societies with a set of limited options, and needlessly high costs that, in many ways, will mirror today's unstable situation. Both now and in the future, the preferred situation is a more diverse, heterogeneous landscape that is more consistent with the historic range of variability, less susceptible to wide-area disturbances, and thus more easily sustainable.

FORESTS AT RISK

The forests at greatest risk are composed of an unsustainable combination of tree species, densities, and age structures that are susceptible to the fire and drought regimes common to the region. Although the situation differs significantly from place to place, the forest areas under the most stress contain too many trees, or too many of the wrong kind of trees, to continue to thrive. As the trees get older and larger, the competition intensifies, stress increases, and the likelihood of catastrophic change goes up accordingly (Covington and Moore 1994; Steele 1994; Blatner et al. 1994).

This is a particular problem in forests where the species mix has shifted away from ponderosa and other long-needled pines and toward firs. This species shift, attributable to a combination of logging, grazing, fire suppression, and related activities over the past century, has been well documented. In a review of Idaho forest data for the period 1952 to 1987, for example, O'Laughlin (1994) found that western white pine and ponderosa pine components had declined 60 percent and 40 percent respectively, while true firs, lodgepole pine, and Douglas-fir had increased 60, 39, and 15 percent.

There are also problems where the species mix is still heavily dominated by pines, but where the lack of fire has contributed to a dense, overcrowded forest. In a study area in Arizona's Coconino National Forest, for example, Northern Arizona University's Wallace Covington and Margaret Moore estimated that stem counts on basalt-derived soils had shifted from a pre-settlement average of 23 trees per acre to a current count of 851. On a study area in the Kaibab National Forest, pre-settlement tree densities on limestone soils averaged 56 per acre, compared to 276 today (Covington and Moore 1994).

Under these altered conditions, competition for moisture and nutrients creates stress that exacts a significant toll in reduced growth, while opening the way for catastrophic outbreaks of insects, disease and wildfire. Wildfires in these ecosystems have gone from a high-frequency, low-intensity regime that sustained the system to numerous high-intensity fires that require costly suppression attempts which often prove futile in the face of overpowering fire intensity. High fuel loads, resulting from the long-time absence of fire and the abundance of dead and dying trees, result in fire intensities that cause enormous damage to soils, watersheds, fisheries, and other ecosystem components.

WILDFIRE THREATS HAVE CHANGED

Change agents and processes, such as drought, pests, and wildfires, are normal components of forest environments. Unfortunately, today's conditions in many Inland West forests allow normal processes to become catastrophic events. Unless the land conditions themselves can be improved, these catastrophic changes seem certain to continue. On the Boise National Forest, for example, wildfire consumed an average of 3,000 acres per year from 1955 to 1985. From 1985 to 1992, the average annual wildfire acreage jumped to 56,000 (Erickson and Toweill 1994). These include large-area, intense, stand-replacing wildfires in ponderosa pine forests, indicating a major shift away from the type of fire regime these forests experienced in the pre-settlement era.

The 1992 Foothills Fire provides an example of the type of events that seem certain to continue. This wildfire, which started on public rangelands east of Boise, Idaho on August 19, 1992, burned more than 257,000 acres, including about 140,000 acres of national forest land (USDA Forest Service 1993). Virtually every tree, including isolated pines and small timbered pockets, was killed throughout the fire's area. The largest ponderosa pine in Idaho–a veteran of dozens of previous fires over its centuries of life–was killed, indicating that this was the most intense wildfire the tree had ever experienced. Trees containing an estimated 300 million board feet of merchantable timber were killed, along with young tree growth across the fire's area. One population of rare bull trout was wiped out and another seriously threatened as intense heat denuded the riparian zone vegetation and heated small streams beyond the fishes' tolerance. Watershed functions were altered through vegetative change and the creation of water-repellent soils by the intense heat (McLean 1993).

Federal costs for fire suppression and emergency restoration work on the national forest totalled $24 million (USDA Forest Service 1993). The need for additional management is far from over; the young ponderosa pine regeneration must be thinned and protected from future fires that could kill the seedlings before they mature sufficiently to produce new seed sources. An intense reburn in the next few years could change areas that have been forested for many decades, if not centuries, to grassland and shrub fields.

Situations similar to what existed on the land impacted by the Foothills Fire are common in many regions of the Inland West. Because these situations reflect conditions outside the historical range of variability characteristic of these sites (Morgan et al. 1994), the authors believe natural forces are unlikely to correct them in ways that satisfy the public's expectations for healthy forests. Change will continue, that is certain, but the result of that change is far from known. The changed forest may be what people consider to be healthy and productive, or it may not. The choice we face is whether to leave the outcome to chance, or to try to guide it toward desirable conditions through what Aldo Leopold described as "intelligent tinkering" with forest ecosystems.

INACTION POSES INCREASED RISK

Though it is readily recognized that scientific knowledge is incomplete, modern ecosystem theory provides a basis for corrective actions that can mitigate current levels of risk and potential damage, and facilitate improved forest health. Through corrective actions and ecosystem management, a balance of forest structures can be achieved across landscapes that will increase the opportunity for maintaining biological diversity and reduce the impact and scale of inevitable disturbances (Oliver et al. 1994).

The question of risk is at the heart of the options facing society in the inland western forests. Any management action–including the option of taking no action at all–has certain costs and uncertain outcomes. It takes courage and leadership, in the face of uncertainty and public cynicism, to take actions that may be inconsistent with past practices (McKetta et al. 1994). That is, however, what is demanded in the current situation facing Inland West forests and forest managers.

The authors conclude that in many Inland West forests, the costs and risks of inaction are greater than the costs and risks of remedial action. Inaction in the face of current forest conditions will likely prove to be the most costly and environmentally destructive option. The judicious control of tree density and species composition through prescribed fire, thinning, and other silvicultural methods is critical to reducing risk and restoring and maintaining forest health. These conclusions may be inconsistent with what citizens, scientists, and policymakers have learned in other forest regions where better ecosystem management is perceived to require less intensive human intervention in the forest.

The current ecological conditions in the forests of the Inland West, however, lead to the conclusion that ecosystem management will demand increased management. This new management will be more intense and cover wider areas of the forest, but it will be different in impact and appearance from what has historically been done, particularly on the federal forests. In most cases it will be a more adaptive form of management, more responsive to local conditions and needs, with results more closely monitored to study effects and continuing changes. The challenge to managers on these threatened forests is to provide preventative treatment as a means both of protecting valuable resources and reducing the effort and cost of such current management activities as fire suppression and emergency restoration.

DETERRENTS TO TIMELY FEDERAL RESPONSE

A concern remains about the capacity of federal land-managing agencies to respond rapidly and adaptively to the actions necessary in the high-risk situations. Federal forest management and environmental laws charge the land-managing agencies (primarily the Forest Service and Bureau of Land Management) with assuring continued ecosystem health and sustainability. Unfortunately, the procedural, regulatory, and judicial framework that has developed in response to that complex array of federal laws imposes time delays that, in many cases, prevent the agency from taking timely action to address fast-changing situations in the forests (Oliver et al. 1994). This is compounded by a lack of public trust that results in appeals and litigation on most proposed actions involving tree removal or logging. The public is also generally unaware of the extreme risks to the forest and adjacent private property inherent in the current situation, so they often fail to see the potential price that will be paid for "doing nothing" (McKetta et al. 1994).

Public understanding, acceptance, and support for management intervention in unhealthy forest situations is unlikely to result from attempts simply to "educate" people or through arguing that "the experts know best." Instead, research needs to provide analytic and illustrative tools that will help concerned citizens and agencies understand forest conditions and related risks. People will need to see the forest with their own eyes, and be able to see how risk ratings and other decision tools have been calculated. They will need to understand not only the likelihood of undesirable changes in forests that are left alone, but what those changes will mean in terms of altered values such as visual quality, wildlife habitat and numbers, fisheries, water quantity and quality, and recreational opportunity. They will have to come to a personal acceptance of the fact that, while all ecosystems undergo constant change, there are ecologically sound methods by which human intervention can guide change in ways that are ecologically, economically, and socially sound (McKetta et al. 1994).

As scientists, land managers, and concerned conservationists, the authors believe it is imperative for national policymakers to recognize the unique and critical situation threatening the forests of the Inland West. Immediate action is needed to encourage and support forest-management programs that maintain healthy forest areas, treat unsustainable conditions, restore forest health where needed, and prevent the widespread ecological setbacks and protect the public and private values that are currently at risk.

REFERENCES

Blatner, K.A., C.E. Keegan III, J. O'Laughlin, and D.L. Adams. 1994. Forest health management and policy: A case study in southwestern Idaho. In: pp. 317-337. Sampson, R.N. and D.L. Adams (Eds.). Assessing Forest Ecosystem Health in the Inland West. Proceedings of the AMERICAN FORESTS workshop, Sun Valley, ID. The Haworth Press, Inc. New York.

Covington, W.W., R.L. Everett, R. Steele, L.L. Irwin, T. A. Daer and A.A.N.

Auclair. 1994. Historical and anticipated changes in forest ecosystems of the Inland West of the United States. In: pp. 13-63. Sampson, R.N. and D.L. Adams (Eds.). Assessing Forest Ecosystem Health in the Inland West. Proceedings of the AMERICAN FORESTS workshop, Sun Valley, ID. The Haworth Press, Inc. New York.

Covington, W.W. and M.M. Moore. 1994. Postsettlement changes in natural fires regimes and forest structure: Ecological restoration of old-growth ponderosa pine forests. In: pp. 153-181. Sampson, R.N. and D.L. Adams (Eds.). Assessing Forest Ecosystem Health in the Inland West. Proceedings of the AMERICAN FORESTS workshop, Sun Valley, ID. The Haworth Press, Inc. New York.

McKetta, Charley, K.A. Blatner, R.T. Graham, J. R. Erickson and S.S. Hamilton. 1994. Human Dimensions of Forest Health Choices. In: pp. 135-149. Sampson, R.N. and D.L. Adams (Eds.). Assessing Forest Ecosystem Health in the Inland West. Proceedings of the AMERICAN FORESTS workshop, Sun Valley, ID. The Haworth Press, Inc. New York.

McLean, Herbert E. 1993. The Boise quickstep. *American Forests* 99(1,2): 11-14.

Morgan, Penelope, G.H. Aplet, J.B.Haufler, H.C.Humpgries, M.M. Moore and W. D. Wilson. 1994. Historical Range of Variability: A Useful Tool for Evaluating Ecosystem Change. In: pp. 87-111. Sampson, R.N. and D.L. Adams (Eds.). Assessing Forest Ecosystem Health in the Inland West. Proceedings of the AMERICAN FORESTS workshop, Sun Valley, ID. The Haworth Press, Inc. New York.

O'Laughlin, Jay. 1994. Assessing forest health conditions in Idaho with forest inventory data. In: pp. 221-247. Sampson, R.N. and D.L. Adams (Eds.). Assessing Forest Ecosystem Health in the Inland West. Proceedings of the AMERICAN FORESTS workshop, Sun Valley, ID. The Haworth Press, Inc. New York.

O'Laughlin, Jay, R.L. Livingston, R. Thier, J.Thorton, D.E. Toweill and Lynnette Morelan. 1994. Defining and measuring forest health. In: pp. 65-85. Sampson, R.N. and D.L. Adams (Eds.). Assessing Forest Ecosystem Health in the Inland West. Proceedings of the *American Forests* workshop, Sun Valley, ID. The Haworth Press, Inc. New York.

Oliver, C.D., D.E. Ferguson, A.E. Harvey, H.Malany, J.M. Mandzak and R. W. Mutch. 1994. Managing Ecosystems for Forest Health: An Approach and the Effects on Uses and Values. In: pp. 113-133. Sampson, R.N. and D.L. Adams (Eds.). Assessing Forest Ecosystem Health in the Inland West. Proceedings of the AMERICAN FORESTS workshop, Sun Valley, ID. The Haworth Press, Inc. New York.

Steele, R. 1994. The role of succession in forest health. In: pp. 183-190. Sampson, R.N. and D.L. Adams (Eds.). Assessing Forest Ecosystem Health in the Inland West. Proceedings of the AMERICAN FORESTS workshop, Sun Valley, ID. The Haworth Press, Inc. New York.

USDA Forest Service. 1993. Burned area report for Foothills Fire. FSH 2509.13 report FS-2500-8. Boise, ID.

SECTION II

The papers presented in this section serve to set the context of the forest health problem in the Inland West, to define the terms and conditions which have led to the current situation, and to offer a synthesis of disciplinary perspectives on the topic.

These papers were written during the workshop in Sun Valley, ID, and were refined by the lead authors and further reviewed by the working group during the weeks following the workshop. They reflect a week of collaboration, discussion, and comment around a set of basic questions posed by the sponsoring committee.

Historical and Anticipated Changes in Forest Ecosystems of the Inland West of the United States

W. Wallace Covington
Richard L. Everett
Robert Steele
Larry L. Irwin
Tom A. Daer
Allan N. D. Auclair

ABSTRACT. Euro-American settlement of the Inland West has altered forest and woodland landscapes, species composition, disturbance regimes, and resource conditions. Public concern over the loss of selected species and unique habitats (e.g., old-growth) has caused us to neglect the more pervasive problem of declining ecosystem health. Population explosions of trees, exotic weed species, insects,

W. Wallace Covington, School of Forestry, Northern Arizona University, Flagstaff, AZ 86011.

Richard L. Everett, Wenatchee Forestry Sciences Laboratory, U.S. Department of Agriculture, Forest Service, Pacific Northwest Research Station, Wenatchee, WA 98801.

Robert Steele, Boise Forestry Sciences Laboratory, U.S. Department of Agriculture, Forest Service, Intermountain Research Station, Ogden, UT 84401.

Larry L. Irwin, National Council of the Paper Industry for Air and Stream Improvement, Inc., Corvallis, OR 97339.

Tom A. Daer, Bureau of Land Management, Missoula, MT 59801.

Allan N. D. Auclair, Science and Policy Associates, Inc., Washington, DC 20005.

[Haworth co-indexing entry note]: "Historical and Anticipated Changes in Forest Ecosystems of the Inland West of the United States." Covington, W. Wallace et al. Co-published simultaneously in the *Journal of Sustainable Forestry* (The Haworth Press, Inc.) Vol. 2, No. 1/2, 1994, pp. 13-63; and: *Assessing Forest Ecosystem Health in the Inland West* (eds: R. Neil Sampson and David L. Adams) The Haworth Press, Inc., 1994, pp. 13-63. Multiple copies of this article/chapter may be purchased from The Haworth Document Delivery Center [1-800-3-HAWORTH; 9:00 a.m. - 5:00 p.m. (EST)].

13

diseases, and humans are stressing natural systems. In particular, fire exclusion, grazing, and timber harvest have created anomalous ecosystem structures, landscape patterns, and disturbance regimes that are not consistent with the evolutionary history of the indigenous biota. Continuation of historical trends of climate change, modified atmospheric chemistry, tree density increases, and catastrophic disturbances seems certain. However, ecosystem management strategies including the initiation of management experiments can facilitate the adaptation of both social and ecological systems to these anticipated changes. A fairly narrow window of opportunity–perhaps 15-30 years–exists for land managers to implement ecological restoration treatments.

INTRODUCTION

Enhanced understanding of historical and anticipated changes in ecological and social systems is central to the design and implementation of mutually beneficial policies for wildland management. Although the ecological and resource management implications of changes in forest conditions since Euro-American settlement have been discussed, in part, in several recent reports (e.g., Caraher et al. 1992; Everett et al. 1993), no such synthesis exists for the Inland West region of the United States (the area west of the Great Plains and east of the Cascade Range in the north and the Sierra Nevada in the south) (Figure 1). We acknowledge that this geographical delineation is somewhat arbitrary from an evolutionary and biogeographical perspective. The biota in the region blends with that to the north, east, south, and west (e.g., Brown 1982). Nonetheless, we feel this area shares several important ecological, cultural, historical, and political features which justify its designation as a region.

Currently, there is widespread agreement among natural resource professionals, the scientific community, and others concerned with forest and woodland management that many land use practices (e.g., overgrazing, high grade logging, fire exclusion) have often inadvertently resulted in serious degradation of many of the region's ecosystems. Although a continuing concern of both professionals and the general public is the loss of old-growth trees from logging (Thomas 1979; Hoover and Wills 1984, Booth 1991; Kaufmann et al. 1992), there is an increasing consensus among natural resource professionals that disruption of natural fire regimes, tree population explosions, and increases in landscape homogeneity are a far greater threat to biological diversity and ecosystem sustainability than is generally realized (Caraher et al. 1992; Kaufmann et al. 1992; Everett et al. 1993; Mutch et al. 1993; Covington and Moore 1994b). Unless concerted actions are taken to reverse ongoing ecosystem degradation, the prospects look grim for the quality of life–not only for the forest and woodland ecosystems of the region, but also for the human populations that rely on these resources (e.g., Caraher et al. 1992; Everett et al. 1993; Mutch et al. 1993; Covington and Moore 1994b).

In fact, public concerns about declining ecosystem health, expanding insect and disease problems, and increasing wildfire size and frequency have led to calls for region-wide ecological treatments designed to restore more natural conditions (Drake et al. 1991; Reynolds et al. 1992; Everett et al. 1993). An essential first step to such restoration is understanding the range of conditions and the natural disturbance regimes present before Euro-American settlement and the mechanisms that underlie undesirable postsettlement changes (Lubchenko et al. 1991; Covington and Moore 1992; Everett et al. 1993). This is particularly true for ecosystems that had a natural disturbance regime characterized by frequent fire, such as those formerly dominated by ponderosa pine (*Pinus ponderosa*) and giant sequoia (*Sequoiadendron gigantea*) (Bonnicksen and Stone 1982; Parsons et al. 1986; Moir and Dieterich 1988; Hessburg et al. 1993; Covington and Moore 1994a).

The purpose of this paper is to present a general synthesis to aid disciplinary specialists in understanding the historical and anticipated changes in forest and woodland conditions for the Inland West region of the United States. An exhaustive literature review is beyond the scope of this paper. Rather, the purpose is to integrate general knowledge with examples from selected research conducted throughout the region and to provide access to references that will lead the reader into more comprehensive and in-depth literature on this important topic.

AN EVOLUTIONARY CONTEXT

Disruption of Natural Mortality

The evolutionary history of the organisms that constitute today's forests and woodlands is characterized by mortality-causing natural disturbance regimes (e.g., fires, predation, disease, insect attack) which have varied in kind, frequency, intensity, and extent (e.g., White 1979; Holling 1981; Kilgore 1981; Pickett et al. 1989; Baker 1992). In fact, transitory phenomena dominate the evolutionary history of organisms, so much so that a near constant state of instability can be said to be characteristic of natural systems (Seastedt and Knapp 1993). As a consequence, many of these species have adaptations that lead to the production of offspring well in excess of the number required to maintain the species through time (e.g.,

Allen 1962; Keeley 1981; Shugart and West 1981; Christensen 1988). In the absence of natural causes of mortality, the population of these species may irrupt to levels never experienced over evolutionary time, sometimes to the point that the natural functioning of ecosystems is severely impaired. If the irrupting species is highly interactive with other members of the ecological community, the consequences can be far-reaching. In fact, a fundamental corollary of the evolutionary postulate of conservation biology is that unnatural irruptions of ecologically dominant or highly interactive species can threaten the persistence of other species and in some cases lead to general system collapse (Soulé 1985).

Perhaps the most familiar ecological irruption process is that of predator control contributing to deer population irruptions that exceed the carrying capacity of the land (Leopold 1949; Flader 1974; Mitchell 1993). Although factors other than predator control are often involved (e.g., Mitchell 1993), the consequence of unchecked deer population growth can be overbrowsing by deer, reduction of ground cover, soil erosion, and eventually massive starvation of the deer herds.

Less generally known is the scenario involving irruption of fir trees following insecticide spraying to prevent spruce budworm outbreaks in eastern Canada (reviewed in Holling 1981). The result was that incrementally larger areas of the landscape reach the stand conditions (fir crown density) that support budworm outbreaks, eventually reaching a point where the susceptible area is too extensive for economical spraying and catastrophically large defoliation may result.

In a similar vein, suppression of natural fires in some ecosystems may lead to tree population explosions and dead fuel accumulation to such an extent that catastrophic adjustments become inevitable. Eventually, catastrophic disturbances such as insect and disease attack and crown fire may cause extensive mortality at a scale never before experienced by the community of organisms.

In sum, although changing ecological conditions are the rule rather than the exception during the course of evolution, the rapidity of postsettlement alterations in the kind, frequency, intensity, and extent of natural disturbances is unprecedented.

A BRIEF BIOGEOGRAPHY OF THE INLAND WEST

The forests of the Inland West are controlled over broad regions by climate (Figure 1) with soils and topography acting as secondary influences. However, in some areas, particularly the drier regions, soils, and especially limestone soils, can have a profound effect on plant community

FIGURE 1. Approximate climatic regions of the Inland West as suggested by distribution of indicative tree species (from Steele and Pfister 1991). I–Core maritime, II–Inland maritime, III–Northern continental, IV–Southern continental.

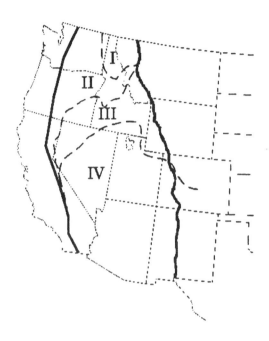

composition. Topography increases diversity of site conditions in a local area thereby increasing diversity of plant and animal communities. Variations in disturbance regimes have also influenced the composition and density of these forests.

Cedar-Hemlock Zone

In the northernmost region, prevailing westerly winds from the Pacific Ocean carry moisture-laden air masses far inland. Extended periods of cloud cover and fog with prolonged cyclonic storms create the dominant weather pattern from fall through late spring. At mid-elevations of the forest zone, this maritime climate supports a rich assemblage of plants and animals characterized by the potential (in the absence of catastrophic disturbance) for succession to lead to western redcedar (*Thuja plicata*) and western hemlock (*Tsuga heterophylla*) dominated forests (Figure 2) (Daubenmire 1956). Historically, this area supported the fire-induced white pine (*Pinus monticola*) forests of the inland northwest; it is among the most productive forest types in the Rocky Mountains (Cooper et al. 1991).

Subalpine Zone

At elevations above the cedar-hemlock zone and at increasingly higher elevations farther south lie extensive forests which, in the absence of disturbance, are dominated by Engelmann spruce (*Picea engelmannii*) and subalpine fir (*Abies lasiocarpa*). These spruce-fir forests are associated with long, cold, snowy winters and cool summers with short growing seasons. They consistently occur at upper elevations of the forest zone ranging from Canada to Mexico (Figure 2). From Canada to central Colorado, lodgepole pine (*Pinus contorta*) is the most prevalent seral dominant. Whitebark pine (*P. albicaulis*) grows on the uppermost ridges as far south as Wyoming and northern Utah (Little 1971). Through much of Colorado to northern New Mexico and Arizona, bristlecone pine (*P. aristata*) and occasionally limber pine (*P. flexilis*) appear as seral components of the spruce-fir forests or as pure stands at the highest elevations and on drier sites (DeVelice et al. 1986). In the Cascade Range, Pacific silver fir (*Abies amabilis*), Shasta red fir (*A. magnifica shastensis*), and mountain hemlock (*Tsuga mertensiana*) are additional associates of these high-elevation forests (Franklin and Dyrness 1973).

Mixed Conifer Zone

South of the cedar-hemlock zone and immediately below the spruce-fir zone is a rather diverse group of forests referred to here as the mixed conifer zone. In central Idaho, western Montana, and eastern Oregon and Washington, grand fir (*Abies grandis*), Douglas-fir (*Pseudotsuga menziesii*), western larch (*Larix occidentalis*) and ponderosa pine are the predominant tree species. They occur here as a mixed conifer forest on the periphery of the maritime climatic regime that supports cedar-hemlock forests (Steele et al. 1981; Cooper et al. 1991). To the south and east, effects of this maritime climate diminish; here western larch, grand fir, and ponderosa pine disappear, leaving only Douglas-fir with occasional limber pine as the predominant tree (Figure 2) over much of eastern Idaho, central Montana, and Wyoming (Pfister et al. 1977; Steele et al. 1983).

A summer continental climatic regime of high altitude convectional storms arising from the southern California coast and Gulf of Mexico carry moisture northward during the growing season. This climatic influ-

FIGURE 2. General elevational distribution of forests of the Inland West.

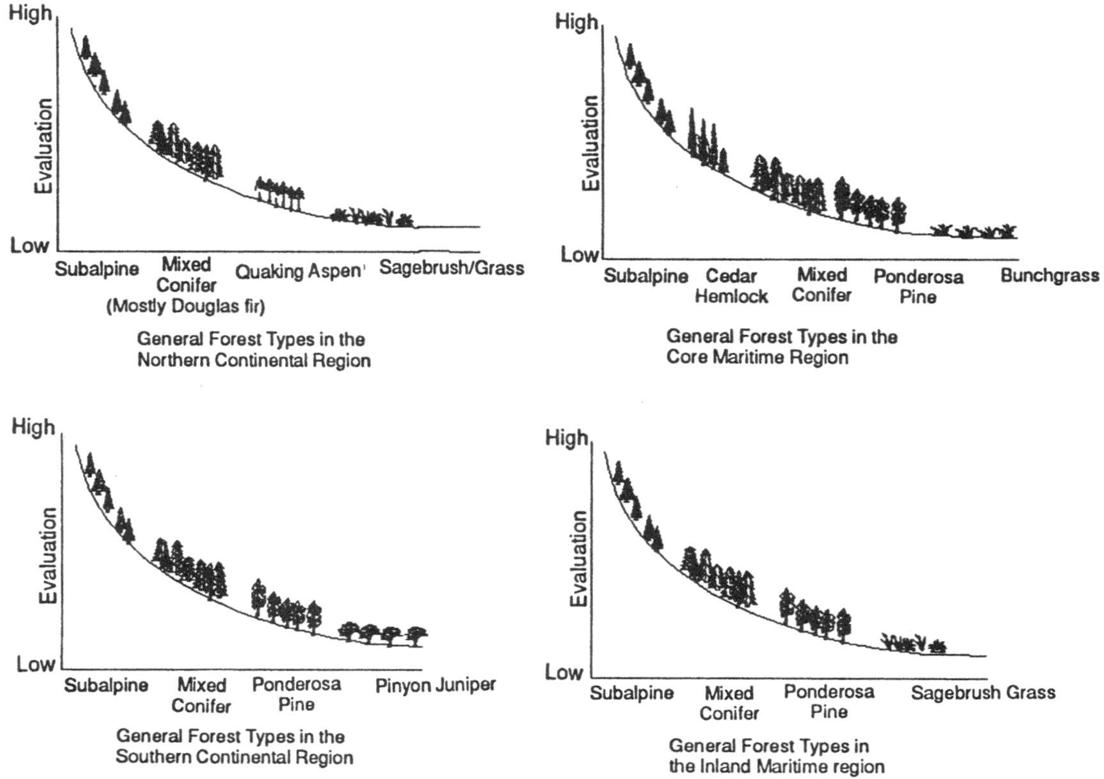

19

ence apparently enables elements of the southern mixed conifer forest such as white fir (*Abies concolor*) to extend northward to northern Utah and central Colorado. In Arizona and New Mexico, southwestern white pine (*Pinus strobiformis*) is a seral component of these forests (Moir and Ludwig 1979) and blue spruce (*Picea pungens*) appears intermittently from Arizona and New Mexico to Wyoming (Little 1971).

Along the eastern slope of the Cascade and Sierra Nevada ranges the mixed conifer zone is more diverse. In addition to white fir, Douglas-fir, and ponderosa pine, one can find incense cedar (*Calocedrus decurrens*), Jeffrey pine (*Pinus jeffreyi*), sugar pine (*P. lambertiana*), and giant sequoia growing either in association or as individual additions to this diverse forest zone (Parsons and DeBenedetti 1979; Paysen et al. 1980).

Ponderosa Pine Zone

At elevations below the mixed conifer forest lies an extensive dry forest zone where ponderosa pine is the climax dominant. Other conifers, except juniper in some areas, are generally sparse, but in Arizona, New Mexico, and the Dakotas, species of oak (*Quercus*) may appear under various soil and topographic conditions (Little 1971; Alexander et al. 1984, 1987; Fitzhugh et al. 1987). The ponderosa pine zone occurs intermittently from the base of the Cascade and Sierra Nevada ranges to South Dakota and Nebraska and from central Montana to Mexico (Steele 1988). The largest continuous ponderosa pine forest occurs in central and northern Arizona (Schubert 1974). But this extensive range is confounded by the lack of ponderosa pine from east-central Idaho to central Wyoming and from southwestern Montana to northern Utah and Nevada (Little 1971). Throughout much of this "pineless zone" the mixed conifer forest is notably reduced to a relatively narrow elevational band composed mainly of Douglas-fir. At their lower limits, these Douglas-fir forests border quaking aspen (*Populus tremuloides*) or sagebrush/grass communities in much of this area but in the southern portion they merge with pinyon-juniper woodlands.

Pinyon-Juniper Zone

Positioned below the lower timberline throughout much of the Inland West is a vast woodland dominated by species of pinyon pine and juniper, either singly or in various associations. This arid woodland composed of relatively small trees is known as the pinyon-juniper zone. It occurs extensively in Arizona, New Mexico, Nevada, Utah, and Colorado. Northern extensions or outliers reach as far north as central Washington in the case of western juniper (*Juniperus occidentalis*) and east-central Idaho and south-central Montana in the case of Utah juniper (*J. osteosperma*) (Little 1971). The pinyon pine element reaches as far north as southern Idaho in the case of singleleaf pinyon (*Pinus monophylla*). Other important elements include Colorado pinyon (*P. edulis*) (the commercial source of pine nuts), Mexican pinyon (*P. cembroides*), alligator juniper (*Juniperus deppeana*) and one-seed juniper (*J. monosperma*). In some areas, shrubby species of oak may associate with the pinyon and juniper.

GLOBAL CLIMATE
AND ATMOSPHERIC CHEMISTRY CHANGES

The evolutionary history of the species that constitute today's ecosystems has been characterized by major climatic fluctuations, especially over the past several million years (Delcourt and Delcourt 1988). These changes in climate have resulted in substantial changes in the geographical and altitudinal range of the biota. During the last major glacial period (20,000-14,000 B.P. [before the present]) continental ice sheets cooled middle latitudes, and steppe vegetation covered extensive areas of the inland Pacific Northwest (Johnson et al. 1993). During this colder period, coniferous forests and woodlands shifted south and down in elevation, only to shift back again with warmer temperatures. Significant expansion of conifers, (maximum representation 10,600 B.P.), and their rapid decline attest to the dynamic flux of terrestrial vegetation over time (Johnson et al. 1993).

Generally, plant species appear to have responded to climate change as individual species, not as plant associations; some species remain unchanged, others appear to have increased, while others decreased and became locally extinct. Western juniper woodlands in eastern Oregon were not regionally important until 4400 B.P. and northern Idaho's hemlock and cedar forests are younger yet (Johnson et al. 1993). Wigand and Nowak (1992) reported singleleaf pinyon arrived only 1700 to 1000 B.P. along the northwest side of the Great Basin, although Utah juniper has maintained a presence for 30,000 years. Ponderosa pine arrived in this area only 2000 years ago. The expansion of juniper woodlands in central Arizona began over 2600 years ago, but has accelerated over time as indicated by its pollen percentages (10%, 2630 yrs. B.P.; 30%, 200 yrs. B.P.; 40% 100 years. B.P.) (Davis 1987). Farther south in Mexico and southern Texas, two-needle pinyon (*Pinus remota*) was contracting in area and Colorado pinyon was expanding during the Holocene (Wells 1987).

Singleleaf pinyon contracted in area to lower elevations of the Mogollon Rim of southern Arizona following its expansion during the Wisconsin glacial.

Species composition continues to change in response to historic and current climatic regimes. Recent warming between the mid-1800s and the mid-1900s caused stress within stands that had established and grown in a previously cooler environment (Johnson et al. 1993.) However, "Change (vegetation) is perpetuated not only by plant responses to climate, but by disturbances that accompany climatic change–and fire is the most obvious of these disturbances" (Johnson et al. 1993, p. 61).

HISTORICAL FOREST DISTURBANCES AS AGENTS OF CHANGE

Fire Regimes

Because of differences in climate, fuel accumulation, and ignition rates (Sando 1978; Heinselman 1981; Kilgore 1981; Agee 1993; Covington and Moore 1994a), natural fire regimes vary widely throughout the Inland West. Fires burn frequently where fuels accumulate rapidly, fire weather is frequent, and ignition sources are common. However, because burns are frequent, fuel accumulation is held in check so that stand replacing fires are rare. If fire weather is rare and/or ignition sources are uncommon, much higher fuel accumulation can occur so that an infrequent, high-intensity natural fire regime develops. Where conditions are intermediate between these two extremes, a frequent, low-intensity fire regime punctuated by infrequent stand-replacing fires can occur (e.g., mesic mixed conifer forests). The relative importance of natural fire regimes in shaping evolutionary processes is greatest in ecosystems that have been characterized by frequent, low-intensity fire regimes and infrequent, high-intensity fire regimes or a combination thereof (Heinselman 1981; Kilgore 1981). For a comprehensive discussion of natural fire regimes, see Mooney et at. (1981).

For millennia before Europeans arrived, Native Americans changed fire regimes in the Inland West (Arno 1985; Gruell 1985; Barrett 1988; Savage and Swetnam 1990, Veblen and Lorenz 1991). The wide variety of Native American cultures used fire in many different ways (Pyne 1982; Phillips 1985). In addition to incidental burning, there is ample evidence that fires were intentionally set for a variety of reasons, including increasing desired plant species, improving desired wildlife habitat, driving game animals, and clearing transportation routes, among others. As a result, Native Americans altered fire frequency and hence successional development in many areas of the region (Gruell 1985). Before Euro-American settlement of the Inland West, some Native American tribes acquired horses which allowed them to range widely over large areas, thus increasing the extent of their influence on landscape dynamics. For example, Fisher et al. (1987) found evidence for a fairly stable ecotone between ponderosa pine forests, savannas, and grasslands around Devil's Tower in Wyoming before 1770, the approximate date of expansion of the Sioux horse culture. From 1770-1900 they found changes in fire frequency and a much more dynamic boundary. Since 1900, fire exclusion has resulted in the expansion of closed pine forests at the expense of pine savannas and grasslands (Fisher et al. 1987).

Insects and Pathogens

Forest ecosystems characterizing much of the Inland West occupy precarious, constantly changing environments limited by moisture, temperature, and/or nutrients. Within these ecosystems, native insects and diseases have historically functioned to regulate biological and physical processes integral to trophic dynamics and nutrient cycling, animal and plant habitat development, and patch and landscape diversity (Hessburg et al. 1993; Harvey, 1994). Although insects and pathogens have played important roles in the forests of the region their extent was often smaller, of shorter duration, and of less intensity than is the case today (Swetnam and Lynch 1989; Wickman 1992; Hessburg et al. 1993). Heterogeneous forest cover, more open cover, and increased representation of insect predators maintained more stable insect populations, although outbreaks were a natural phenomena of inland forests wherever favored tree hosts grew in abundance (Mason and Wickman 1988, 1993; Wickman 1992). Endemic levels of root diseases (e.g., annosus, tomentosus, armillaria, and laminated root rot) were present and individual or group mortality caused increased biodiversity within stands (Hessburg et al. 1993).

Native insects and diseases rarely caused widespread mortality in the presettlement subalpine, western hemlock, or western redcedar forests (Hessburg et al. 1993). In mixed conifer forests, western pine beetle (*Dendroctonus brevicomis*) attacked stressed ponderosa pine and mountain pine beetle (*D. ponderosae*) attacked lodgepole pine but extent of the attacks was limited (Hessburg et al. 1993).

In the ponderosa pine climax forests, western pine beetle attacks were continuous but localized except in the tension zone between forest and shrub steppe where trees were subjected to increased stress (Hessburg et al. 1993). Mountain pine beetle historically attacked lodgepole and ponderosa pine stands with sufficient number of trees greater than 9″ in diameter. Root diseases were present but limited in extent. Mistletoes were present but kept in check by recurring fire. Outbreaks of western spruce budworm (*Choristoneura occidentalis*) and Douglas-fir tussock moth (*Orgyia pseudotsugata*) were common in presettlement forests of the northwest. For example, between 1775 and 1809 five major outbreaks of western spruce budworm occurred in the Blue Mountains of Oregon (Wickman et al. 1993). Pine butterfly (*Neophasia menapia*) attacked over 130,000 acres of ponderosa pine near Spokane, Washington from 1883 to 1905 (Oliver et al. 1993).

Flooding

Flooding played a major role in sculpting the land's surface, controlling species distribution patterns, and re-initiating successional processes in the Inland West. Pleistocene lakes covered vast areas of the Northwest and the Great Basin, leaving remnants of historical beaches several hundred meters above the valley floor (Johnson et al. 1993). These lakes replaced previous plant communities and impacted species migration, but also increased heterogeneity of habitat for more mesic species. Periodic failures of ice-dams that held these lakes created massive floods that scoured soil and vegetation from vast areas as seen in the scablands of eastern Washington (Waitt 1985). On a continuing basis, major storms periodically flushed sediment from stream systems, re-initiated riparian succession, and provided nutrients to adjacent lowlands.

CHANGES IN ECOSYSTEM CONDITIONS AND RESOURCES DURING EURO-AMERICAN SETTLEMENT

Human Population Irruption

After initial declines of indigenous human populations due, in large part, to introduced diseases such as smallpox, a human population explosion has continued to occur in the Inland West. The current boom began with the arrival of ever-increasing numbers of Euro-Americans in the 1800s. Possessed with cultural, social, and economic attributes in many ways different from Native American cultures, the newcomers initiated dramatic ecological modifications that continue to the present day (Robbins and Wolf 1993).

Within decades the population of the Inland West had increased dramatically. Population levels rapidly exceeded the natural carrying capacity of the land (under indigenous technologies) leading to the initiation of new agricultural practices (farming/irrigation, ranching, timber harvesting) to meet subsistence and economic needs. For example, the Euro-American population increased from 13,000 in 1850 to over 930,000 by 1900 in the northwest portion of the region (Robbins and Wolf 1993). Since Euro-American settlement, resource utilization has increased exponentially, not only for use within the region but also for export. As a consequence, resource demand has exceeded resource renewal capability leading to unsustainable land use practices such as the excessive fur trapping of beaver from the early 1800s, the overutilization of forage by domestic livestock dating from the mid-1800s, and the depletion of old-growth trees and old forest types in this century (Oliver et al. 1993; Covington and Moore 1994a). Mining activity was extensive and unregulated, leading to degradation of upland slopes, riparian areas, and stream systems. Fisheries were significantly impacted by hydroelectric dams, overharvesting of fish stocks, and loss of stream habitat. Also, streams were dewatered to fulfill irrigation needs (Wissmar et al. 1993).

Unprecedented Herbivory of Forest and Rangeland Ecosystems

Grass and shrub communities of the intermountain portions of the Inland West evolved largely in the absence of large hoofed herbivores (Mack 1989). With the introduction of domestic livestock, unprecedented herbivory began on forest and rangeland ecosystems of the entire region. The acquisition of horses by Native Americans in the 1700s and resulting environmental impacts was but a small prelude to the hundreds of thousands of domestic livestock brought by the early Euro-American settlers in the mid- to late-1800s (Irwin et al. 1993). The rapid increase in domestic animal numbers (Figure 3) was tantamount to a population explosion of an introduced species with no population control.

Livestock grazing differed significantly from that of native herbivores in intensity, season, and dispersal of use across the landscape. No longer did the grazing animal follow the seasons up the hill slopes; rather, domestic livestock remained in the lowlands unless forced otherwise. Rapid shifts in types, class, and numbers of grazing animals occurred in response to economic returns rather than ecological reasoning.

Sheep numbers dramatically increased on forest and rangelands of the region in the 1890s, but this rapidly shifted to cattle (Figure 3) by the 1940s (e.g., Clary 1975; Oliver et al. 1993). Each type of livestock has an

FIGURE 3. Cattle, calf, sheep and lamb numbers over time in representative counties of eastern Washington (Yakima) and Oregon (Umatilla). Trends in livestock numbers compiled from the U.S. Department of Agriculture 1890-1990). Similar shifts occurred throughout the Inland West.

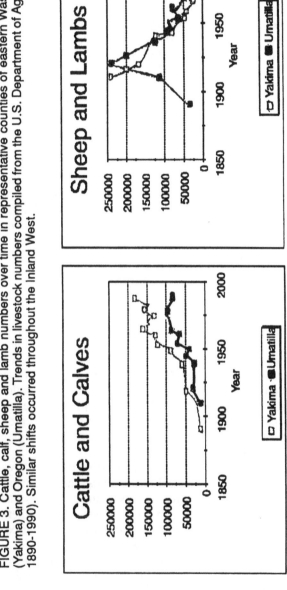

innate preference for certain plants (Heady 1964). Thus, a succession of plant species were selectively depleted from grazed sites as class of livestock changed. However, livestock are also adaptable and if preferred forage is unavailable other forage is taken (Provenza and Balph 1987) causing a decline in all palatable species when grazing pressure is extreme. Also, domestic animals have been a vector for the introduction and establishment of exotic plant species that have altered native community structure and processes.

Grazing may have depleted original forage species as early as the 1870s in eastern Oregon and Washington (Gordon et al. 1883) and by the early 1900s in the rest of the region. For example, depletion of forage production was seen as a serious problem on a majority of range sites in the region by 1935 (U.S. Senate 1936). Overgrazing sets into motion a chain of ecosystem degradation processes which can spread rapidly across the landscape. Loss of plant and litter cover exposes the soil surface to erosion; thus, continued grazing after loss of ground cover can cause the loss of productive top soil and the lowering of site potential (e.g., Platts 1984; Irwin et al. 1993). Reduction in forage and litter production alters fire intensity and extent, and contributes to the shift from high-frequency, low-severity fire regimes to those of longer duration but more intensity (e.g., Arno and Ottmar 1993; Covington and Moore 1994a).

Altered fire regimes, in conjunction with grazing, provided opportunities for the encroachment of woodland and forest tree species into rangeland plant communities (Leopold 1924; Bock and Bock 1983; Fisher et al. 1987). The spread of juniper and pinyon tree species into adjacent shrub and grassland communities is well documented (Johnsen 1962; Drivas and Everett 1988) and encroachment of conifers into forest meadows is a significant concern throughout the region (e.g., Arno and Gruell 1986; Allen 1989; USDA Forest Service 1993).

Because of their frequently degraded conditions, today's rangelands are not meeting the multitude of demands placed upon them. In 1987 approximately 15 percent of rangelands administered by the Forest Service and 18 percent administered by the Bureau of Land Management were considered to be in poor condition (U.S. General Accounting Office 1988). However, conditions are much worse in some locales. For example, non-federal lands in Oregon had over 40 percent of the area in poor condition (USDA Soil Conservation Service 1985). Riparian areas are often utilized to a greater degree than upslope areas. As a consequence, riparian zones are disproportionately impacted; for example, 47 percent of riparian areas managed by the Forest Service are not meeting management objectives (U.S. General Accounting Office 1991). Overgrazing has contributed to the loss of riparian vegetation, altered channel morphology, lowered ground water tables, and reduced fisheries and wildlife diversity (Platts 1984; Elmore and Beschta 1987; Elmore 1992).

Brief History of Industrial Timber Harvesting in the Inland West

Euro-American settlement and associated activities created major changes in forest composition and density. Throughout much of the Inland West, the first Euro-American settlement activities were centered around mining which placed heavy demands on local forests for mining props and building materials, especially large dimension timbers for mill construction (Oliver et al. 1993). In all cases, the local pine species, and occasionally Douglas-fir, were preferred for their greater strength and durability as building materials and their greater heat content as fuelwood. For these same reasons, western larch was also a preferred species where it occurred. In many areas, these preferred species occurred in mixtures of grand fir, white fir, subalpine fir, and Engelmann spruce which have inferior wood quality and greater vulnerability to fire, insects, and disease. As the large seed-bearing pines were removed, the remaining true firs, Douglas-fir, and other species left behind began to increase their numbers wherever adequate soil moisture permitted. Logging slash was generally left where it fell creating abnormal fuel loads in these dry forests. Thus, the potential for future epidemics of insects and disease and more destructive fires was becoming established.

In the late 1800s, the first inland timber harvests for commercial production of lumber for export to other regions occurred in locations accessible by railroad (Pearson 1950; Schubert 1974). Harvest techniques involved whip-saws and teams of oxen and horses, along with steam donkey-engines. In many areas, only large, high-grade ponderosa pines were cut. During this period, many railroad grades and skid trails were located in small stream channels and draws, with no consideration for riparian habitats or water quality. In some instances, temporary splash-dams were constructed, and the resultant reservoirs filled with logs, which were later sent downstream as debris torrents when the dams were dynamited (Oliver et al. 1993). Extensive railroad logging throughout the region created dangerous levels of dry logging slash resulting in extreme fire hazard. For example, in 1931 on the Boise National Forest, stand-destroying wildfire burned 62,000 acres of ponderosa pine/Douglas-fir forest that had previously experienced only surface fires for, at the least, 300 years (Steele et al. 1986). Observers attributed the intensity of this fire to the large volume of slash created by railroad logging in the late 1920s (Smith 1983).

Beginning in the 1920s, new technology including chainsaws, bulldozers, and logging trucks allowed for the rapid harvesting of steeper slopes than was possible with previous technologies. Dead trees were often felled as well. In the Depression years, the Civil Conservation Corps (CCCs) cut large snags from forests (many of which had burned in the late 1800s), because of concern that snags were the source of lightning-caused wildfires. This snag-felling practice has continued until the present day in some areas because OSHA (Occupational Safety and Health Act) regulations mandate removal of snags for safety reasons. In the 1940s, species harvested included medium-sized ponderosa pine, larger high-grade trees, and higher-quality Douglas-fir trees, at least in Idaho (e.g., Smith 1983).

In the 1950s and 1960s, a dramatic increase in harvesting and road building occurred (Schubert 1974; Oliver et al. 1993), and timber harvest on National Forests was increasingly influenced by principles that favored even-aged management. An "agricultural" model of forest management became institutionalized. Shelterwood, seedtree, and clearcut timber harvests prevailed, and the harvest of shade-tolerant species such as grand fir began. By then, thickets of young trees, especially the more shade-tolerant firs, had begun to develop. With few exceptions, little thought was given to the potential hazards these dense thickets would pose for fire, insects, and disease. By now the more accessible forests had changed from open stands of predominantly large pines, Douglas-fir, and western larch to closed stands of smaller trees, often with thickets of saplings, and a much larger component of true firs and spruce which can develop particularly dense stand conditions. The combination of these logging practices and fire exclusion accelerated this trend with ever-increasing areas changing from low-density stands composed mainly of large pine to dense stands of small diameter true fir, Douglas-fir, and spruce.

Timber management practices on National Forests in the 1970s and 1980s continued to emphasize intensive, even-aged management, despite concerns of many resource professionals and an increasingly vociferous public that even-aged management negatively impacted visual quality, wildlife, riparian zones, and water quality (Burk 1970). During this period the use of cable yarding systems exacerbated the problem because they allowed harvests on steep slopes and fragile soils. In addition, favorable economic conditions allowed for increased harvests of grand fir, white fir, spruce, and lodgepole pine. By the 1980s rising economic conditions allowed for some commercial thinning operations and some harvesting of true fir species.

Also during this period, harvesting trends on many state and industrial forestlands shifted toward selective harvesting, creating stands of well-spaced and often thrifty trees. In some States, forest practices acts were strengthened by new standards for road construction, tree stocking rates, wildlife habitat, riparian zones, and water quality.

After passage of the National Environmental Policy Act (NEPA) in 1969, and the Federal Land Policy and Management Act (FLPMA) and National Forest Management Act, both in 1976, new regulations which restrict many management activities were applied to federally-administered forests. Also, concerns for conserving genetic diversity were brought into policy decisions. For the most part, private, non-industrial forestlands continued to be managed on an economic-harvest basis.

The net legacy of past timber management practices, in association with fire exclusion policies, has been the removal of most of the large old-growth trees and their replacement with dense stands of young trees. In forest associations that include true firs and Douglas-fir at climax, these dense stands contain high percentages of shade-tolerant, climax species that are highly susceptible to insects and disease.

DISRUPTION OF HISTORICAL DISTURBANCE REGIMES

Changes in Fire Regimes

A general framework for discussing postsettlement changes in fire regimes is presented by Covington and Moore (1994a). The basic principles of this perspective are that:

- attempted fire exclusion in forest and woodland types which had infrequent crown fires results in increasingly large crown fires; and
- attempted fire exclusion in forest and woodland types which had frequent surface fires results in a shift to infrequent crown fires and then to increasingly large crown fires.

The end result of fire exclusion in fire-prone forests and woodlands is increasingly synchronous landscapes dominated by large, catastrophic disturbance regimes. These changes are in process throughout the Inland West.

Because of the differences in fire history, fuel accumulation rates, inherent flammability, climate, and land use history, different ecosystems of the Inland West are at different stages of these shifts toward landscape-level crown fires. Central to understanding this variation is knowledge of the presettlement fire regimes throughout the region. However, a comprehensive review of primary references is beyond the scope of this paper. A summary table constructed from secondary sources (Kilgore 1981; Arno 1985; Swetnam 1990; Agee 1993) should be useful (Table 1) to readers who want more detail.

By the early 1900s, fire exclusion had started to alter forest structure and fire regimes in the region (Agee 1990, 1993; Kauffman 1990). Initially, fire suppression had its greatest effects on high-frequency, low-intensity fire regimes and lesser effects on sites with lower fire frequency (Agee 1993; Arno and Ottmar 1993; Covington and Moore 1994a). Thus, high-elevation subalpine forests that historically burned at intervals greater than 100 years have been altered to a lesser extent than open-pine stands at lower elevations. Open-pine stands that once burned at 2 to 10 year intervals with low intensity now burn much less frequently but with high-intensity, stand-replacing fires (Hessburg et al. 1993; Covington and Moore 1994a). Areas with increased fuel loadings from abnormally high levels of forest floor and dead or live understory pine, true firs, or Douglas-fir are more continuous across the landscape, causing today's fires to be larger and more catastrophic.

Changes in Tree Insect and Pathogen Dynamics

Every forest ecosystem has its own unique association of native insects and pathogens. When ecosystem attributes such as species composition, tree density, horizonal and vertical diversity, and above ground biomass change, so do insect and pathogen complexes (Hessburg et al. 1993). Fire exclusion and selective harvesting have accelerated forest succession in all major forest ecosystems throughout the region, creating unstable community structures characterized by high stem density and above ground biomass and nutrient reservoirs; increasing dominance of shade-tolerant, pest-intolerant, climax species; and unprecedented build-up of continuous fuels and high-risk host coverage across the landscapes of the Inland West. Ecosystems under stress commonly exhibit increased tree insect and pathogen activity. This could partially explain the recent extensive tree mortality over millions of acres not only in the Inland West, but also throughout the United States (Haack and Byler 1993) (Table 2). This "unnatural" insect and pathogen damage may be symptomatic of declining forest health rather than the cause.

Insect and pathogen attacks on subalpine fir, western hemlock, and western redcedar do not appear to have changed significantly since settlement (Hessburg et al. 1993). In the grand fir/Douglas-fir forests, western pine beetle is presently a threat, but mountain pine beetle is having the greatest impact in over-stocked ponderosa pine and lodgepole pine stands. The fir engraver (*Scolytus ventralis*) is increasing in extent and intensity as its host species become weakened in overstocked stands (Wright et al. 1984). Western spruce budworm and Douglas-fir tussock moth also are increasing in extent as host species become continuous across the land-

TABLE 1. Presettlement fire regimes and postsettlement vegetation of forests and woodlands in the Inland West[1].

Potential Climax Vegetation	Location[2]	Presettlement Fire Freq. (yr)	Vegetation After Fire Exclusion	Reference[3]
Pinyon Pine and/or Juniper				
grassland	region-wide	15-30	closed woodland	Kilgore 1981
grassland/savanna	region-wide	15-90	closed woodland	Arno 1985
grassland/sage	NV, CA, ID	7-25	closed woodland	Agee 1993
Gambel Oak	AZ, CO, NM	20-30	closed woodland	Arno 1985
Ponderosa Pine				
park-like stands	AZ, NM	2-5	closed pine forest	Swetnam 1990
and savannas	region-wide	5-25	closed pine forest	Arno 1985
	OR	3-38	closed pine forest	Agee 1993
Douglas-fir				
open canopy and/or young forest	CO, MT, WY	25-60	closed Douglas-fir	Arno 1985
mountain grassland	OR, WA	20-30	closed Douglas-fir	Arno 1985
Mixed Conifer				
park-like ponderosa pine	AZ, NM, TX	6-10	closed mixed conifer	Swetnam 1990
open canopy mixed conifer	AZ, NM, TX	7-9	closed mixed conifer	Swetnam 1990

Forest type	Location	Years	Cover type	Reference
Douglas-fir, larch, grand fir	ID, WA, OR, MT	150	closed mixed conifer	Kilgore 1981
dry site Douglas-fir	N. Rocky Mts.	7-20	closed mixed conifer	Kilgore 1981
wet site Douglas-fir	MT	117-146	closed mixed conifer	Kilgore 1981
Cedar-hemlock western white pine larch, lodgepole	ID, MT	60-350	cedar-hemlock	Arno 1985
cedar-hemlock	ID, MT	>400	cedar-hemlock	Kilgore 1981
Lodgepole pine	region-wide	25-150	lodgepole-fir-spruce	Kilgore 1981
	region-wide	25-75	lodgepole-fir-spruce	Kilgore 1981
	region-wide	60-500	lodgepole-fir-spruce	Arno 1985
Subalpine fir and/or spruce young, open	region-wide	50-300	closed spruce-fir	Arno 1985
spruce-fir	region-wide	>150	closed spruce-fir	Kilgore 1981
lodgepole pine	region-wide	350	closed spruce-fir	Kilgore 1981
spruce-fir	MT, WA	25-300	closed spruce	Agee 1993

[1] Although some of this information is based on expert opinion most of it is based on quantitative research.

[2] Two letter abbreviations are used to denote States within the USA (e.g., OR = Oregon, AZ = Arizona).

[3] These references are synthesis papers which provide more detailed discussions as well as primary literature citations.

TABLE 2. Average annual acreage of US commercial forests affected by major insects and pathogens, 1979-83.

Insect or patbogen	Millions of acres	Native or exotic	Region affected
Dwarf mistletoe	22.6	Native	Western States
Root pathogens	16.8	Native	Primarily western States
Fusiform rust	15.3	Native	Southeastern States
Southern pine beetle	9.3	Native	Southeastern States
Western spruce budworm	6.8	Native	Western States
Gypsy moth	5.8	Exotic	Northeastern and Great Lakes States
Eastern spruce budworm	5.7	Native	Northeastern and Great Lakes States
Mountain pine beetle	4.3	Native	Western States

Source: Haack and Byler 1993

scape (Hessburg et al. 1993). All major tree-killing root diseases, except P-group annosum, are widespread. Dwarf mistletoe is also on the increase. For example, forty-three percent of the Douglas-fir stands in eastern Oregon and Washington are infected with dwarf mistletoe.

In the ponderosa pine forests, western pine beetle and pandora moth (*Coloradia pandora*) outbreaks are occurring on severely stressed sites. Mountain pine beetle is attacking lodgepole pine in previously harvested sites. Root diseases and western dwarf mistletoe (*Arceuthobium campylopodum*) are increasing as well (Hessburg et al. 1993). Thus, insect and pathogen disturbances which were previously limited by fire-induced landscape heterogeneity, now have expanded.

Changes in Hydrologic Regimes

Euro-American settlement altered hydrologic regimes throughout the Inland West (Leopold 1924; Clary 1975; Wissmar et al. 1993). Fire exclusion and the resultant increase in understory trees and forest floor accumulations increased interception and evapotranspiration in coniferous forests, thus reducing surface, subsurface, and instream flows and water availability in upslope soils. Grazing, timber harvest, agricultural practices, and road construction have degraded riparian habitat throughout the region. Riparian vegetation has been lost, the riparian zone has been reduced, and erosion has increased (Gregory and Ashkenas 1990; Wissmar et al. 1993).

Excessive grazing by domestic livestock reduced plant and litter cover of soils, increased surface erosion, and destabilized riparian areas. Grazing impacts are pervasive. On high-elevation summer ranges, sheet erosion has been shown to increase when grazing reduces combined plant, litter, and rock cover to below 70 percent (Meeuwig 1970). At lower elevations grazing and fire exclusion acted as catalysts in the encroachment of pinyon and juniper woodlands into adjacent grasslands and shrublands. These woodlands have significantly greater soil erosion than grasslands, shrublands, or coniferous forests (Buckhouse and Gaither 1982). Throughout much of the region, high grazing pressure not only from domestic livestock but also from unusually high elk and deer herds continue to degrade meadow and riparian environments (Barton and Fosburgh 1986). Furthermore, many current recreational uses of meadows and riparian areas are causing soil compaction and devegetation.

On sites where excessive tree harvest occurred, surface runoff and subsurface flows temporarily increased, causing high peak flows and accelerated in-stream erosion (Wissmar et al. 1993). Removal of tree overstory reduces interception and evapotranspiration, and provides for increased soil moisture that can decrease hillslope stability. Mass wasting as a result of unstable soils can be triggered by tree harvest and road construction. Timber harvest reduces hillslope stability by reducing root reinforcement and temporarily increasing water inputs and soil moisture (Sidle et al. 1985). Road-related slides can exceed those due to vegetation removal alone. Road construction can disrupt subsurface and surface water flow patterns, steepen the slope surface, create a load in the embankment fill, remove support of the cut slope, and channel water onto the road surface (Megahan et al. 1978; Sidle et al. 1985). Roading can cause changes in quality and quantity of stream flow and impair aquatic habitat. Accelerated sedimentation is the most common cause of deteriorated water quality from road construction (Megahan et al. 1993). Accelerated sediment loading, changed channel morphology, and altered riparian conditions can result from road-related effects and adversely impact all freshwater stages of salmonids (Furniss et al. 1991) and most other aquatic organisms.

Recent large-scale fires have increased amounts of water, sediment, and debris delivered to stream channels causing scarring of streams and loss of fish habitat. For example, on the Grande Ronde watershed in Oregon, which has been severely defoliated by tree harvest and insect outbreaks, base flow of streams has increased and peak discharge has shifted to one month earlier (Wissmar et al. 1993). Flood control actions have straightened stream channels and reduced the amount of sediment entering into and passing through stream systems; the replacement of sand and gravel in stream beds is required for fish rearing habitat (Johnson et al. 1993). Stream diversions for irrigation, stream channelization for flood control, and removal of beavers and their dams all contribute to reduced upstream storage and reduced low flows (Wissmar et al. 1993).

Permanent blockages of river systems by hydroelectric dams inundated rivers, destroyed spawning and rearing habitat for fish, and continue to cause mortality of juvenile fish moving downstream and adults moving upstream (Wissmar et al. 1993). For example, dam-related mortality rates have been estimated at between 77 and 96 percent for juveniles and 37 to 51 percent for adults in the Columbia River (Wissmar et al. 1993).

ECOSYSTEM RESPONSE
TO ALTERED DISTURBANCE REGIMES

Tree Population Explosions

As Euro-American settlement expanded, the effects of timber harvesting, grazing, and fire control created major changes in forests and woodlands of the Inland West. Pinyon-juniper woodlands expanded into adjacent grasslands from the rocky refugia that protected these fire-sensitive trees from frequent wildfire (Burkhardt and Tisdale 1976). The ponderosa pine zone increased in density as fire exclusion permitted more pine seedlings to develop in the understory (Cooper 1960). The once open park-like stands of ponderosa pine now contain thickets of postsettlement trees that serve as fuel ladders to carry fire into the overstory of large old pine. The ecological and resource implications of postsettlement changes in forests where ponderosa pine is the climax species have been severe (Covington and Moore 1994a, 1994b).

Changes have also been dramatic in mixed conifer forests where open, fire-maintained stands of large diameter ponderosa pine, Jeffrey pine, and western larch have been removed by logging. In their place is a mixture of small diameter pine and western larch, and a large component of small Douglas-fir and true firs which are more vulnerable to fire as well as a broad variety of insects and diseases (Wickman 1992). These "new forests" grow quite dense and thus competition for moisture and nutrients is intense. The severe competition results in tree stress that increases vulnerability to insect and disease attack (Vasechko 1983; McDonald 1991). As a result, many acres of mixed conifer forest are now in a dead or dying condition. Much of the mixed conifer zone, especially in the northern Rocky Mountains receives a high incidence of lightning storms. Ultimately, wildfire occurs and these dense, highly-flammable stands, dead or alive, create catastrophic fires that cause severe ecological damage and cost millions of dollars to suppress. These extensive high-intensity fires can destroy conifer seed sources over large areas and stimulate shrubfield development, making it difficult for shade-intolerant pines to establish naturally. Large scale planting efforts generally are carried out after these fires, creating additional expense, and reinforcing landscape homogeneity.

The subalpine zone has had changes similar to those of the mixed conifer zone but to a lesser degree. In the northern Rocky Mountains, vast forests of fire-induced lodgepole pine once dominated this zone and in many areas still do. But the mosaic of different tree sizes and densities created by natural fire regimes has diminished across the landscape due to fire control (Tande 1979). In their place are more uniform and denser stands of lodgepole pine often with a layer of subalpine fir and Engelmann spruce developing in the understory. These stands have greater vulnerability to bark beetles and stand-destroying fire, such as the Yellowstone fires of 1988, than the mosaic stand condition (Romme and Despain 1989). Logging practices have also altered many subalpine stands by creating large clearcut areas that regenerate to uniform and dense stands.

Logging, wildfires, and introduced disease have severely impacted forests of the cedar-hemlock zone. Most stands of large western white pine were harvested during the first half of this century. In 1910, wildfires burned over 2.5 million acres in northern Idaho and western Montana (Wellner 1970). Much of the burned area created seedbeds for new white pine stands. Unfortunately, the introduced white pine blister rust (*Cronartium ribicola*) has decimated most of these and other young white pine forests. The result has been a major shift in seral species composition toward Douglas-fir and grand fir, both of which are vulnerable to root rots which can create large patches of dead and dying trees scattered through the forest (Haack and Byler 1993). The climax species of western redcedar and western hemlock occur throughout most understory layers and can become prevalent in the overstory. Because fire return intervals may reach 200 years or more in this type (Arno and Davis 1980), these climax species can sometimes achieve large size.

Effects of Increased Conifer Populations on Herbaceous, Shrubby and Deciduous Woody Species

Overgrazing and increased conifer canopy closure have reduced density and biomass of herbaceous and shrubby understory species in forests and woodlands of the Inland West (Clary 1975; Clary 1988). The loss of understory diversity and production from canopy closure is due to decreased light levels, buildup of litter on the forest floor, and allelopathic inhibition of understory seed germination (Anderson et al. 1969; Clary 1975; Kelsey and Everett, in press). The decline in herbaceous and woody plants from increased tree cover is readily apparent, but the simultaneous decline in soil seed reserves of these species is less obvious (Koniak and Everett 1982). The long-term impacts of increased tree populations and extended rotations on understory species is unknown.

Increases in conifer populations have lead to loss of aspen stands throughout the Inland West (Schier 1975; DeByle 1985). Further losses have occurred because aspen parklands have been converted to meadows for livestock and others have been degraded from continuous and intense recreational use (DeByle 1985). In some areas, losses have been dramatic. For example, in Arizona and New Mexico alone, the area dominated by aspen has decreased by 222,000 acres or 46 percent (USDA Forest Service 1993).

The encroachment of conifers into meadows is widespread in the Inland West and has occurred from subalpine meadows (Franklin et al. 1971) to the low-elevation ecotones between forests and the grasslands (Strang and Parminter 1980). Almost all meadows in the mixed conifer zone of the

Southwest show evidence of conifer invasion (USDA Forest Service 1993). A major cause of this phenomenon is fire exclusion which stopped the fires which once killed invading conifer seedlings and periodically expanded meadows into the surrounding forest (Arno and Gruell 1986).

Wildlife Population Changes and Implications

In the 150 years since Euro-American settlement, significant changes have occurred in wildlife populations in the Inland West. Because of the highly varied environmental gradients and disturbance regimes, wildlife communities were diverse before settlement. In general, some species have been extirpated, some have increased in abundance, and some new species have been introduced. The history of changes in fish and wildlife populations includes a period of excessive exploitation and a period of protection/restoration which was followed by the rise of scientific management (Peek 1986). Some species have been persecuted throughout these periods, yet continue to survive (e.g., coyote [*Canis latrans*]). Following is a brief account of the history of major representative groups of fish and wildlife.

Large predators were persecuted until as recently as 1960; gray wolves (*Canis lupus*), mountain lions (*Felis concolor*), and grizzly bears (*Ursus arctos*) were largely extirpated from most western states. For example, government hunters and trappers from the U.S. Bureau of Biological Survey, established in 1915, were charged with the task of eliminating such predators from the western states. Despite the heavy persecution, populations of grizzlies survived in refuges in the regions containing Yellowstone and Glacier National Parks, and a few have been reported in northern Idaho and Washington. Wolves have recently re-established small populations in Washington, Idaho, and Montana, and a few have been reported in Wyoming. Mountain lion populations have increased in most western states since 1970. Black bears (*Ursus americanus*) and mountain lions are managed through regulated hunts in many states.

Thomas and Toweill (1982) reported the history of elk (*Cervus elaphus*) populations. Unregulated hunting, including market hunting, reduced elk populations to low numbers by the 1890s. Following a 4-decade period of low or no legal hunting, elk populations began to recover. Thousands of elk were translocated from the Yellowstone region after 1892, establishing new populations and supplementing existing populations in some areas.

Today, elk numbers in some areas appear higher than during any period in recorded history (Irwin et al. 1993). There is limited evidence that combined herbivory by large game and livestock have influenced wildfire regimes in ponderosa pine zones and affected soil nutrient capital with potential concomitant effects on growth rates of conifers (Weigand et al., in press). Further, the dense populations may be affecting forage nutritional regimes such that survival rates of juvenile elk are reduced, although the relative abundance of adult bull elk may be an influence as well (Irwin et al. 1993).

Workman and Low (1976), Wallmo (1981), and Mitchell and Freeman (1993) have documented the history of mule deer (*Odocoileus hemionious*) herds in the Inland West. Deer numbers in some areas, notably the north Kaibab deer herd in northern Arizona, experienced irruptions as early as the 1920s (Rasmussen 1941). Causes for the irruptions were varied and probably interdependent (Caughley 1976; Pengelly 1976; Mitchell and Freeman 1993), ranging from predator control and protection from hunting to logging and cattle grazing that favored shrubs that deer eat. The subsequent collapse of the Kaibab deer herd was also multicausal, including overgrazing by deer, drought conditions in 1924-1925, and disruption of the natural fire regime which accelerated losses of available browse (Mitchell and Freeman 1993). Most western states experienced significant irruptions during the period from mid-1950 through the early 1970s, in which large populations were observed that subsequently crashed to low numbers. Mule deer populations have rebounded in most western states (Schommer 1991).

Native fish populations were influenced by management practices that included poisoning of "rough" fish and introductions of popular game fish, such as rainbow (*Oncorhynchus mykiss*), brown (*Salmo trutta*), and eastern brook trout (*Salvelinus fontinalis*). For example, the brown trout is a strong competitor that has changed the distribution and abundance of bull trout (*S. confluentus*) (Robinson and Bolen 1984). As previously discussed, aquatic communities have been changed as a result of agricultural practices, range management, and road building that affected spawning sites, in conjunction with several other factors that reduced populations of both resident fish and anadromous fish (Kaczynski and Palmisano 1993).

Little is known of the history of nongame species, particularly amphibians and reptiles. Because relatively more is known about birds, they have often been used as indicators of biodiversity responses of other wildlife to management (e.g., Williams and Marcot 1991; Hansen and Urban 1992; Reynolds et al. 1992). Despite the dearth of data there is a general consensus that as resource management intensified, natural populations of many animals were reduced because ecosystems were simplified and habitats were fragmented and reduced (Harris 1984). It seems clear, also, that the distribution and abundance of many snag-dependent species (e.g., woodpeckers [*Picoides* sp.], flying squirrels [*Glaucomys sabrinus*]) were changed by snag-removal practices.

Assessing wildlife responses to land use is complicated because environmental gradients constrain the effects of spatial patterning of stands (Turner 1990). Therefore, a quantified understanding of the influences of landforms and soils on animal/plant interactions is required (Haufler and Irwin, in press; Irwin 1994). An ecologically-based landscape classification system (e.g., Wertz and Arnold 1972; Jones and Floyd 1993) can provide a basis for predicting the biological potential of ecosystems, which subsequently will determine the range of management options that affect wildlife.

Exotic Species Introductions

Exotic species introduced primarily from Europe and Asia have colonized disturbed habitats and invaded undisturbed native communities with deleterious effects on native plants, animals, and watersheds throughout the Inland West. Natural population control mechanisms for the exotics, such as predators or competitors, are usually absent, which is a major successional advantage to the exotics over native species.

Exotic Plants–Introduction of exotic weedy plants has altered biodiversity, site productivity, and economic resource values. Exotic plant species usually colonize following disturbances (Branson 1985) such as overgrazing, timber harvest and road construction, cropland abandonment, or high-intensity fires. Competitive advantages of introduced exotic plants over native species stems from a variety of mechanisms including earlier and more rapid growth (e.g., cheatgrass, Bromus tectorum) that depletes soil moisture, chemical inhibition (e.g., knapweed, Centaurea sp.) of other plants, and reduced palatability (e.g., leafy spurge, (Euphorbia esula)) relative to associated species (Harris 1989; Messersmith 1989; Kelsey and Everett, in press). A variety of reproductive systems, high fecundity, efficient dispersal mechanisms, and variation in germination requirements allow many exotic plants to outcompete native vegetation (Harrod et al., in press). Unconstrained population expansion of exotics such as knapweed/ yellow starthistle, cheatgrass, halogeton (Halogeton glomeratus), Dalmation toadflax (Linaria dalmatica), and leafy spurge suppresses indigenous species and reduces biodiversity on invaded sites (Belcher and Wilson 1989). Loss of native plant diversity is typically associated with loss in wildlife habitat (animal species diversity), as in the Clark Fork Basin of Montana where knapweed and other noxious weeds pose a significant threat to big game winter range (Lesica 1993).

The introduction of cheatgrass (an annual) serves as a case example of the immigration and exploitation of postsettlement-disturbed environments by exotics in the Inland West. Cheatgrass originated in the arid steppes of Eurasia where it had evolved in association with large herbivores (Mack 1989). Cheatgrass was introduced around 1889 into a bunchgrass environment that had an open niche between grass and shrub clumps. Further sites for cheatgrass establishment were made available by overgrazing. By the 1930s, cheatgrass had taken over much of the disturbed rangelands and open forestlands within its range of adaptability (Harris 1989). Cheatgrass can alter fire regimes because it grows in dense stands and produces flashy fuels which enhance the potential for ignition and fire spread. The combination of competition and recurring fire can exclude indigenous perennial species resulting in burn sites dominated by cheatgrass (Plummer et al. 1968).

The development of extensive exotic plant communities can result in long-term soil productivity loss. Accelerated erosion can occur due to the shallow fibrous or tap-rooted nature of many exotics compared to the deep-spreading rooting nature of native bunchgrass communities. Potential nutrient losses can occur when nonmycorrhizal weeds cause fungal populations to decline in communities originally dominated by mycorrhizal native bunchgrasses (Goodwin 1992).

Because many exotic plants evolved under grazing pressure in their native lands, they have developed chemical (e.g., oxalates in halogeton) or physical (e.g., awns in cheatgrass) defense mechanisms which can present significant threats to native grazers and domestic livestock, reducing range resource values (Kingsbury 1964). As an example, annual forage loss in Montana to spotted knapweed alone in 1984 was estimated at 4.5 million dollars (Bucher 1984) and if it were to occupy all potential habitat the loss could increase to 156 million (Lacey 1989). This estimate is for livestock production and does not include losses of food webs due to competition between native plants and exotics.

Exotic Animals–One of the most far-reaching exotic species introductions in the Inland West was the introduction of domestic livestock, and more specifically, the unregulated overgrazing by cattle, sheep, and goats. Some of these animals became feral. For example, by the 1700s, wild horse and burro populations occurred in much of the western United States; they are now protected by the Wild Horses and Burros Protection Act of 1971. These wild ungulates graze rangeland and woodland species that did not evolve under their intense form of grazing pressure (Mack 1989; Johnson et al. 1993). Current population levels are such that herd size must be carefully controlled to minimize environmental damage.

Conflicts have arisen because forage allocated to wild horses and burros reduces that available for other animals and domestic livestock.

Species introductions are not restricted to terrestrial ecosystems. For example, native bull trout population declines in Montana has been partially attributed to crosses with introduced brook trout which produce sterile hybrids (Holton 1990). In addition, the decline of bull trout may also be attributed in part to altered fire regimes in mixed conifer forests adjacent to spawning tributaries, where extensive, high-intensity fires can accelerate erosion and sedimentation far above levels before settlement.

Exotic Insects and Pathogens–Exotic insects and pathogens have dramatically altered forest ecosystem diversity, function, and productivity. More than 20 exotic fungal pathogens and 360 exotic insects now attack woody trees and shrubs in North America (Haack and Byler 1993). A classic example is that of white pine blister rust, introduced from Europe in 1909, which eliminated over 90 percent of western white pine forests throughout its historical range within the Inland West (Monnig and Byler 1992). Before settlement, low-intensity, frequent ground fires in concert with native insect and disease mortality provided the historical disturbances necessary to allow vast western white pine communities to persist and develop into seral old-growth forests. By the turn of the century, the introduction of blister rust, selective harvest, and fire suppression shifted the successional advantage to less well-adapted shade tolerant species. Subsequent tree mortality from epidemic populations of native root disease pathogens, bark beetles, and defoliators soared. This ultimately reduced timber productivity by over 50 percent in some areas and increased the risk of catastrophic wildfire (Monnig and Byler 1992). Much of the old-growth western white pine forests have been replaced by brushfields to the benefit of some wildlife species (such as deer and elk) and to the detriment of historic old-growth dependent species.

The gypsy moth (*Lymantria dispar*) is probably the most notorious exotic insect in North America. Since its introduction in 1869, millions of acres of eastern hardwood forests have been defoliated annually (Haack and Byler 1993). European and Asian strains of gypsy moth have recently been discovered in forests of the Inland West (e.g., in forests of the Pacific Northwest and riparian communities in Utah). Potential damage from this pest is host-related, with hardwood trees and shrubs, western larch, and Douglas-fir posing the greatest risk (personal communication, R.L. Livingston, Idaho Department of Lands). This suggests that riparian, big game winter range, and mixed conifer communities are at risk from this pest within the Inland West.

Improved Air Quality at the Expense of Ecosystem Sustainability

Air quality in ecosystems influenced by Native Americans was low in summer months when large tracts of forest were underburned and grasslands were ignited to enhance hunting and gathering of roots and berries (Robbins and Wolf 1993). Notes from explorers William Clark, Peter Ogden, John Fremont, John Townsend, Benjamin Bonneville, and early travelers of the Oregon Trail provide copious comments on Native Americans setting fire to the grass and the "smoky weather" in the northern portions of the Inland West. Cooper (1960) found similar descriptions in accounts from early explorers and Euro-American settlers in the southern portion of the Inland West. Low air quality was the price for maintaining open park-like forests and lush grasslands to meet the needs of indigenous hunter-gather cultures.

As previously discussed, fires caused by Native Americans and ignitions from lightning strikes maintained low fuel loadings on the landscape. Euro-American settlement interrupted this frequent, low-intensity burning, but weather conditions that historically provided the ignitions for the high-frequency, low-intensity fire regime continued, or perhaps intensified. Continual fire exclusion and an epidemic of understory trees has created an anomalous build up of fuels (Arno and Ottmar 1993; Covington and Moore 1994a). High fuel loadings that are continuous over the landscape provide the potential for extreme wildfire events (Mutch et al. 1993) and tremendous volumes of smoke emissions as seen in the recent Yellowstone fires. Yellowstone is not an isolated case. Nearly 100 years earlier, wildfires irrupted over much of the northern Rocky Mountains, causing one Idaho newspaper to note that, "For more than a month the city (Boise) has been enveloped in a dense cloud of smoke (from a wildfire)" (Idaho Statesman 1889). The potential for episodic low air quality exists in the future if these fuel-laden systems readjust through catastrophic crown fire.

FUTURE FORESTS OF THE INLAND WEST

New Environmental Stresses

Over the next century, radical changes in climate are expected to have a marked and lasting impact on forests of the Inland West. In the absence of vigorous and versatile management responses, it is likely that the regional forests will be strongly impacted and undergo substantial changes in their species makeup, structure, and function.

Conventional forest management strategies have implicitly assumed that climate remains constant over time. Recent evidence of marked global temperature increases this century and model projections of continued warming under scenarios of doubling of atmospheric CO_2 have changed this perception. General circulation models (GCM) predict an increase of 1.5-2.5°C in mean annual temperature within the next 30 to 50 years (Mitchell et al. 1990). Even at current rates of warming (approximately 1°C/century), forest species in some regions are perceived to be out of synchrony with the prevailing climatic conditions, and may be unable to acclimate with sufficient rapidity to survive (Davis 1989). Actual meteorological records for the 1980 decade and for the century indicate the Inland West region is particularly vulnerable to global warming and to extreme moisture stress.

Climatic variability is likely to be less at the future, higher temperature equilibrium. In the interval of rapid change, however, extreme variability is probable, including increases in the incidence of climatic discontinuities or "jumps" (Ebbesmeyer et al. 1991), the intensity and frequency of storms, the frequency of lightning, and the incidence of extreme weather fluctuations. Changes in the frequency and severity of extreme temperature and precipitation fluctuations are expected to increase during the transition period. The implications of this are that all types of weather damage and mortality to forests will increase in the coming decades. Of particular note are cavitation-type injuries on trees caused by winter thaw-freezes or prolonged summer drought (Auclair et al. 1990). These injuries result in hyper-sensitivity to water stresses and extensive areas of dieback and mortality, particularly among moisture-demanding species such as true firs (Tyree and Sperry 1989).

The concentration of CO_2 in the atmosphere is expected to increase at an accelerating rate. Current levels (359 ppm) are 28% above pre-industrial concentrations, and may reach 550 ppm by 2030. A positive effect on tree growth rate may be apparent at the higher CO_2 levels of the next century. Moreover, trees and shrubs with the C3 metabolic pathway are expected to respond more vigorously than some grasses (C4 pathway) with the consequence being accelerated tree and shrub invasion of grassland communities.

Although CO_2 is by far the most abundant greenhouse gas, a variety of other trace gases are biologically active and will assume increased importance over the coming decades. The continued destruction of stratospheric ozone by CFC compounds will lead to increased damage to forests by ultraviolet-B radiation. Tropospheric ozone and nitrogen oxides (NOx), despite regulations to limit their levels, have been increasing rapidly and are expected to continuously increase well into the next century. A "nitrate fertilization effect," resulting from atmospheric nitrogen oxides, although a potentially positive stimulus to vegetation growth, has yet to be conclusively demonstrated in the field (Graybill et al. 1992).

Ecological systems are highly dynamic entities. As climate and atmospheric chemistry continue to change over the decades ahead, forest ecosystems will respond, often in dramatic and unexpected ways. Some of the major, anticipated changes include:

- *Acceleration of Tree Growth Rate.* The average increase in tree growth among all major forest biomes in North America has been estimated to have been 2.25 fold between 1890 and 1990 (Auclair and Bedford 1993). The largest increases have been in the more northern biomes suggesting a response to warming rather than CO_2 per se, although the latter possibility is not excluded. The largest tree growth acceleration (3.2 fold) was in the western Canadian boreal forest (Jozsa and Powell 1987); the acceleration in the western U.S. temperate conifer forest was estimated to be 1.7 fold (Auclair and Andrasko 1991); a progressive acceleration of growth rate has also been evident in bristlecone pines at high elevations in California (Graybill and Idso 1993).
- *Increases in Water and Temperature Stresses.* These will, to some extent, counteract the positive effects of accelerated tree growth rates through greater frequency of wildfire, dieback and growth rate decline, pest-kill, and the invasion of exotic pest species. More rapid tree growth and the potential for greater tree densities and larger tree size suggest that forests may be more vulnerable to overstocking and resultant moisture and other stresses.
- *Increases of Within-Stand Competition and Mortality.* Accelerated tree growth will result in higher levels of competition within stands, and a higher rate of tree mortality. The potential benefits of increases in coarse woody debris and soil organic matter will contrast with the increased risk of catastrophic fire due to the accumulation of ground fuels.
- *Changes in Species Composition.* Physiological responses will vary among species (e.g., Polley et al. 1993), causing cascading shifts in species composition at all trophic levels.
- *Changes in Ecosystem Carbon Balance.* As an integrated indicator of the complex of energy and nutrient processes in forest ecosystems, the net carbon balance represents a useful measure of long-term sustainability. Marked imbalances in carbon storage or loss suggest corrective changes may be necessary. Regional, national, and global es-

timates of carbon balances are possible using forest inventory statistics (Auclair and Bedford 1993). The global estimates of carbon (CO_2) flux in the biosphere indicate a five decade trend toward increased carbon storage in forests (1925-1975), followed by an abrupt reversal of this trend toward carbon release (Sarmiento et al. 1992). The reversal correlated with the surge of wildfires, dieback, pests, and harvesting in the late 1970s and 1980s (Auclair and Carter 1993). It is probable that on a large-scale this net release will continue for several decades or until a balance (i.e., increased health and sustainability) is achieved.

Open, vigorous forests are more likely to be able to absorb these impacts without catastrophic readjustment. What do these anticipated changes imply for the Inland West? Although a quantitative region-wide analysis has yet to be done, we can draw some inferences from a systems analysis of forest responses to a greenhouse climate in western Montana (Running and Nemani 1991). Using ecological simulation they concluded that:

1. annual photosynthesis of forest ecosystems may increase by 20 to 30%,
2. growing seasons may lengthen by two months,
3. snowpack may decrease by two months, and
4. as a consequence of higher temperatures and less snowpack, hydrologic outflow may decrease by 30%.

Continued Tree Irruptions and Catastrophic Disturbances

In the absence of a major shift in the way society and the natural resource professions approach forest and woodland management, many of the current trends in forest and woodland deterioration will continue. In areas where forest density exceeds the carrying capacity of the land, drought, insect and disease epidemics, and wildfire will occur at scales and intensities never experienced. In areas which escape (temporarily) these large-scale disturbances or areas which have not yet crossed critical thresholds, tree densities will continue to increase with far-reaching consequences. For example, Covington and Moore (1994a, 1994b) used simulation analysis to forecast the consequences of continued increases of dominance by pine trees in ponderosa pine/bunchgrass ecosystems for two areas in Arizona. Their analyses predict: (1) increased density of post-settlement trees at the expense of old-growth trees and understory vegetation, (2) increased fuel loading, (3) decreased aesthetic values, and (4) a continuing shift away from wildlife populations that depend upon herbaceous and shrub production for forage (e.g., bluebirds [*Sialia* sp.], elk, turkey [*Meleagris gallopavo*], rabbits [*Sylvilagus* sp.]) toward wildlife which rely upon ponderosa pine as a basis for their food webs (e.g., tassle-eared squirrels [*Sciurus alberti*], porcupines [*Erethrizon dorsatum*], woodpeckers).

We are limited in our estimation of the character of future forests and woodlands by uncertainties regarding the effects of global climate change, but the immediate future can be estimated from current trends. Successional processes in the pinyon-juniper zone will continue to manifest in the expansion of pinyon-juniper communities into adjacent grasslands and shrublands. As woodland communities increase in canopy cover, herbaceous understories will continue to decline and wildlife and livestock grazing opportunities will diminish. In the ponderosa pine zone, seedlings and saplings of dense pine thickets will continue to develop and act as fuel ladders, allowing surface fires to become increasingly destructive crown fires. Pine needles and bark will continue to accumulate around the base of the large pines so that even infrequent surface fires will generate enough heat to kill the large pines which previously survived frequent fires for centuries. In the mixed-conifer zone, succession will continue toward nearly pure stands of true firs and Douglas-fir. Stand densities will continue to increase and exacerbate tree stress and vulnerability to insect and disease outbreaks. These dense stands are extremely flammable, creating fire conditions that can imperil not only resource values, but also human lives and property. In the cedar-hemlock zone, long fire-free intervals followed by catastrophic fires are the normal pattern and will likely continue. Western white pine is gradually developing resistance to the blister rust (Hoff and McDonald 1993), but this process could take millennia for white pine to regain its role as a major seral dominant. The subalpine zone will continue to advance successionally toward dense layers of climax Engelmann spruce and subalpine fir, increasing the flammability and vulnerability of these forests to insects and disease. Where lodgepole pine occurs, the mosaics of varying size and density will continue to diminish, leaving vast uniform stands which are more vulnerable to destructive fire, bark beetles, mistletoes, and other disease problems. The ultimate consequence is that throughout most of the Inland West, the present trends will leave us with seriously degraded ecosystems with reduced aesthetic, economic, and ecological value (e.g., Mutch et al. 1993; Covington and Moore 1994b).

Adaptation of Biosocial Systems to Anticipated Changes

Continuation of historical trends of dramatic changes in climate and atmospheric chemistry, coupled with interference of natural disturbance regimes, present both ecosystems and dependent human social systems with formidable challenges, and possibly with exciting opportunities. It seems highly unlikely that ecological systems will be able to absorb these environmental changes without major disruptions of existing ecosystem structures (species composition and demography, biomass and nutrient storage, and soils) and processes (carbon assimilation, nutrient cycling, trophic dynamics, and successional and landscape dynamics). In a similar vein, human social systems will be hard-pressed to respond rapidly and on the scale at which changes are occurring. Hence, remedial actions must be taken.

Our assessment of historical and likely future changes in climate, atmospheric chemistry, and disturbance processes in the Inland West leads us to the conclusion that the risk of inaction seems far greater than that associated with taking reasoned remedial actions. Nonetheless, we concur with concern by some environmental interest groups that increased publicity and alarm cries about ecosystem health may lead to ill-conceived and haphazard solutions, with the possible result that the "cure" may be worse than the "disease" (O'Laughlin et al. 1994). Fortunately, recent calls for applied systems approaches for dealing with undesirable environmental changes (e.g., Walters 1986; Walters and Holling 1990; Bonnicksen 1991; Lubchenko et al. 1991) are beginning to be heeded by both resource interest groups and governmental organizations (e.g., U.S. Department of Agriculture Forest Service's recent policy change toward ecosystem management and the U.S. Department of the Interior's shift toward ecosystem approaches to conservation of biological diversity). As a result, circumspect solutions are more likely than before.

Also on an optimistic note, within limits, periods of change can be beneficial for both biotic and social systems. Evidence is mounting that maximum rates of productivity and biodiversity appear to occur during periods of transition at a broad variety of scales in ecological systems (Seastedt and Knapp 1993). Similarly, human social systems often respond to periods of change with advances in both the intellectual and economic quality of life (e.g., Kuhn 1962; Devall and Sessions 1985; Bonnicksen 1992).

CONCLUSIONS

The evolutionary history of the organisms which constitute today's Inland West ecosystems has been characterized by natural disturbance regimes (e.g., fires, predation, defoliation), which have varied in kind, frequency, intensity, and extent. Natural fire regimes were particularly important in shaping the communities present at the time of Euro-American settlement. Exclusion of natural fires in the forests and woodlands of the Inland West, coupled with global climatic fluctuations and changes in atmospheric chemistry, has led to tree population explosions, dead fuel accumulations, and landscape level fuel continuity to such an extent that the niches of some species of plants, animals, and microbes have become threatened. Parallel declines in resource conditions for humans have occurred to a greater or lesser extent in all types.

In many cases, the natural functioning (e.g., successional processes, recycling processes) of these ecosystems has been severely impaired. As a result of increased tree densities in the ponderosa pine type, the increase in late successional species in the mixed conifer climax type, and the increasing landscape homogeneity in all types, catastrophic resetting of these systems by either large crown fires or large insect and disease epidemics is certain.

These changes, in concert with ongoing global changes in atmospheric CO_2 and climate, imply the need for extensive ecosystem restoration and management to prevent wide-scale collapse of existing ecological systems. This is especially true for wilderness areas and nature reserves where unnatural tree densities often exceed that of surrounding wildlands.

While some might quibble over the exact magnitude of the changes, the general trajectory seems unequivocal. Continued climate changes are likely to lead to increased tree seedling establishment (especially of shade-tolerant species), intensified competition among established trees, further deterioration of tree vigor, and increased tree mortality from insects, disease, and drought. Thus, we anticipate an acceleration of historical changes in the Inland West including increased fuel accumulations, lengthened fire seasons, and intensified burning conditions, all contributing to larger and more catastrophic wildfires. A fairly narrow window of opportunity–perhaps 15-30 years–exists for land managers to implement ecosystem management treatments to restore more nearly natural and robust ecosystem structures and processes.

For many areas it may already be too late. However, for most of the region there is still time for remedial actions, both to minimize the negative environmental consequences of these changes and to capitalize on the positive. The consequences of inaction far exceed those of action.

REFERENCES

Agee, J.K. 1990. The historical role of fire in Pacific Northwest forests. In: pp. 25-38, Chapter 3. Walstad, J.D., S. R. Radosevich, D.V. Sandberg (Eds.). Natural and prescribed fire in Pacific Northwest forests. Oregon State University Press. Corvallis, OR.

Agee, J.K. 1993. Fire and weather disturbances in terrestrial ecosystems of the eastern Cascades. In: pp. 359-414. Hessburg, P.F. (Comp.). Eastside forest ecosystem health assessment–Volume III: Assessment. USDA Forest Service, Pacific Northwest Research Station. Portland, OR. 750 p.

Alexander, B.G., F. Ronco, Jr., E.L. Fitzhugh and J.A. Ludwig. 1984. A classification of forest habitat types on the Lincoln National Forest, New Mexico. USDA Forest Service, Rocky Mountain Forest and Range Experiment Station. General Technical Report RM-104. 29 p.

Alexander, B.G., E.L. Fitzhugh, F. Ronco, Jr. and J.A. Ludwig. 1987. A classification of forest habitat types of the northern portion of the Cibola National Forest, New Mexico. USDA Forest Service, Rocky Mountain Forest and Range Experiment Station. General Technical Report RM-143. 35 p.

Allen, C.D. 1989. Changes in the landscape of the Jemez Mountains, New Mexico. Ph.D. dissertation, University of California, Berkeley. 346 p.

Allen, D. 1962. Our wildlife legacy. Fitzhenry and Whiteside, Ltd. Toronto, Canada. 422 p.

Anderson, R.L., O.L. Loucks and A.M. Swain. 1969. Herbaceous response to canopy cover, light intensity, and throughfall precipitation in coniferous forests. *Ecology* 50:225-263.

Arno, S.F. 1985. Ecological effects and management implications of Indian fires. In: pp. 81-86. Proceedings–Symposium and workshop on wilderness fire. November 15-18, 1983. Missoula, MT. USDA Forest Service, Intermountain Forest and Range Experiment Station. General Technical Report INT-182.

Arno, S.F. and D.H. Davis. 1980. Fire history of western redcedar/hemlock forests in northern Idaho. In: pp. 21-26. Stokes, M.A. and J.H. Dieterich (Technical Coordinators). Proceedings of the Fire History Workshop. 20-24 Oct. 1980. Tucson, AZ. USDA Forest Service, Rocky Mountain Forest and Range Experiment Station. General Technical Report RM-81.

Arno, S.F. and G.E. Gruell. 1986. Douglas-fir encroachment into mountain grasslands of southwestern Montana. *Journal of Range Management* 39:272-275.

Arno, S.F. and R.D. Ottmar. 1993. Reducing hazard for catastrophic fire. In: pp. 17-18. Everett, R. (Comp.). Eastside forest ecosystem health assessment–Volume IV: Restoration of stressed sites and processes. Portland, OR: USDA Forest Service, Pacific Northwest Research Station. 114 p.

Auclair, A.N.D. and K.J. Andrasko. 1991. Net CO_2 flux in temperate and boreal forests in response to climate changes this century: a case study. In: Proceedings of the NATO workshop on the biological implications of Global Climate Change. Clemson, SC.

Auclair, A.N.D. and J.A. Bedford. 1993. Forest depletion and accrual dataset: area and volume estimates. Documentation file, Version 2. Office of Research and Development, U.S. Environmental Protection Agency. Washington, DC.

Auclair, A.N.D. and T.B. Carter. 1993. Forest wildfires as a recent source of CO_2 at northern latitudes. *Canadian Journal of Forest Research* 23:1528-1536.

Auclair, A.N.D., H.C. Martin and S.L. Walker. 1990. A case study of forest decline in western Canada and the adjacent United States. *Water, Air, and Soil Pollution* 53:13-31.

Baker, W.L. 1992. The landscape ecology of large disturbances in the design and management of nature reserves. *Landscape Ecology* 7:181-194.

Barrett, S.W. 1988. Fire suppression's effects on forest succession within a central Idaho wilderness. *Western Journal of Applied Forestry* 3:76-80.

Barton, K. and W. Fosburgh. 1986. The U.S. Forest Service. In: Eno, A., R. Disilvestro, and W. Chandler (Eds.). Audubon wildlife report 1986. New York. The National Audubon Society. 158 p.

Belcher, J.W. and S.D. Wilson. 1989. Leafy spurge and the species composition of a mixed-grass prairie. *Journal of Range Management* 42:172-175.

Bock, J.H. and C.E. Bock. 1983. Effect of fires on woody vegetation in the pine-grassland ecotone of the southern Black Hills. *American Midland Naturalist* 112:35-42.

Booth, D.E. 1991. Estimating prelogging old-growth in the Pacific Northwest. *Journal of Forestry* 89(10):25-29.

Bonnicksen, T.M. 1991. Managing biosocial systems: a framework to organize society-environment relationships. *Journal of Forestry* 89(10):10-15.

Bonnicksen, T.M. and E.C. Stone. 1982. Reconstruction of a presettlement giant sequoia-mixed conifer community using the aggregation approach. *Ecology* 63:1134-1148.

Branson, F.A. 1985. Vegetation changes for western rangelands. Society for Range Management. Denver, CO. 76 p.

Brown, D.E. (Ed.). 1982. Biotic communities of the American Southwest–United States and Mexico. *Desert Plants* 4(1-4):1-342.

Bucher, R.F. 1984. Potential spread and loss of spotted knapweed on range. Bulletin 1316. Montana Cooperative Extension Service.

Buckhouse, J.C. and R.E. Gaither. 1982. Potential sediment production within vegetative communities in Oregon's Blue Mountains. *Journal of Soil and Water Conservation* 37(2):120-122.

Burk, D.A. 1970. The clearcut crisis: controversy in the Bitterroot. Jursnick Printing. Great Falls, MT. 152 p.

Burkhardt, J.W. and E.W. Tisdale. 1976. Causes of juniper invasion in southwestern Idaho. *Ecology* 57:472-484.

Caraher, D.L., J. Henshaw, F. Hall et al. 1992. Restoring ecosystems in the Blue Mountains–a report to the Regional Forester and the Forest Supervisors of the Blue Mountains Forests. USDA Forest Service, Pacific Northwest Region. Portland, OR. 15 p.

Caughley, G. 1976. Wildlife management and the dynamics of ungulate populations. *Applied Biology* 1:183-246.

Christensen, N.L. 1988. Succession and natural disturbance: paradigms, problems, and preservation of natural ecosystems. In: pp. 62-88. Agee, J.K. and D.R. Johnson (Eds.). Ecosystem management for parks and wilderness. University of Washington Press. Seattle.

Clary, W.P. 1975. Range management and its ecological basis in the ponderosa pine type of Arizona: the status of our knowledge. USDA Forest Service. Research Paper RM-158. 35 p.

Clary, W.P. 1988. Silvicultural systems for forage production in ponderosa pine forests. In: pp. 185-191. Baumgartner, D.M. and J.E. Lotan (Eds.). Ponderosa pine: the species and its management. Proceedings of the symposium held Sept. 29-Oct. 1, 1987, Spokane, WA. Office of Conferences and Institutes, Washington State University. Pullman. 281 p.

Cooper, C.F. 1960. Changes in vegetation, structure, and growth of southwestern pine forests since white settlement. *Ecological Monographs* 30:129-164.

Cooper, S.V., K.E. Neiman and D.W. Roberts. 1991. Forest habitat types of northern Idaho: a second approximation. USDA Forest Service, Intermountain Research Station. General Technical Report INT-236. 143 p.

Covington, W.W. and M.M. Moore. 1992. Postsettlement changes in natural fire regimes: implications for restoration of old-growth ponderosa pine forests. In: pp. 81-99. Kaufmann, M.R. and W. H. Moir (Technical Coordinators). Old-growth forest in the Southwest and Rocky Mountain Regions. USDA Forest Service, Rocky Mountain Forest and Range Experiment Station. General Technical Report RM-213.

Covington, W.W. and M.M. Moore. 1994a. Postsettlement changes in natural fire regimes and forest structure: ecological restoration of old-growth ponderosa pine forests. In: pp. 153-181. Sampson, R.N. and D. Adams (Eds.). Assessing Forest Ecosystem Health in the Inland West. Proceedings of the American Forests scientific workshop, November 14-20, 1993. Sun Valley, ID. The Haworth Press, Inc. New York.

Covington, W.W. and M.M. Moore. 1994b. Southwestern ponderosa forest structure and resource conditions: changes since Euro-American settlement. *Journal of Forestry* 92(1):39-47.

Daubenmire, R. 1956. Climate as a determinant of vegetation distribution in eastern Washington and northern Idaho. *Ecological Monographs* 26:131-154.

Davis, M.B. 1989. Lags in vegetation response to greenhouse warming. *Climate Change* 15:75-82.

Davis, O.K. 1987. Palynological evidence for historic juniper invasion in central Arizona: A late-Quaternary perspective. In: Proceedings: Pinyon-Juniper Conference. 13-15 January, 1986, Reno, NV. USDA Forest Service, Intermountain Research Station. Ogden, UT. General Technical Report INT-215. 581 p.

DeByle, N.V. 1985. Management for esthetics and recreation, forage, water, and wildlife. In: pp. 223-232. Aspen: ecology and management in the western United States. USDA Forest Service, Rocky Mountain Research Station. General Technical Report RM-119.

Delcourt, H.R. and P.A. Delcourt. 1988. Quarternary ecology: relevant scales in time and space. *Landscape Ecology* 2:23-44. ~

Devall, B. and G. Sessions. 1985. Deep ecology: living as if nature mattered. Gibbs Smith, Inc. Peregrine Smith Books. Salt Lake City, UT. 267 p.

DeVelice, R.L., J.A. Ludwig, W.H. Moir and F. Ronco Jr. 1986. A classification of forest habitat types of northern New Mexico and southern Colorado. USDA Forest Service, Rocky Mountain Forest and Range Experiment Station. General Technical Report RM-131. 59 p.

Drake, B., L.D. Garrett, D. Gill et al. 1991. Findings and recommendations on the Arizona Blue Ribbon Task Force. USDA Forest Service, Southwestern Region. 35 p. + 10 appendices.

Drivas, E.P. and R.L. Everett. 1988. Water relations characteristics of competing singleleaf pinyon seedlings and sagebrush nurse plants. *Forest Ecology and Management* 23:27-37.

Ebbesmeyer, C.C., D.R. Cayan, D.R. McLain, F.H. Nichols, D.H. Peterson and K.T. Redmond. 1991. 1976 step in the Pacific climate: Forty environmental changes between 1968-1975 and 1977-1984. In: pp. 115-126. Betancourt, J.L. and V.L. Tharp (Eds.). Proceedings, 7th Annual Pacific Climate (PACLIM) Workshop, April 1990. Interagency Ecological Studies Program Technical Report 26. California Department of Water Resources. San Francisco, CA.

Elmore, W. 1992. Riparian responses to grazing practices. In: pp. 442-457. Naiman, R.J. (Ed.). Watershed management. Springer-Verlag. New York.

Elmore, W. and R.L. Beschta. 1987. Riparian areas: perceptions in management. *Rangelands* 9:260-265.

Everett, R.L., P.F. Hessburg, M.E. Jensen and B.T. Bormann. 1993. Eastside forest ecosystem health assessment, Volume I, Executive Summary. USDA Forest Service, Pacific Northwest Research Station. 57 p.

Fisher, R.F., M.J. Jenkins and J.W. Fisher. 1987. Fire and the prairie-forest mosaic of Devil's Tower National Monument. *The American Midland Naturalist* 117:250-257.

Fitzhugh, E.L., W.H. Moir, J.A. Ludwig and F. Ronco Jr. 1987. Forest habitat types in the Apache, Gila, and part of the Cibola National Forests, Arizona and New Mexico. USDA Forest Service, Rocky Mountain Forest and Range Experiment Station. General Technical Report RM-145. 116 p.

Flader, S.L. 1974. Thinking like a mountain: Aldo Leopold and the evolution of an ecological attitude toward deer, wolves, and forests. University of Missouri Press. Columbia. 284 p.

Franklin, J.F. and C.T. Dyrness. 1973. Natural vegetation of Oregon and Washington. USDA Forest Service, Pacific Northwest Forest and Range Experiment Station. General Technical Report PNW-8. 417 p.

Franklin, J.F., W.H. Moir, G.W. Douglas and C. Wiberg. 1971. Invasion of subalpine meadows by trees in the Cascade Range, Washington and Oregon. *Arctic and Alpine Research* 3:215-224.

Furniss M.J., T.D. Roelofs and C.S. Yee. 1991. Road construction and maintenance. In: pp. 297-323. Meehan, W. R. (Ed.). Influences of forest and rangeland management on salmonid fishes and their habitats. Special Pub. 19. American Fisheries Society.

Goodwin, J. 1992. The role of mycorrhizal fungi in competitive interactions among native bunchgrasses and alien weeds: a review and synthesis. *Northwest Science* 66(4):251-260.

Gordon, C., J. McCoy et al. 1883. Report on cattle, sheep and swine; supplementary to enumeration of livestock on farms in 1880. U. S. Department of the Interior, Census Office. 3:953-1116.

Graybill, D.A. and S.B. Idso. 1993. Detecting the aerial fertilization effect of atmospheric CO_2 enrichment in tree-ring chronologies. *Global Biogeochemical Cycles* 7(1):81-95.

Graybill, D.A., D.L. Peterson and M.J. Arbaugh. 1992. Coniferous forests of the Colorado front range. In: pp. 365-401. (Chapter 9). Olson, R., D. Binkley, and M. Bohm (Eds.). Effects of air pollution on coniferous forests of the western United States. Springer-Verlag. New York.

Gregory, S.V. and L. Ashkenas. 1990. Riparian management guide. Willamette National Forest, USDA Forest Service, Pacific Northwest Region. Eugene, OR.

Gruell, G.E. 1985. Indian fires in the Interior West: a widespread influence. In: pp. 68-74. Proceedings–Symposium and workshop on wilderness fire. November 15-18, 1983. Missoula, MT. USDA Forest Service, Intermountain Forest and Range Experiment Station. General Technical Report INT-182.

Haack, R.A. and J.W. Byler. 1993. Insects and pathogens: regulators of forest ecosystems. *Journal of Forestry* 91(9):32-37.

Hansen, A.J. and D.L. Urban. 1992. Avian response to landscape pattern: the role of species life histories. *Landscape Ecology* 7:163-180.

Harrod, R.J., R.J. Taylor, W.L. Gaines et al. [In press]. Noxious weeds in the Blue Mountains. In: Jaindle, R. and T. Quigley (Eds.). Search for a solution–sustaining the land, people, and economy of the Blue Mountains, a synthesis of our knowledge. USDA Forest Service, Blue Mountains Natural Resource Institute.

Harris, G.A. 1989. Cheatgrass: invasion potential and managerial implications. In: pp 5-11. Roche, B. and C. Talbot Roche (Eds.). Range Weeds Revisited: Symposium Proceedings, January 24-26, 1989. Spokane, WA. Washington State University. Pullman.

Harris, L. 1984. The fragmented forest. Island biogeography theory and the preservation of biotic diversity. University of Chicago Press. Chicago, IL. 211 p.

Harvey, A.E. 1994. Integrated roles for insects, diseases, and decomposers in fire dominated forests of the Inland Western United States. In: pp. 211-220. Sampson, R.N. and D. Adams (Eds.). Assessing Forest Ecosystem Health in the Inland West. Proceedings of the American Forests Scientific Workshop, November, 14-20, 1993. Sun Valley, ID. The Haworth Press, Inc. New York.

Haufler, J.B. and L.L. Irwin. [In press]. An ecological basis for planning for biodiversity and resource use. Proceedings, International Union Game Biol. Halifax, Nova Scotia.

Heady, H.F. 1964. Palatability of herbage and animal preference. *Journal of Range Management* 17:76-82.

Heinselman, M.L. 1981. Fire intensity and frequency as factors in the distribution and structure of northern ecosystems. In: pp. 7-57. Fire regimes and ecosystem properties. USDA Forest Service. General Technical Report WO-26. 594 p.

Hessburg, P.F., R.G. Mitchell and G.M. Flip. 1993. Historical and current roles of insects and pathogens in eastern Oregon and Washington forest landscapes. In: pp. 485-536. Hessburg, P.F. (Comp.). Eastside forest ecosystem health assessment–Volume III: Assessment. USDA Forest Service, Pacific Northwest Research Station.

Hoff, R.J. and G.I. McDonald. 1993. Variation of virulence of white pine blister rust. *European Journal of Forest Pathology* 23:103-109.

Holling, C.S. 1981. Forest insects, forest fires, and resilience. In: pp. 445-464. Fire regimes and ecosystem properties. Proceedings of the conference held in Honolulu, Hawaii, December 11-15, 1978. USDA Forest Service. General Technical Report WO-26. 594 p.

Holton, G.D. 1990. A field guide to Montana fishes. Montana Department of Fish, Wildlife and Parks. Helena, MT.

Hoover, R.L. and D.L. Wills (Eds.). 1984. Managing forested lands for wildlife. Colorado Division of Wildlife. Denver. 459 p.

Idaho Statesman. 1889. August 24th edition. Boise, ID.

Irwin, L.L. 1994. A process for improving wildlife habitat models for assessing forest ecosystem health. In: pp. 293-306. Sampson, R.N. and D. Adams (Eds.). Assessing Forest Ecosystem Health in the Inland West. Proceedings of the American Forests Scientific Workshop, November, 14-20, 1993. Sun Valley, ID. The Haworth Press, Inc. New York.

Irwin, L.L., J.G. Cook, R.A. Riggs and J.M. Skovlin. 1993. Effects of long-term grazing by big game and livestock in the Blue Mountains forest ecosystems. In: pp. 537-588. Hessburg, P.F. (Comp.). Eastside forest ecosystem health assessment–Volume III: Assessment. USDA Forest Service, Pacific Northwest Research Station. 750 p.

Johnsen, T.N. 1962. One-seed juniper invasion of northern Arizona grasslands. *Ecological Monographs* 32:187-207.

Johnson, C.G., R.R. Clausnitzer, P.J. Mehringer and C.D. Oliver. 1993. Biotic and abiotic processes of eastside ecosystems: the effects of management on plant community ecology, and on stand and landscape vegetation dynamics. In: pp. 35-99. Hessburg, P.F. (Comp.). Eastside forest ecosystem health assessment–Volume III: Assessment. USDA Forest Service, Pacific Northwest Research Station. 750 p.

Jones, S.M. and F.T. Floyd. 1993. Landscape classification: the first step toward ecosystem management in the southeastern United States. In: pp. 181-201. Aplet, G.H., N. Johnson, J.T. Olson and V.A. Sample (Eds.). Defining sustainable forestry. Island Press. Washington, DC.

Jozsa, L.A. and J.M. Powell. 1987. Some climatic aspects of biomass productivity of white spruce stem wood. *Canadian Journal of Forest Research* 17:1075-1079.

Kaczynski, V.W. and J.F. Palmisano (with assistance from J.E. Levin). 1993. Oregon's wild salmon and steelhead trout: a review of the impact of management and environmental factors. Oregon Forest Industries Council. Salem, OR. 328 p.

Kaufmann, J.B. 1990. Ecological relationships of vegetation and fire in Pacific Northwest forests. In: Chapter 4. Walstad, J.D., S. R. Radosevich and D.V. Sandberg (Eds.). Natural and prescribed fire in Pacific Northwest forests. Oregon State University Press. Corvallis.

Kaufmann, M.R., W.H. Moir and W.W. Covington. 1992. The status of knowledge of old-growth forest ecology and management in the central and southern

Rocky Mountains and Southwest. In: pp. 1-11. Kaufmann, M.R. and W.H. Moir (Technical Coordinators). Old-growth forest in the Southwest and Rocky Mountain Regions. USDA Forest Service, Rocky Mountain Forest and Range Experiment Station. General Technical Report RM-213.

Keeley, J.E. 1981. Reproductive cycles and fire regimes. In: pp. 231-277. Mooney, H.A., T.M. Bonnicksen, N.L. Christensen, J.E. Lotan and W.A. Reiners (Technical Eds.). Fire regimes and ecosystem properties. Proceedings of the conference held December 11-15, 1978. Honolulu, HI. USDA Forest Service. General Technical Report WO-26. 594 p.

Kelsey, R.G. and R.L. Everett. [In press]. Allelopathy.

Kilgore, B.M. 1981. Fire in ecosystem distribution and structure: western forests and scrublands. In: pp. 58-89. Mooney, H.A., T.M. Bonnicksen, N.L. Christensen, J.E. Lotan, and W.A. Reiners (Technical Eds.). Fire regimes and ecosystem properties. Proceedings of the conference held December 11-15, 1978. Honolulu, HI. USDA Forest Service. General Technical Report WO-26. 594 p.

Kingsbury, J.M. 1964. Poisonous plants of the United States and Canada. Prentice-Hall, Inc. Englewood Cliffs, NJ.

Koniak, S. and R.L. Everett. 1982. Seed reserves in soils of successional stages of pinyon woodlands. *American Midland Naturalist* 108:295-303.

Kuhn, T.S. 1962. The structure of scientific revolutions. University of Chicago Press Chicago. 210 p.

Lacey, C. 1989. Knapweeds–the situation in Montana. In: Roche, B. and C. Talbot Roche (Eds.). Range weeds revisited: Symposium proceedings, January 24-26 1989. Spokane, WA. Washington State University, Pullman.

Leopold, A. 1949. A Sand County almanac. Oxford University Press. New York. 295 p.

Lesica, P. and G. Hirschenberger. 1993. Ovando Valley (Blackfoot River, Montana) ecological report. Unpublished report for the Nature Conservancy. Helena, MT.

Little, E.L., Jr. 1971. Atlas of United States trees. Volume I: Conifers and important hardwoods. USDA Forest Service, Miscellaneous Publications No. 1146. 8 p. + 200 maps.

Lubchenko, J. et al. 1991. The sustainable biosphere initiative: an ecological research agenda. *Ecology* 72:371-412.

Mack R.N. 1989. Invaders at home on the range. In: pp 1-4. Roche, B. and C. Talbot Roche (Eds.). Range Weeds Revisited: Symposium Proceedings. January. 24-26, 1989. Spokane, WA. Washington State University, Pullman.

Mason, R.R. and B.E. Wickman. 1988. The Douglas-fir tussock moth in the interior Pacific Northwest. In: pp. 179-209. Berryman, A.A. (Ed.). Dynamics of forest insect populations. Plenum Press. New York.

Mason, R.R. and B.E. Wickman. 1993. Procedures to reduce landscape hazard from insect outbreaks. In: pp. 19-20. Everett, R.L. (Comp.). Eastside forest ecosystem health assessment–Volume IV: Restoration of stressed sites, and processes. USDA Forest Service, Pacific Northwest Research Station. 114 p.

McDonald, G.I. 1991. Connecting forest productivity to behavior of soil-borne diseases. In: pp. 129-144. Harvey, A.E. and L.F. Neuenschwander (Comps.). Proceedings–Management and productivity of western-montane forest soils. April 10-12, 1990. Boise, ID. USDA Forest Service, Intermountain Research Station. General Technical Report INT-280. 254 p.

Meeuwig, R.O. 1970. Sheet erosion on intermountain summer ranges. USDA Forest Service, Intermountain Research Station. Research Paper INT-85. 25 p.

Megahan, W.F., N.F. Day and T.M. Bliss. 1978. Landslide occurrence in the western and central Northern Rocky Mountains physiographic province in Idaho. In: pp. 116-139. Proceedings of the 5th North American Forests soils conference. Colorado State University. Fort Collins.

Megahan, W.F., L.L. Irwin and L.L. LaCabe. 1993. Forest roads and forest health. In: pp. 91-92. Everett, R. (Comp.) Eastside forest ecosystem health assessment–Volume IV: Restoration of stressed sites, and processes. USDA Forest Service, Pacific Northwest Research Station. 114 p.

Messersmith, C.G. 1989. Leafy spurge morphology and ecology. In: pp. 47-51. Roche, B. and C. Talbot Roche (Eds.). Range weeds revisited: Symposium proceedings, January 24-26, 1989. Spokane, WA. Washington State University, Pullman.

Mitchell, J.E. and D.R. Freeman. 1993. Wildlife-livestock-fire interactions on the North Kaibab: A historical review. USDA Forest Service. General Technical Report RM-222. 12 p.

Mitchell, J.F.B., S. Manabe, T. Tokioka and V. Meleshko. 1990. Equilibrium climate change. In: pp. 131-172. J.T. Houghton, G.J. Jenkins and J.J. Ephraums (Eds.). Climate Change: The IPCC Scientific Assessment. Cambridge University Press. Cambridge, UK.

Moir, W.H. and J.H. Dieterich. 1988. Old-growth ponderosa pine from succession in pine-bunchgrass forests in Arizona and New Mexico. *Natural Areas Journal* 8(1):17-24.

Moir, W.H. and J.A. Ludwig. 1979. A classification of spruce-fir and mixed conifer habitat types of Arizona and New Mexico. USDA Forest Service, Rocky Mountain Forest and Range Experiment Station. Research Paper RM-207. 47 p.

Monnig, E.M. and J.W. Byler. 1992. Forest health and ecological integrity in the Northern Rockies. USDA Forest Service, FPM Report 92-7. 18 p.

Mooney, H.A., T.M. Bonnicksen, N.L. Christensen, J.E. Lotan and W.A. Reiners (Technical Coordinators). Fire regimes and ecosystem properties. Proceedings of the conference held December 11-15, 1978. Honolulu, HI. USDA Forest Service, General Technical Report WO-26. 594 p.

Mutch, R.W., S.F. Arno, J.K. Brown et al. 1993. Forest health in the Blue Mountains: a management strategy for fire-adapted ecosystems. USDA Forest Service, Pacific Northwest Research Station. General Technical Report PNW-GTR-310. 14 p.

O'Laughlin, J. 1994. Assessing forest health conditions in Idaho with forest inventory data. In: pp. 221-247. Sampson, R.N. and D. Adams (Eds.). Assessing Forest Ecosystem Health in the Inland West. Proceedings of the American

Forests Scientific Workshop, November, 14-20, 1993. Sun Valley, ID. The Haworth Press, Inc. New York.

Oliver, C.D., L.L. Irwin and W.H. Knapp. 1993. Eastside forest management practices: historical overview, extent of their application, and their effects on sustainability of ecosystems. In: pp. 217-290. Hessburg, P.F. (Comp.). Eastside forest ecosystem health assessment–Volume III: Assessment. USDA Forest Service, Pacific Northwest Research Station. Portland, OR. 750 p.

Parson, D.J., D.M Graber, J.K. Agee and J.W. van Wagtendonk. 1986. Natural fire management in national parks. *Environmental Management* 19(1):21-24.

Parsons, D.J. and S.H. DeBenedetti. 1979. Impact of fire suppression on a mixed-conifer forest. *Forest Ecology and Management* 2:21-33.

Paysen, T.E., J.A. Derby, H. Black, Jr., V.C. Bleich and J.W. Mincks. 1980. A vegetation classification system applied to southern California. USDA Forest Service, Pacific Southwest Forest and Range Experiment Station. General Technical Report PSW-45. 33 p.

Pearson, G.A. 1950. Management of ponderosa pine in the Southwest. USDA Forest Service, Agriculture Monograph 6. 218 p.

Peek, J.M. 1986. A review of wildlife management. Prentice-Hall. Englewood Cliffs, NJ. 486 p.

Pengelly, W.L. 1976. Probable causes of the recent decline of mule deer in western U.S.–a summary. In: pp. 129-134. Workman, G. and J.B. Low (Eds.). Mule deer decline in the west–a symposium. Utah State University Agriculture Experiment Station. Logan.

Pfister, R.D., B.L. Kovalchik, S.F. Arno and R.C. Presby. 1977. Forest habitat types of Montana. USDA Forest Service, Intermountain Forest and Range Experiment Station. General Technical Report INT-34. 174 p.

Phillips, C.B. 1985. The relevance of past Indian fires to current fire management programs. In: pp. 87-92. Proceedings–Symposium and workshop on wilderness fire. November 15-18, 1983. Missoula, MT. USDA Forest Service, Intermountain Forest and Range Experiment Station. General Technical Report INT-182.

Pickett, S.T.A., J. Kolsa, J.J. Armesto and S.L. Collins. 1989. The ecological concept of disturbance and its expression at various hierarchical levels. *Oikos* 54:129-136.

Platts, W.S. 1984. Compatibility of livestock grazing strategies with riparian-stream systems. In: pp. 67-74. Range watershed riparian zones and economics: interrelationships in management and use short course. Oregon State University. Corvallis.

Plummer, A.P., D.R. Christensen and S.B. Monsen. 1968. Restoring big game range in Utah. Utah Division of Fish and Game, Publication No. 68-3. Publishers Press. Salt Lake City, UT. 183 p.

Polley, H.W., H.B. Johnson, B.D. Marino and H.S. Mayeux. 1993. Increase in C3 plant water use efficiency and biomass over glacial to present CO_2 concentrations. *Nature* 361:61-64.

Provenza F.D. and D.F. Balph. 1987. Diet training: Behavioral concepts and

management objectives. In: Proceedings-Symposium on plant-herbivore interactions. USDA Forest Service, Intermountain Research Station. General Technical Report INT-222.

Pyne, S.J. 1982. Fire in America: a cultural history of wildland and rural fire.
Princeton University Press. Princeton, NJ. 654 p.

Rasmussen, D.I. 1941. Biotic communities of the Kaibab Plateau, Arizona. *Ecological Monograph* 11:229-275.

Reynolds, R.T., R.T. Graham, M.H. Reiser, R.L. Bassett, P.L. Kennedy, D.A.
Boyce, Jr., G. Goodwin, R. Smith and E.L. Fisher. 1992. Management recom-
mendations for the northern goshawk in the southwestern United States.
USDA Forest Service. General Technical Report RM-217. 90 p.

Robbins, W.G. and D.W. Wolf. 1993. Landscape and the intermontane Northwest:
an environmental history. In: pp. 1-34. Hessburg, P.F. (Comp.). Eastside forest
ecosystem health assessment–Volume III: Assessment. USDA Forest Service,
Pacific Northwest Research Station. 750 p.

Robinson, W.L. and E.G. Bolen. 1984. Wildlife ecology and management. Macmillan Publication Co. New York.

Romme, W.H. and D.G. Despain. 1989. Historical perspective on the Yellowstone
fires of 1988. *BioScience* 39:695-699.

Running, S.W. and R.R. Nemani. 1991. Regional hydrologic and carbon balance
responses of forests resulting from potential climate change. *Climatic Change*
19:349-368.

Sando, R.W. 1978. Natural fire regimes and fire management–foundations for
direction. *Western Wildlands* 4(4):34-44.

Sarmiento, J.L., J.C. Orr and U. Siegenthaler. 1992. A perturbation simulation of
CO_2 uptake in an ocean general circulation model. *Journal of Geophysical
Research* 97:3621-3645.

Savage, M. and T.W. Swetnam. 1990. Early 19th-century fire decline following
sheep pasturing in a Navajo ponderosa pine forest. *Ecology* 71:2374-2378.

Schier, G.A. 1975. Deterioration of aspen clones in the middle Rocky Mountains.
USDA Forest Service, Intermountain Research Station. Research Paper
INT-170. 14 p.

Schommer, T. 1991. Analysis of big game statistics, 1965-90. Wallowa-Whitman
National Forest. Part II: deer. USDA Forest Service, Wallowa-Whitman National Forest. Baker City, OR. 59 p.

Schubert, G.H. 1974. Silviculture of southwestern ponderosa pine: the status of
our knowledge. USDA Forest Service. Research Paper RM-123. 71 p.

Seastedt, T.R. and A.K. Knapp. 1993. Consequences of non-equilibrium resource
availability across multiple time scales: the transient maxima hypothesis.
American Naturalist 141:621-633.

Shugart, H.H., Jr. and D.C. West. 1981. Long-term dynamics of forest ecosystems. *American Scientist* 69:647-652.

Sidle, R.C., A.J. Pearce and C.L. O'Loughlin. 1985. Hillslope stability and lands
use. American Geophysical Union. Washington, DC. Water Resource Monograph 11. 140 p.

Smith, E.M. 1983. History of the Boise National Forest 1905-1976. Idaho State Historical Society. Boise. 166 p.

Soulé, M.E. 1985. What is conservation biology. *BioScience* 35:727-734.

Steele, R. 1988. Ecological relationships of ponderosa pine. In: pp. 71-76 Baumgartner, D.A. and J.E. Lotan (Eds.). Ponderosa pine, the species and its management. Cooperative Extension, Washington State University. Pullman.

Steele, R., S.F. Arno and K. Geier-Hayes. 1986. Wildfire patterns change in central Idaho's ponderosa pine–Douglas-fir forest. *Western Journal of Applied Forestry* 1(1):16-18.

Steele, R., S.V. Cooper, D.M. Ondov, D.W. Roberts and R.D. Pfister. 1983. Forest habitat types of eastern Idaho-western Wyoming. USDA Forest Service, Intermountain Forest and Range Experiment Station. General Technical Report INT-114. 138 p.

Steele, R. and R.D. Pfister. 1991. Western-montane plant communities and forest ecosystems perspectives. In: Harvey, A.E. and L.F. Neuenschwander (Comp.). Proceedings–management and productivity of western-montane forest soils. April 10-12, 1990. Boise, ID. USDA Forest Service, Intermountain Research Station. General Technical Report INT-280. 254 p.

Steele, R., R.D. Pfister, R.A. Ryker and J.A. Kittams. 1981. Forest habitat types of central Idaho. USDA Forest Service, Intermountain Forest and Range Experiment Station. General Technical Report INT-114. 138 p.

Strang, R.M. and J.V. Parminter. 1980. Conifer encroachment on the Chilcotin grasslands of British Columbia. *The Forestry Chronicle* 2:13-18.

Swetnam, Thomas W. Fire history and climate in the Southwestern United States. In: pp.6-17. Effects of fire management of Southwestern natural resources. USDA Forest Service General Technical Report RM-191.

Swetnam, T.W. and J.H. Dieterich. 1985. Fire history of ponderosa pine forests in the Gila Wilderness, New Mexico. In: pp. 390-397. Lotan, J.E., B.M. Kilgore, W.C. Fischer and R.M. Mutch (Technical Coordinators). Proceedings–Symposium and workshop on wilderness fire. USDA Forest Service. Intermountain Research Station. General Technical Report INT-182. 434 p.

Swetnam, T.W. and A.M. Lynch. 1989. A tree-ring reconstruction of western spruce budworm in the southern Rocky Mountains. *Forest Science* 35(4): 962-986.

Tande, G.F. 1979. Fire history and vegetation pattern of coniferous forests in Jasper National Park, Alberta. *Canadian Journal of Botany* 57:1912-1931.

Thomas, J.W. (Ed.). 1979. Wildlife habitats in managed forests–the Blue Mountains of Oregon and Washington. USDA Forest Service. Agriculture Handbook No. 553. 512 p.

Thomas, J.W. and D.A. Toweill (Eds.). 1982. Elk of North America–history and management. Stackpole Books, Harrisburg, PA.

Turner, M.G. 1990. Spatial and temporal analysis of landscape patterns. *Landscape Ecology* 4:21-30.

Tyree, M.T. and J.S. Sperry. 1989. Vulnerability of xylem to cavitation and embolism. *Annals of Plant Physiology and Molecular Biology* 40:19-38. U.S. Department of Agriculture. 1890-1990. Census of Agriculture. Washington DC.

U.S. Department of Agriculture. 1985. Oregon's soil: a resource condition report. Soil Conservation Service. Portland, OR. 26 p.

U.S. Department of Agriculture. 1993. Changing conditions in Southwestern forests and implications on land stewardship. Forest Service, Southwestern Region. Albuquerque, NM. 8 p.

U.S. General Accounting Office. 1988. Rangeland management. GAO/RCED-88-80. Washington, DC.

U.S. General Accounting Office. 1991. Rangeland Management: Forest Service not performing needed monitoring of grazing allotments. GAO/RCED-91-148. 8 p.

U.S. Senate. 1936. The western range: a great but neglected natural resource. Executive Document 199, 74th Congress, 2nd Session. 620 p.

Vasechko, G.I. 1983. An ecological approach to forest protection. *Forest Ecology and Management* 5:133-168.

Veblen, T.T. and D.C. Lorenz. 1991. The Colorado front range: a century of ecological change. University of Utah Press. Salt Lake City. 186 p.

Waitt, R.G., Jr. 1985. Case for periodic colossal jokulhlaups from Pleistocene Glacial Lake, Missoula. *Geological Society of America Bulletin* 96(10): 1271-1286.

Wallmo, O.C. (Ed.). 1981. Mule and black-tailed deer of North America. University of Nebraska Press. Lincoln. 605 p.

Walters, C.D. 1986. Adaptive management of renewable resources. MacMillan Publishing Co. New York. 374 p.

Walters, C.J. and C.S. Holling. 1990. Large-scale management experiments and learning by doing. *Ecology* 71:2060-2068.

Weaver, H. 1959. Ecological changes in the ponderosa pine forest of Cedar Valley in southern Washington. *Journal of Forestry* 57:12-20.

Weigand, J.F., R. Haynes, A.R. Tiedemann and R.A. Riggs. [In press]. Ecological and economic impacts of ungulate herbivory on commercial stand development in forests of eastern Oregon and Washington. *Forest Ecology and Management.*

Wellner, C.A. 1970. Fire history in the northern Rocky Mountains. In: pp 42-64. The role of fire in the intermountain west. Symposium proceedings. October 27-29, 1970. Missoula, MT. Intermountain Fire Research Council and University of Montana, Missoula.

Wells, P.V. 1987. Systematics and distribution of pinyons in the late Quaternary. In: Proceedings: Pinyon-Juniper Conference. January 13-15, 1986, Reno, NV. USDA Forest Service, Intermountain Research Station. General Technical Report INT-215. 581 p.

Wertz, W. and J.A. Arnold. 1972. Land systems inventory. USDA Forest Service, Intermountain Regional Office.

White, P.S. 1979. Pattern, process, and natural disturbance in vegetation. *Botanical Review* 45:229-297.

Wickman, B.E. 1992. Forest health in the Blue Mountains: the influence of insects and disease. USDA Forest Service, Pacific Northwest Research Station. General Technical Report PNW-GTR-295. 15 p.

Wickman, B.E., R.R. Mason and T.W. Swetnam. 1993. Searching for long-term patterns of forest insect outbreaks. Proceedings of a conference on individuals, populations, and patterns, September 7-10, 1992. Norwich, UK.

Wigand, P.E. and C.L. Nowak. 1992. Dynamics of northwest Nevada plant communities during the past 30,000 years. In: pp. 40-62. Hall, E.A., Jr., V. Doyle-Jones and B. Widawski (Eds.). The history of water: eastern Sierra Nevada, Owens Valley, White-Inyo Mountains. White Mountain Research Station Symposium, Volume 4. White Mountain Research Station, University of California. Los Angeles.

Williams, B.L. and B.G. Marcot. 1991. Use of biodiversity indicators for analyzing and managing forest landscapes. *Transactions of the North American Wildlife Natural Resources Conference* 56:613-627.

Wissmar, R.C., Smith, J.E., McIntosh, B.A. et al. 1993. Ecological health of river basins in forested regions of eastern Washington and Oregon. In: pp. 415-484. Hessburg, P.F. (Comp.). Eastside forest ecosystem health assessment–Volume III: Assessment. USDA Forest Service, Pacific Northwest Research Station. 750 p.

Workman, G.W. and J.B. Low (Eds.). 1976. Mule deer decline in the west–a symposium. Utah State University Agricultural Experiment Station. Provo. 134 p.

Wright, L.C., A.A. Berryman and B.E. Wickman. 1984. Abundance of the fir engraver, *Scolytus ventralis* and the Douglas-fir beetle, *Dendroctonus pseudotsugae,* following tree defoliation by the Douglas-fir tussock moth, *Orgyia pseudotsugata. Canadian Entomologist* 116:293-305.

Defining and Measuring Forest Health

Jay O'Laughlin
R. Ladd Livingston
Ralph Thier
John Thornton
Dale E. Toweill
Lyn Morelan

ABSTRACT. If forest health is to be approached scientifically, it must be defined and measured. Forest health is a condition of forest ecosystems that sustains their complexity while providing for human needs. We developed this broad definition because a widely acceptable definition is lacking, and forest health is a focal point in discussions of how to sustain forest ecosystems in the United States. Steps for measuring forest health are: (1) select a representative set of indicators for a particular ecosystem; (2) establish baseline data, such as a historical range of variability; (3) develop standards against

Jay O'Laughlin is Director, Policy Analysis Group, College of Forestry, Wildlife and Range Sciences, University of Idaho, Moscow, ID 83844-1134.

R. Ladd Livingston is Supervisor, Insect and Disease Section, and Forest Entomologist, Idaho Department of Lands, P.O. Box 670, Coeur d'Alene, ID 83816-0670.

Ralph Thier is Forest Entomologist, USDA Forest Service, Forest Pest Management, Boise, ID 83702.

John Thornton is Forest Hydrologist, Boise National Forest, Boise, ID 83702.

Dale E. Toweill is Wildlife Program Coordinator, Natural Resources Policy Bureau, Idaho Department of Fish and Game, Boise, ID 83707.

Lyn Morelan is Forest Health Coordinator, Boise National Forest, Boise, ID 83702.

[Haworth co-indexing entry note]: "Defining and Measuring Forest Health." O'Laughlin, Jay et al. Co-published simultaneously in the *Journal of Sustainable Forestry* (The Haworth Press, Inc.) Vol. 2, No. 1/2, 1994, pp. 65-85; and: *Assessing Forest Ecosystem Health in the Inland West West* (eds: R. Neil Sampson and David L. Adams) The Haworth Press, Inc., 1994, pp. 65-85. Multiple copies of this article/chapter may be purchased from The Haworth Document Delivery Center [1-800-3-HAWORTH; 9:00 a.m. - 5:00 p.m. (EST)].

65

which to compare current conditions; and (4) establish a monitoring program to assess current conditions and modify baseline data as new trends develop.

INTRODUCTION

Aldo Leopold (1949) described the need to develop a science of land health, and, in a general fashion, how to begin the task: "Health is the capacity of the land for self-renewal. Conservation is our effort to understand and preserve this capacity . . . A science of land health needs, first of all, a base datum of normality, a picture of how healthy land maintains itself as an organism." Implementing this concept is difficult because Leopold used a nontechnical notion of land health not tied to a specific ecological theory (Hargrove 1992). Few if any scientists recognize land or land-based ecological systems as organisms (Burns 1992), but Leopold's challenge is nonetheless compelling.

It is important to develop a scientific concept of forest health because the term has become part of forest policy dialogue at the national level. At the conclusion of the historic Forest Conference in Portland, Oregon, held on April 2, 1993, President Bill Clinton said " . . . as we craft a [forest management] plan, we need to protect the long-term health of our forests, our wildlife and our waterways."

Can forest health be measured? Although components or elements of a forest ecosystem can be measured, forest health defies precise definition, and is not now a measurable concept. The comparison of ecosystem health to human health can provide useful analogies, as human health is similarly undefinable. We suggest some preliminary approaches that may, in the future, allow scientists to make judgments about forest ecosystem health condition. Until the appropriate indicators of forest ecosystem health have been selected, measured, and evaluated in a social context, objective judgments about forest health cannot be made.

WHAT IS FOREST HEALTH?

By extension of Aldo Leopold's definition, forest health is the capacity of the forest for self-renewal, and forest conservation is our effort to understand and preserve this capacity. Forest health has appeared only lately in forestry literature; Waring (1980) and Smith (1985, 1990) were some of the first forest scientists to use the term. It generally refers to forest decline phenomena that became a subject of scientific inquiry in the 1980s.

Today, in discussions of forest resource management and policy throughout the United States, forest health is frequently mentioned as a component of what some people now refer to as sustainable ecosystems. Ecosystem health concepts are embodied in the term sustainability (Riitters et al. 1990; Costanza 1992). Indeed, in the Eastside Forest Ecosystem Health Assessment (Everett et al. 1993) a healthy forest is defined as a sustainable forest ecosystem. The terms health, ecosystem management, and sustainability defy precise definition–they tend to mean whatever people want them to mean.

Various national policies and programs have been designed to monitor the condition of forest ecosystems, but none have been fully implemented. One is the National Forest Health Monitoring Program of the USDA Forest Service (1992). Another is the U.S. Environmental Protection Agency's Environmental Monitoring and Assessment Program, or EMAP (Hunsaker and Carpenter 1990). EMAP scientists said "No widely accepted definition of forest health exists" (Riitters et al. 1990). Smith (1990) said definitions are critical in any assessment of forest health. Recent literature was reviewed for existing definitions (O'Laughlin et al. 1993) and revealed a general lack of agreement on an acceptable definition. We therefore developed one.

The concept of forest health, however defined, is a useful communication device for relating forest conditions to something people understand, thus attracting their attention to forest ecosystem management problems and inspiring their imagination toward socially desirable solutions. Forest health focuses attention on (a) the *prevention* of socially undesirable forest conditions by integrating the various concerns of protecting the forest from insects, disease, and wildfire in an ecological framework; and (b) the *restoration* of socially desired forest conditions as needed.

Forest Ecosystem Health

The attempt to define forest health necessarily begins by defining the three words in the phrase *forest ecosystem health*. Verbatim definitions selected for their precision and brevity are presented below.

Forest–A forest is an ecosystem characterized by a more or less dense and extensive tree cover. More particularly, a plant community predominantly of trees and other woody vegetation, growing more or less closely together (Society of American Foresters 1983).

Ecosystem–Many definitions of an ecosystem exist. We selected these: (1) any complex of living organisms with their environment, that we isolate mentally for the purposes of study (Society of American Foresters 1983); (2) a set of interacting species and their local, nonbiological environment, functioning together to sustain life (Botkin 1990); and (3) the complex of biotic and abiotic elements interacting over time and space (Everett et al. 1993).

Health–The Random House dictionary (1971) provides four definitions of health. All but one applies to the general condition of the human body and mind. The fourth and broadest definition of health is: "vigor; vitality: *economic health.*"

Forest Ecosystem Health–Taken together, the three preceding definitions lead one to define forest health as the vigor or vitality of interacting biotic and abiotic elements of a system characterized by extensive tree cover that function together to sustain life and are isolated mentally for human purposes. This definition is inclusive, but rather cumbersome. We therefore developed the shorter definition presented in the abstract and at the conclusion of this section.

Is Forest Health Comparable to Human Health?

The health of the human body offers a somewhat useful but limited analogy to forest ecosystem health conditions. Managing for health involves not only treatment of the symptoms but also preventative actions to reduce future disturbances or disruptions of system function. But, as Kimmins (1992) pointed out, unlike the human system that inevitably must die, a forest ecosystem will not die as long as its recovery processes are not destroyed.

The purpose of drawing analogies is usually to further an understanding of the unfamiliar by reference to features of the familiar. The health of organisms and populations of organisms can be understood objectively, but similar reasoning cannot be applied to ecosystems (Calow 1992). The concepts of stress and health were developed for organisms. The benefit of the analogy is to move the almost incomprehensibly complex idea of ecosystem conditions into an understandable context.

Once again the ideas of Aldo Leopold (1949) have been instrumental in developing the concept of forest health as related to human health:

> In general, the trend of evidence indicates that in land, just as in the human body, the symptoms may lie in one organ and the cause in another. The practices we now call conservation are, to a large extent, local alleviations of biotic pain. They are necessary but they must not be confused with cures. The art of land doctoring is being practiced with vigor, but the science of land health is yet to be born.

As Leopold drew the analogy, he urged the development of a new science of land health. Forty years later, his call is now being answered, and existing knowledge is being viewed in a different and broader context.

Both the human body and forests are complex systems, composed of many parts carrying on various functions that are essential to the well-being of the whole. When these systems are threatened, we must administer aid to protect the potential for long and productive lives (Waring 1980). This implies some type of management.

Health Management Analogy

Health may be thought of as the absence of conditions that result from disease or other known stresses (Palmer et al. 1992; Rapport 1992). The modern thinking on human health is to prevent problems before they occur by adopting a healthful lifestyle, rather than diagnosing and treating problems as they occur. This approach is currently being applied in forest pest management–development and maintenance of healthy forests by managing them on sound ecological principles and preventing problems before they occur (Space 1992).

A goal of forest health management is to avoid conditions that promote potentially extreme insect and disease outbreaks and wildfire damages. Because insects, diseases, and wildfire are natural components of forest ecosystems, the avoidance of unhealthy conditions is perhaps a more practical management approach than the attainment of healthy conditions.

Forest scientists lack complete knowledge of how forest ecosystems function, just as physicians lack complete knowledge of how the human body functions. Yet when we become ill we expect treatment based on what medical science does know. Because we lack complete knowledge, should we do nothing until we know everything about a particular ailment or health situation? Some people think so when it comes to managing forests. They believe nature knows best.

Some representatives of environmental groups have expressed doubts that forest health is an appropriate management objective, particularly if it involves logging, even if trees are dead or dying. Their concern is that the cure may be worse than the disease and that nature knows best how to cure ailing forests (Blatner et al., 1994). In medical science, a similar idea is called therapeutic nihilism. This was a form of medical practice in the mid-19th century where doctors concluded that the limits of medical knowledge had been reached and that treatment of a wide range of illnesses actually reduced patients' quality of life and sometimes seriously endangered their lives–that is, the cure was worse than the disease (Hargrove 1992).

The notion of therapeutic nihilism found its way into professional environmental management when problems with the scientific management of natural systems arose in the 1930s and 1940s. One example was the deer irruptions on the Kaibab plateau that led Aldo Leopold to develop some of his ecological ideas. Therapeutic nihilism, or nature knows best, is embodied in the National Park Service policy of natural regulation. It is also a belief commonly held by most environmentalists and the general public (Hargrove 1992).

Does nature know best? Nature and natural regulation provide what some people desire from a forest, but not necessarily everyone. In this context, ecologist Daniel Botkin (1993a) said, "When you do nothing, you'll get something you didn't expect." That unexpected something may or may not be what you want.

Two Approaches to Forest Health

What do people want from forests? It is difficult to determine what mix of goods and services forests should provide. Because forests can provide many things, debates over what they should provide are at the core of all forest management policy debates (Cubbage et al. 1993). Everyone desires healthy forests. How can that be accomplished in light of conflicting human needs?

As a starting point, two approaches are available for considering whether or not a forest is healthy. Both are necessary considerations, leading to the development of important ideas that need to be captured in a complete definition of forest health. The first approach is to focus on forest management objectives, the second is to focus on forest ecosystem function. Judgments about forest health need to consider both ecosystem function and management objectives (Monnig and Byler 1992).

Objective-Oriented Approach–A management objective-oriented approach has led the U.S. Forest Service (1988, 1993a) to this definition: "an unhealthy forest inhibits managers from achieving objectives; a healthy forest does not pose such obstacles." It follows that a healthy forest may not be insect-free or pathogen-free, but sufficiently free of pest damage to meet management objectives (Byler and Zimmer-Grove 1991). Furthermore, a forest can be maintained in such condition that it will meet the objectives of future generations, which may be different than those of today and require maintaining various options.

There are two challenges in this definition. First, management to achieve objectives requires a clear and explicit statement of those objectives so the managers know whether they are on target to meet the objectives. Much of the forest policy debate about public forest management stems from disagreement over management objectives. Second, objectives must reflect limitations posed by ecosystem characteristics or properties.

Ecosystem-Oriented Approach–Ecosystems are comprised of various components. Some components might be in a healthy or sustainable condition while others may not be. Ecosystems are dynamic. It is therefore conceptually difficult to think of components being in some kind of equilibrium or balance. A more feasible approach is to protect desirable ecosystem properties. Resistance and resilience are properties that enable the system to persist in many different states or successional stages. Resilience is the ability of the ecosystem to respond to disturbances. The concept of resilience as it is commonly used incorporates the idea of resistance to disturbance as well as the period of time it takes a system to respond. Botkin (1993b) has suggested persistence as a better term to describe this ecosystem property, but it is perhaps too late to discard the widely-used notion of resilience. Whatever it is called, this ecosystem property needs to be protected to ensure ecosystem sustainability.

Forest Health Management

Selecting a preferred ecosystem state as a management objective for a particular forest is a subjective undertaking, and difficult (perhaps impossible) because change is a natural property of ecosystems. To produce a healthy forest, managers need to expand their vision.

Professionals involved with forest resources are driven by a variety of concerns implied by the terms health, ecosystem management, and sustainability, to move from stand management to a broader view that manages across the landscape. These concerns include current issues such as deforestation, habitat loss, air and water pollution, global climate change, damage from a variety of forest insects and diseases, and management practices.

The change in resource management philosophy (some call it a revolution) is a shift from sustained yield–usually expressed in terms of outputs–to sustainability–often expressed in terms of a forest condition or outcomes. Management focus is shifting to a more inclusive view of what remains in the forest ecosystem after management activities, rather than primarily on what goods and services are produced by those activities. Forest health enters the discussion to the extent that ecosystem management is designed to leave a "healthy" forest. The dividing line between healthy and unhealthy, however, remains elusive.

Forest Health is a Complex Subject

In its strategic plan for forest health, the USDA Forest Service (1988) stated,

Forest health is a complex subject with both real and perceived problems which can arouse strong emotions. The actual problems are the product of events occurring over a long period of time. The perceived problems reflect an incomplete understanding of forest ecosystems, the biological processes operating within them, and alternative views of the purposes to be served by the forest.

When the concept of health is imposed on a forest, the complexity of the system must be considered. Otherwise, important ecosystem aspects may be overlooked. Aspects that must be considered include the spatial and temporal settings which are reflected in the diversity of the forest. Some spatial aspects, such as distribution of old-growth forest vegetation, may apply to the entire forest but others, such as distribution of riparian areas, can be extremely limited. Likewise, a forest changes over time. For example, a forest supports different vegetation in an early successional stage than it does later when it reaches a climax condition.

Characteristics of the forest ecosystem can be scientifically measured and compared to indicate some tangible aspects of forest health, but forest health is a human perception; hence, the concept must also consider human values. This idea was expressed in the National Forest Health Monitoring Plan (USDA Forest Service 1992): "Although forest condition can be specified and measured objectively, forest health carries an element of subjectivity, as it is a value judgement."

Because forest health is a human perception, the concept defies a definition acceptable to everyone. This is exemplified by the multitude of definitions in use (O'Laughlin et al. 1993). We offer the following definition, which reflects the tangible aspects of forest ecosystems and recognizes human needs:

Forest health is a condition of forest ecosystems that sustains their complexity while providing for human needs.

CAN FOREST HEALTH BE MEASURED?

We will not know whether the health of forest ecosystems is improving, stable, or declining until we become serious about long-term forest health assessment (Smith 1990). But can we do the assessment? Science has yet to devise methods for measuring ecosystem health, and perhaps never will. Ecosystems do not die in the sense that an individual organism or a population of organisms do; rather, ecosystem characteristics and properties change. Although objective indicators of forest ecosystem characteristics can be specified and measured, forest health is subjective as it is based on value judgements. Some problems associated with these value judgments can be overcome by selecting representative ecosystem parameters to measure.

The measurement of forest ecosystem health depends first on the selection of indicators. They are defined and described in the next section. Measures of indicators can be used to develop baseline data for the forest ecosystem under consideration, allowing the use of such data to develop standards. Monitoring can then be employed to collect data on current conditions for comparison against the standard.

Based on Aldo Leopold's writings, Monnig and Byler (1992) said forest health is best measured by how its patterns and rates of change compare to historic patterns. Specific components of a forest ecosystem can be selected and monitored with standard sampling procedures. Through time, a set of baseline data can be developed that will illustrate presence/absence, relative abundance, distribution, and patterns of change for these indicators. This information, coupled with historical information from other sources, might be viewed as representing an expected range of variability. By comparing current data with the patterns established through time, it is possible to see if they are within or outside the historic range of variability.

By analyzing information gathered from monitoring specific indicators, it becomes possible to make judgments about the status of the health of the forests. These judgements are necessarily subjective, as each participant in the decision making process comes endowed with his or her own specific set of needs and objectives. For example, a site where trees are killed by root disease and replaced with shrub species is seen as unhealthy by those whose objective is wood fiber production or protection from wildfire, while it may be judged healthy by those interested in having cover and browse for deer, or in increasing biodiversity. Indicators of conditions, such as the presence/absence of root disease, can be measured and monitored, but the inherent subjectivity of judgments about health will remain until a representative set of indicators has been selected and standards established.

CHOICE OF INDICATORS

The first step in measuring ecosystem health is identification of relevant indicators and acceptable ranges of parameters. Forty years of ecological research have identified many such variables and parameters, but there are several problems in selecting an appropriate set of them. First, each ecosystem has its own set of indicators and endpoints, and each must be assessed separately. Second, indicators and variables must be sufficiently dynamic to measure ecosystem change. Third, each scientist evaluating the ecosystem will choose variables depending on his or her specific interest and expertise (Haskell et al. 1992).

Many different categories of ecosystem components need to be considered, depending on the composition of a particular forest ecosystem. Selection of a representative set of indicators depends on three key factors. Indicators should be (1) closely related to parameters of interest, (2) effectively measurable, and (3) little affected by other parameters. While it is difficult to generalize about indicator selection because of the complex nature of ecosystems, indicators discussed in this section have been selected from hundreds of candidates and widely applied. Riitters et al. (1990) suggested that someday we may be able to identify a set of indicators that relates to everyone's perception of forests.

Environmental Monitoring and Assessment Program (EMAP)

The Environmental Protection Agency has designed an Environmental Monitoring and Assessment Program (EMAP) to monitor status and trends in the condition of major ecological resources throughout the nation (Norton and Slonecker 1990). EMAP scientists have identified seven ecological resource classes for monitoring. Among them are forests and inland surface waters. An almost unlimited number of potential measurements of forests could be considered for monitoring. Numeric criteria based on experience and expert judgement are available for forests, but these are difficult to justify objectively. The EMAP program will use "response indicators" to quantify and classify the condition of forests. These indicators may be used to identify forests in "subnominal" condition.

An indicator is defined by EMAP scientists (Hunsaker and Carpenter 1990) as a characteristic of the environment that, when measured, quantifies the magnitude of stress, habitat characteristics, degree of exposure to the stressor, or degree of ecological response to the exposure. Indicators that represent ecosystem characteristics can be identified and quantitatively measured. The EPA recognizes three broad categories of indicators:

1. *Response Indicator*–A characteristic of the environment measured to provide evidence of the biological condition of a resource at the organism, population, community, or ecosystem process level of organization.

2. *Exposure and Habitat Indicators*–Diagnostic indicators measured in conjunction with response indicators.

- *Exposure Indicator*–A characteristic of the environment measured to provide evidence of the occurrence or magnitude of a response indicator's contact with a physical, chemical, or biological stress.
- *Habitat Indicator*–A physical attribute measured to characterize conditions necessary to support an organism, population, or community in the absence of pollutants.

3. *Stressor Indicator*–A characteristic measured to quantify a natural process, an environmental hazard, or a management action that effects changes in exposure and habitat.

Certain indicators are appropriate to represent the forest ecosystem in the context of forest health. EPA scientists recognize three response indicators as having high priority status in forests: (1) visual symptoms of foliar damage, (2) tree growth efficiency, and (3) relative abundance of animals. It is important to note that none of them are ready for implementation as health criteria (Riitters et al. 1990).

Indicators for forest ecosystems can be subdivided into the seven categories described below. In some categories, individual indicators are identified. In others, a general description is provided.

Soil

Soil is a medium on the Earth's surface which has both abiotic and biotic characteristics and features. The interaction of the soil with other physical and biological processes results in the abundance and diversity of plants and animals within an ecosystem. Three indicators follow.

Microbial Activity–Soil microorganisms play a vital role in the retention and release of nutrients and energy in forested soils. Nutrient and energy fluxes influence the biological and chemical activity of an ecosystem and are sensitive to structural changes in that system (Riitters et al. 1990).

Litter Dynamics–Large quantities of nutrients circulate within a forest ecosystem. Although part of the annual requirements of plants can be met by reabsorption before the loss of tissues, the remaining nutrients must be obtained by uptake from the soil. Most soil nutrients are derived from the decomposition of organic litter, including woody material, insect frass, and fallen leaves. Thus litter dynamics, such as the rate and pathways of decomposition, are important determinants of ecosystem productivity and condition (Riitters et al. 1990).

Soil Productivity Index–Soil productivity is usually defined as the capacity of a given volume of soil to produce a vegetative response under a specified system of management (Riitters et al. 1990).

Water

Water is ubiquitous and essential in all life forms while being spatially and temporally limiting in many ecosystems. The quality of water flowing through the landscape is a barometer of ecosystem health. Three indicators follow.

Macroinvertebrate Counts–The presence/absence and the abundance of certain aquatic invertebrates can provide a measure of health and complexity. This measure can be used to compare similar aquatic habitat in an ecological region. These organisms are sensitive to physical, chemical, and biological stressors; i.e., they can be used to assess taxonomic and trophic groups (including sensitive and tolerant species), and data can be aggregated to integrate species composition data into an index useful for interpretation of community abundance and condition and ecosystem health (Hughes and Paulsen 1990).

Routine Water Chemistry–This refers to identification and evaluation of a number of conventional chemicals and nutrients, particularly those of importance to aquatic life. Parameters are used to calculate trophic state and to assess such processes as eutrophication, acidification, and salinity (Hughes and Paulsen 1990).

Physical Habitat Features–Hydrological and physical changes including sediment, large woody debris, temperature, streambank stability, and in-stream flows can be measured to assess both morphological condition of stream or lake beds and banks and to assess suitability of these areas for spawning, rearing, and feeding by biota (Hughes and Paulsen 1990).

Vegetation

Woody plant vegetation is the defining characteristic of a forest ecosystem. Two vegetation indicators have been identified as high priority response indicators of forests–visual symptoms of foliar damage and tree growth efficiency. A third indicator, understory vegetation, is a lower priority response indicator (Riitters et al. 1990).

Visual Symptoms of Foliar Damage–These are measures of the health of individual trees and populations in terms of pathological conditions, and they are measures of aesthetic quality. Foliar damage may range from discoloration to complete loss of foliage resulting from insects, disease, fire, or other disturbances that either weaken or kill the tree (Riitters et al. 1990). Foliar damage indicators, however, do not address shifting patterns of vegetative species composition or density, both of which are related to forest health.

It is also difficult to relate the number of trees killed or the extent of a defoliated area to the actual impact on forest health. This has led environmental groups to question the severity of insect damage. For example, Gast et al. (1991) reported that 53% of the forested area in the Blue Mountains national forests had been affected by defoliating insects, and provided no other measure of impact. Aplet (1992), forest ecologist with the Wilderness Society, expressed difficulty interpreting what this statement implied about forest health management. Indeed, the Forest Service subsequently reported (Associated Press 1993) that precipitation has enabled portions of the Blue Mountain forests to recover, and defoliation during drought conditions may have spared many trees from death by causing them to become dormant during a period of stress.

When do visual symptoms of foliar damage suggest that a forest health problem exists? There is not yet any standard classification of crown density, crown vigor, and foliage for a tree or forest stand primarily because the visual method of crown ratings is quite subjective (Anderson and Belanger 1986). Applying crown vigor or other measure of canopy density to indicate forest health cannot be implemented now (Riitters et al. 1990). But given time and effort these problems can be overcome.

Tree Growth Efficiency–This is a measure of the overall ability of trees to maintain themselves in an ecosystem, which is an obvious but sometimes overlooked condition for the perpetuation of forests. Measurements of periodic tree dimensional or biomass growth can be used with an index for the capacity of growth to construct a vegetative index. Literature indicates that certain levels of tree growth efficiency are associated with the probability of insect attack and mortality. These starting points can be refined with additional experience or research (Riitters et al. 1990).

Given adequate resources, tree growth efficiency can be measured, either at the individual tree level or at the stand level. For forest-wide health estimates, a more practical approach may be to use the growth and mortality data collected by the USDA Forest Service. Tree growth and mortality can be dramatically influenced by endemic and epidemic pest populations. Norris et al. (1993) suggested the relationship of mortality to growth as an indicator of forest health. O'Laughlin (this volume) used existing tree growth and mortality data to assess forest health conditions in Idaho.

Understory Vegetation–This can be a sensitive indicator of forest responses to environmental stresses. Measures of the amount and distribution of various life-forms can be summarized into useful indices. Two general parameters, species composition and abundance, should be measured to develop a quantitative assessment of ecological status. The exact methods will depend on the community type that is to be sampled (Riitters et al. 1990).

Animals

Vertebrate and invertebrate animals typically occupy the higher trophic levels of forest ecosystems, making them useful indicators of forest ecosystem condition. A major concern in using animals as indicators lies in correctly matching the organism to the landscape scale of interest, since virtually all animals are mobile to a greater or lesser degree. Some species, such as neotropical birds, may range so widely that they may be used as indicators of entire ecosystems, although they may also prove to be useful indicators at the stand level within selected timeframes.

A particular value in using animals as indicators lies in their widespread appeal to many segments of the public, resulting in a large volume of data collected in the same manner across broad geographic areas.

Relative Abundance–The most direct and easily understood application using animals as indicators is a measurement of relative abundance. This indicator ranges from presence/absence to precise census of entire populations within some defined area. Most appropriately, confirmation of presence plus some index to number of individuals per unit area over a defined period of time is employed to measure relative abundance. Values for individual species can be aggregated or manipulated in many ways to derive estimates relating to forest ecosystem characteristics, such as site-specific measures of diversity, species richness, species evenness, and others (Patton 1992).

Demographics–Population demographics provide an indicator of potential sustainability of ecosystems. Mobile species often require particular habitat features for species life functions (i.e., feeding, security, reproduction). Measurements of population productivity or other demographic indicators (e.g., age structure) may indicate the abundance and/or distribution of such critical habitat features, and with sufficient base-line data, may indicate changes in these features through time.

Morphological Asymmetry–Most animal species are constructed on a body framework of bilateral symmetry. Asymmetry in body form is often an expression of concentration of toxins or pollutants at high levels in the trophic chain (Riitters et al. 1990).

Physical Habitat Features–Physical habitat features, especially those known to provide critical habitat for particular life stages of specific organisms, provide a stationary indirect index to organism distribution. Physical habitat features should be assessed both in terms of presence/absence, number of sites, and distribution across the landscape (Patton 1992). Distribution and relative proximity of these features to each other and the presence of corridors allowing animals to move between separated sites (Harris 1984; Patton 1992; e.g., Erickson and Toweill 1994) must be scaled to the specific organism(s) under consideration. Physical habitat features can include topographical features (e.g., Irwin 1994).

Ecosystem Cycling

The cycling process of chemical elements within an ecosystem is fundamental to the maintenance of its components. Organisms depend on the availability of some 20 elements of which all are required for life processes. The effects on accelerated removal or addition of contaminants on nutrient availability ultimately translates into effects on many other aspects of ecosystem structure and function (Campbell et al. 1990).

Landscape Pattern

Landscape pattern addresses a larger area than some of the other categories or indicators. The indicators under landscape pattern by definition addresses a broad landscape which could range from thousands of acres to over several hundred thousands of acres.

Habitat Proportions (Cover Types)–Determining proportions of various vegetation cover types in a landscape is a basic measurement when considering both extent and change in vegetation and associated animal composition and diversity (Harris 1984; Carpenter et al. 1990). Specific vegetation cover types or characteristics, such as riparian areas and wetlands, need to be considered at the landscape level as they provide special habitats associated with animal composition and diversity. Other indicators mentioned by the EPA that relate to habitat proportions are habitat patchiness and patch size (Carpenter et al. 1990).

Landform–Landform is generally described in a hierarchical framework including physical characteristics (e.g., mountains, valleys, plateaus, etc.) and their formative processes. Information on parent materials and their durability is a consideration. Landform directly influences other indicators, such as water, vegetation, and animals.

Non-Native Plants and Animals

The introduction, even in small numbers, of non-native, introduced animal or plant organisms can pose a threat to forests. They usually arrive without their normal complement of biocontrol agents. Without these agents, the introduced species may spread rapidly. If they impact native species, the resulting damage can be extensive. The presence of any introduced species should be evaluated carefully, as it may result in an unhealthy forest condition.

TOWARD COHERENT MONITORING

A monitoring program is needed to identify changes in conditions. After key parameters or indicators have been identified, their historic range should be determined. This will provide a reference point for the establishment of comparison standards. Medical practitioners routinely consult published tables giving normal values for key physiological attributes. Ecologists are hindered by the scarcity of such information, but the concept of historic range of variability (Morgan et al. 1994) can be used to assist in the development of standards or indices for assessing forest health.

In their review of forest health monitoring strategies, Everett et al. (1993) pointed out that current monitoring is often piecemeal rather than part of a coherent strategy for all species, processes, and ecosystems. Because forest ecosystems depend on a complex relationship of abiotic and biotic elements, monitoring should integrate selected elements or indicators with societal values to assess effectiveness in conserving biological diversity, long-term site productivity, and the sustainability of producing resources for human needs (Everett et al. 1993).

Hunter (1990) proposed a coarse filter approach to ecosystem monitoring. That approach promotes biodiversity by maintaining ecosystem structure and components as well as associated species and processes within the landscape. To supplement the coarse filter strategy, Everett et al. (1993) recommended a three-part fine-filter monitoring strategy focused on: (1) landscape structure and composition; (2) threatened, endangered, and sensitive species and unique habitats; and (3) disturbance regimes, disturbance effects, and hazards.

This monitoring approach could allow for more efficiency and effectiveness in data collection, analysis, and interpretation when based on a hierarchical landscape system such as the National Hierarchy of Ecological Units For Ecosystem Classification (USDA Forest Service 1993b). The various scales of ecological landscape units can be nested into such groups as: potential vegetation, soils, landform, hydrology, biological life and disturbance patterns. The physical and biological processes and responses of nested units to disturbance are similar. This allows extrapolation of data and assists in determining effects on societal values.

Forest inventories have historically been conducted at an extensive scale for the purpose of measuring overall conditions and trends of timber resources, but do little to reflect the needs of ecosystem management. More comprehensive and intensive inventories are necessary to include commodity and non-commodity values and estimates of changes in forest health, biodiversity, and sustainability. With detailed ecosystem characterization of stands, streams, habitats, and landscapes, improved forest planning and decisions will result (Norris et al. 1993).

Because ecosystems are complex, one recommendation of Everett et al. (1993) has much merit: a short list of important monitoring variables or indicators should be identified to accomplish representative and cost-effective monitoring. Until that is done, monitoring will continue to be piecemeal. Without a coherent monitoring program we will not know whether the health of forest ecosystems is improving, stable, or declining.

CONCLUSION

The concept of forest health is elusive yet useful. Forest health provides a medium for discussion of forest conditions relative to human needs and a framework for measurement of ecosystem indicators. Individuals may come to similar conclusions about the condition of one ecosystem component based on an objectively measured indicator, but the value-based aspects of forest health, which cannot be objectively measured, will be repeatedly debated.

We conclude with this summary of major points:

- Forest health is a condition of forest ecosystems that sustains their complexity while providing for human needs. It is a useful communications device in helping people understand the current condition of the forest.
- Forest scientists and managers, working with their customers, can identify, define, and determine ranges of desired conditions for a set of measurable characteristics in each forest ecosystem that can be useful in helping evaluate the condition of the forest at any time, in relation to those desired conditions.
- Objective indicators of forest ecosystem condition can be specified and measured, but forest health assessments contain subjective value judgments which must be clearly recognized.
- Comprehensive and intensive inventories of a short list of indicators representing commodity and non-commodity values will improve

forest health assessments, as well as forest planning and management decisions, by enabling understanding of ecosystem characteristics of stands, habitats, streams, and landscapes.

* As is true in other "health" contexts, it may be easier to identify when a forest is experiencing an "unhealthy" condition in one or more aspects than it is to define exactly what "healthy" means.

REFERENCES

Anderson, R.L. and R.P. Belanger. 1986. A crown rating method for assessing tree vigor of loblolly and shortleaf pines. In: pp. 538-543. Proceedings, Fourth Biennial Silvicultural Research Conference. USDA Forest Service General Technical Report SE-42. Asheville, NC.

Aplet, G. 1992. Forest health: Ecological crisis or timber driven hype? *Forest Watch* 13(2): 19-22. Cascade Holistic Economic Consultants. Portland, OR.

Associated Press. 1993. [Blue Mountains] forests recovering from drought, budworms. *Lewiston Morning Tribune,* (December 8): 7C. Lewiston, ID.

Blatner, K.A., C.E. Keegan III, J. O'Laughlin and D.L. Adams. 1994. Forest health management and policy: A case study in southwestern Idaho. In: pp. 317-337. Sampson, R. N. and D. L. Adams (Eds.). Assessing Forest Ecosystem Health in the Inland West West. Proceedings of the American Forests scientific workshop, November 14-20, 1993. Sun Valley, ID. The Haworth Press, Inc. New York.

Botkin, D.B. 1990. Discordant Harmonies: A New Ecology for the Twenty-first Century. Oxford University Press. New York. 241 p.

Botkin, D.B. 1993a. Quoted in interview by W. Kaufmann. How nature really works: A gaggle of "new ecologists" is debunking the sacred commandments of environmentalism and working to bring order to nature's chaos. *American Forests* 99(3/4): 17-19, 59-61.

Botkin, D.B. 1993b. Fire and stability in forest ecosystems. Keynote address, Assessing Forest Ecosystem Health in the Inland West: A Scientific Workshop. November 14-20, 1993. Sun Valley, ID.

Burns, T.P. 1992. Ecosystem: A powerful concept and paradigm for ecology. *Bulletin of the Ecological Society of America* 73(1): 39-43.

Byler, J.W. and S. Zimmer-Grove. 1991. A forest health perspective on interior Douglas-fir management. In: pp. 103-108. Baumgartner, D.M. and J.V. Lotan (Eds.). Interior Douglas-fir and its Management. Washington State University Cooperative Extension. Pullman.

Calow, P. 1992. Can ecosystems be healthy? Critical consideration of concepts. *Journal of Aquatic Ecosystem Health* 1(1): 1-5.

Campbell, C.L., W. Heck and T. Moser. 1990. Indicator strategy for agroecosystems. In: pp. (8)1-11. Hunsaker, C.T. and D.E. Carpenter (Eds.). Ecological indicators for the Environmental Monitoring and Assessment Program. U.S. Environmental Protection Agency, EPA 600/3-90/060, Office of Research and Development. Research Triangle Park, NC.

Carpenter, D.E., C.T. Hunsaker and R.F. Noss. 1990. Appendix G-1: Indicator fact sheets for animals. In: pp. (G)1-5. Hunsaker, C.T. and D.E. Carpenter (Eds.). Ecological indicators for the Environmental Monitoring and Assessment Program. U.S. Environmental Protection Agency, EPA 600/3-90/060, Office of Research and Development. Research Triangle Park, NC.

Costanza, R. 1992. Toward an operational definition of ecosystem health. In: pp. 239-256. Costanza, R. et al. (Eds.). Ecosystem Health: New Goals for Environment Management. Island Press. Washington, DC.

Cubbage, F.W., J. O'Laughlin and C.S. Bullock, III. 1993. Forest Resource Policy. John Wiley & Sons. New York. 562 p.

Erickson, J.R. and D.E. Toweill. 1994. Forest health and wildlife habitat management on the Boise National Forest. In: pp. 389-409. Sampson, R.N. and D.L. Adams (Eds.). Assessing Forest Ecosystem Health in the Inland West West. Proceedings of the American Forests scientific workshop, November 14-20, 1993. Sun Valley, ID. The Haworth Press, Inc. New York.

Everett, R., P. Hessburg, M. Jensen, B. Bormann, P.S. Bourgeron, R.W. Haynes, W.C. Krueger, J.F. Lehmkuhl, C.D. Oliver, R.C. Wissmar and A.P. Youngblood. 1993. Eastside forest ecosystem health assessment, Volume I: Executive summary. USDA Forest Service unnumbered report. National Forest System and Forest Service Research. Portland, OR. 57 p.

Gast, W.R., Jr., D.W. Scott, C. Schmitt, D. Clemens, S. Howes, C.G. Johnson, R. Mason, F. Mohr and R.A. Clapp, Jr. 1991. Blue Mountains forest health report: New perspectives in forest health. USDA Forest Service unnumbered publication, Pacific Northwest Region, Malheur, Umatilla, and the Wallowa-Whitman National Forests.

Hargrove, E.C. 1992. Environmental therapeutic nihilism. In: pp. 124-131. Costanza, R. et al. (Eds.). Ecosystem Health: New Goals for Environment Management. editors). Island Press. Washington, DC.

Harris, L.D. 1984. The Fragmented Forest: Island Biogeography Theory and the Preservation of Biotic Diversity. University of Chicago Press, Chicago, IL. 211 p.

Haskell, B.D., B.G. Norton and R. Costanza. 1992. What is ecosystem health and why should we worry about it? In: pp. 3-19. Costanza, R. et al., (Eds.). Ecosystem Health: New Goals for Environmental Management. Island Press. Washington, DC.

Hughes, R.M. and S.G. Paulsen. 1990. Indicator strategy for inland surface waters. In: pp. (4)1-20. Hunsaker, C.T. and D.E. Carpenter (Eds.). Ecological indicators for the Environmental Monitoring and Assessment Program. U.S. Environmental Protection Agency, EPA 600/3-90/060, Office of Research and Development. Research Triangle Park, NC.

Hunsaker, C.T. and D.E. Carpenter. (Eds.). 1990. Ecological indicators for the Environmental Monitoring and Assessment Program. U.S. Environmental Protection Agency, EPA 600/3-90/060, Office of Research and Development. Research Triangle Park, NC.

Hunter, M.L., Jr. 1990. Wildlife, Forests, and Forestry: Principles of Managing Forests for Biological Diversity. Prentice Hall. Englewood Cliffs, NJ. 370 p.

Irwin, L.L. 1994. A process for improving wildlife habitat models for assessing forest ecosystem health. In: pp. 293-306. Sampson, R.N. and D.L. Adams (Eds.). Assessing Forest Ecosystem Health in the Inland West West. Proceedings of the American Forests scientific workshop, November 14-20, 1993. Sun Valley, ID. The Haworth Press, Inc. New York.

Kimmins, J.P. 1992. Balancing Act: Environmental Issues in Forestry. University of British Columbia Press. Vancouver. 244 p.

Leopold, A.S. 1949. A Sand County Almanac, and Sketches Here and There. Oxford University Press. New York. 226 p.

Monnig, E. and J. Byler. 1992. Forest health and ecological integrity in the northern Rockies. USDA Forest Service FPM Report 92-7, second edition. R1-92-130, Northern Region. Missoula, MT. 18 p.

Morgan, P. et al. 1994. Historic range of variability: A useful tool for evaluating ecosystem change. In: pp. 87-111. Sampson, R.N. and D.L. Adams (Eds.). Assessing Forest Ecosystem Health in the Inland West West. Proceedings of the American Forests scientific workshop, November 14-20, 1993. Sun Valley, ID. The Haworth Press, Inc. New York.

Norris, L.A., H. Cortner, M.R. Cutler, S.G. Haines, J.E. Hubbard, M.A. Kerrick, W.B. Kessler, J.C. Nelson, R. Stone and J.M. Sweeney. 1993. Sustaining long-term forest health and productivity. Task force report, Society of American Foresters. Bethesda, MD. 83 p.

Norton, D.J. and E.T. Slonecker. 1990. The ecological geography of EMAP. *GeoInfo Systems* (November/December): 33-43.

O'Laughlin, J. 1994. Assessing forest health conditions in Idaho with forest inventory data. In: pp. 221-247. Sampson, R.N. and D.L. Adams (Eds.). Assessing Forest Ecosystem Health in the Inland West West. Proceedings of the American Forests scientific workshop, November 14-20, 1993. Sun Valley, ID. The Haworth Press, Inc. New York.

O'Laughlin, J., J.M. MacCracken, D.L. Adams, S.C. Bunting, K.A. Blatner and C.E. Keegan, III. 1993. Forest health conditions in Idaho. Report no. 11, Idaho Forest, Wildlife and Range Policy Analysis Group, University of Idaho. Moscow.

Palmer, C.J., K.H. Riitters, T. Strickland, D.L. Cassell, G.E. Byers, M.L. Papp and C.I. Liff. 1992. Monitoring and research strategy for forests—Environmental Monitoring and Assessment Program (EMAP). U.S. Environmental Protection Agency, EPA/600/4-91/012. Washington, DC.

Patton, D.R. 1992. Wildlife Habitat Relationships in Forested Ecosystems. Timber Press. Eugene, OR. 392 p.

Random House. 1971. Dictionary of the English Language, unabridged edition. Random House. New York. 2059 p.

Rapport, D.J. 1992. What is clinical ecology? In: pp. 144-156. Costanza, R. et al. (Eds.). Ecosystem Health: New Goals for Environmental Management. Island Press. Washington, DC.

Riitters, K.H., B. Law, R. Kucera, A. Gallant, R. DeVelice and C. Palmer. 1990. Indicator strategy for forests. In: pp. (6)1-13. Hunsaker, C.T. and D.E. Carpenter (Eds.). Ecological indicators for the Environmental Monitoring and Assessment Program. U.S. Environmental Protection Agency, EPA 600/3-90/060, Office of Research and Development. Research Triangle Park, NC.

Smith, W.H. 1985. Forest quality and air quality. *Journal of Forestry* 83(2): 83-92.

Smith, W.H. 1990. The health of North American forests: Stress and risk assessment. *Journal of Forestry* 88(1): 32-35.

Society of American Foresters. 1983. Terminology of forest science, technology, practice, and products. Society of American Foresters, Bethesda, MD. 370 p.

Space, J.C. 1992. The future of forest pest management. In: pp. 14-16. Allen, D.C. and L.P. Abrahamson (Eds.). Proceedings, North American Forest Insect Work Conference. USDA Forest Service General Technical Report PNW-GTR-294. Portland, OR.

USDA Forest Service. 1988. Forest health through silviculture and integrated pest management–A strategic plan. Unnumbered publication. Washington, DC. 26 p.

USDA Forest Service. 1992. National plan: Forest pest management and associated state component, National Forest Health Monitoring Program. USDA Forest Service mimeo. Washington, DC. 14 p.

USDA Forest Service. 1993a. Healthy forests for America's future: A strategic plan. MP-1513. Washington, DC. 58 p.

USDA Forest Service. 1993b. National hierarchical framework of ecological units for ecosystem classification. Memorandum, June 29. Washington, DC. 3 p. + enclosures.

Waring, R.H. 1980. Vital signs of forest ecosystems. In: pp. 131-136. Waring, R.H. (Eds.). Forests: Fresh perspectives from ecosystem analysis. Proceedings, 40th Annual Biology Colloquium. Oregon State University Press. Corvallis.

Historical Range of Variability:
A Useful Tool
for Evaluating Ecosystem Change

Penelope Morgan
Gregory H. Aplet
Jonathan B. Haufler
Hope C. Humphries
Margaret M. Moore
W. Dale Wilson

ABSTRACT. The concept of historical range of variability in eco-system structure or process is valuable in understanding and illustrating the dynamic nature of ecosystems; the processes that sustain and change ecosystems, especially disturbances; the current state of

---

Penelope Morgan is Associate Professor, College of Forestry, Wildlife and Range Sciences, University of Idaho, Moscow, ID 83843.

Gregory H. Aplet is Forest Ecologist, The Wilderness Society, 900 17th St., Washington DC 20006.

Jonathan B. Haufler is Manager, Wildlife and Ecology, Boise Cascade Corporation, P. O. Box 50, Boise, ID 83728.

Hope C. Humphries is Ecological Modeler, The Nature Conservancy, 2060 Broadway, Suite 230, Boulder, CO 80302.

Margaret M. Moore is Associate Professor, School of Forestry, Northern Arizona University, Flagstaff, AZ 86011.

W. Dale Wilson is Forest Soil Scientist, Clearwater National Forest, USDA Forest Service, Orofino, ID 83544.

The authors thank the following individuals for their review and comments on this paper: Patrick Bourgeron, Arthur Zack, Lee Harry, and Leon Neuenswander.

[Haworth co-indexing entry note]: "Historical Range of Variability: A Useful Tool for Evaluating Ecosystem Change." Morgan, Penelope et al. Co-published simultaneously in the *Journal of Sustainable Forestry* (The Haworth Press, Inc.) Vol. 2, No. 1/2, 1994, pp. 87-111; and: *Assessing Forest Ecosystem Heatlh in the Inland West* (eds: R. Neil Sampson and David L. Adams) The Haworth Press, Inc., 1994, pp. 87-111. Multiple copies of this article/chapter may be purchased from The Haworth Document Delivery Center [1-800-3-HAWORTH; 9:00 a.m. - 5:00 p.m. (EST)].

87

the system in relationship to the past; and the possible ranges of conditions that are feasible to maintain. Because ecosystems are structured hierarchically, historical range of variability must be characterized at multiple spatial scales and relevant time scales. Historical range of variability is a useful reference for determining a range of desired future conditions, and for establishing the limits of acceptable change for ecosystem components and processes.

KEYWORDS. Natural range of variability, hierarchy, presettlement, scale, sustainability, forest health, ecosystem management.

INTRODUCTION

The historical range of variability characterizes fluctuations in ecosystem conditions or processes over time (Figure 1). It can describe variations in diverse characteristics, such as tree density, vertebrate population size, water temperature, frequency of disturbance or rates of change, and it can be applied at multiple spatial scales from the site to regions comprising millions of acres or more (Swanson et al. 1993). The essential function of a description of historical range of variability is to define the bounds of system behavior that remain relatively consistent over time.

The concept of historical range of variability has been developed to describe the dynamics of ecosystems undergoing continual change. Pristine forests were not undisturbed, but were subject to recurrent disturbances. In fact, recurrent disturbances are necessary to maintain the diversity of ecosystem structures and functions that have supported the variety of life and processes that we call biodiversity (Keystone Center 1991). Many species are adapted not only to recurrent disturbance (Noble and Slatyer 1980), but to particular disturbance regimes (Mutch 1970). Accumulating evidence of the pervasiveness of change has led ecologists to reject the concept of an equilibrium reached by ecological systems (Botkin 1990), although nonequilibrium dynamics at one scale can be incorporated within steady-state or equilibrium dynamics at a broader scale (Urban et al. 1987).

The rate of change in ecosystem characteristics is as important to the concept of historical range of variability as is the magnitude of historical fluctuations. The status of an ecosystem variable, such as the amount of late successional forest in the landscape, may have varied dramatically over time, but it did so at characteristic rates that reflect important ecosystem processes. The rate of change affects the ability of species to adapt to

FIGURE 1. Historical range of variability in ecosystem processes or structure over time.

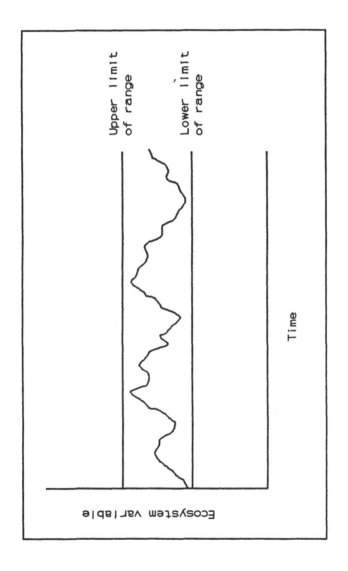

new conditions. Thus, the rate of change is likely to have as great an influence on biodiversity as the ecosystem conditions themselves.

The range of variability in ecosystem conditions and processes has been described using terms such as "historical," "natural," and "presettlement" (Kilgore 1987; Brooks and Grant 1992a, 1992b). Each of these terms conveys different meanings to different people. To some, "historical" refers only to written or oral history; to others, "natural" and "presettlement" leave people out of the ecosystem. Still others are concerned that none of these terms are bounded in time. We use "historical" more broadly to describe dynamics over a time frame relevant to understanding the behavior of contemporary ecosystems and the implications for management. This period does not need to be on the scale of evolutionary time, but it should reflect the adaptation of species to their dynamic environment.

In this paper we will address the following questions: What is the value of the concept of historical range of variability? How can this concept be applied across multiple temporal and spatial scales? What are the methods for describing historical range of variability? What are the implications for management? We conclude with a brief set of recommendations to guide its use in the maintenance of forest health.

VALUE OF THE CONCEPT

One of the most difficult challenges facing resource management professionals is the task of maintaining biological diversity, the components of the ecosystems we manage. One reason this is so difficult is that managers lack tangible models of sustainable ecosystems. As the human population has exploded throughout the world, humans have altered ecosystems in ways and at rates that are unprecedented in the evolutionary history of these systems. The concept of historical range of variability provides a window for understanding the set of conditions and processes that sustained ecosystems prior to their recent alterations by humans.

Historical range of variability can serve as a useful tool for understanding the causes and consequences of change in ecosystem characteristics over time. It provides a context for interpreting natural processes, especially disturbance, and it allows variability in patterns and processes to be understood in terms of a dynamic system. Study of past ecosystem behavior can provide the framework for understanding the structure and behavior of contemporary ecosystems, and is the basis for predicting future conditions.

There is evidence for the relationship between historical ecosystem behavior, especially the process of disturbance, and the maintenance of biological diversity. For instance, large areas of wilderness in interior Alaska are likely within the bounds of historical disturbance regimes. As a result, these areas remain largely intact, having suffered few recent species extinctions. Remote wilderness is not the only example in which maintenance of natural disturbance regimes sustains ecosystems. One of the most dynamic systems in the world, the rocky intertidal zone, experiences frequent severe disturbances during which fresh substrate is exposed to colonization by new individuals. In the absence of unnatural disruptions, such as oil spills, the dominant disturbance regime maintains the biodiversity of the ecosystem (Sousa 1979).

The effect of maintaining natural disturbance regimes extends to the managed landscape as well. Prescribed fires are being used to restore many native plant communities. Similar effects can be achieved accidentally. At Fort Bragg, North Carolina, fires ignited by artillery shelling have maintained a historical disturbance regime of frequent, low-intensity fires that supports a phenomenal diversity of herbaceous plants beneath the pine overstory. In the same pine type where fires have been excluded, shrubs have displaced many herbaceous species, resulting in the loss of understory species diversity.

Deleterious effects on biological diversity and ecosystem function are often observed where ecosystem conditions and processes have been altered significantly from the historical range of variability. The exclusion of low-intensity surface fires from the ponderosa pine forests of the Inland West has so altered ecosystem structure and function that catastrophic fire now threatens the sustainability of the forest ecosystem in some places (Covington and Moore 1994a, 1994b). Barrows (1978) and Swetnam (1990) documented that fires occurring in ponderosa pine forests of the southwestern United States are now 3 to 6 times larger than those occurring prior to settlement by Euro-Americans. Habitat for animals such as the flammulated owl, northern goshawk and other species may be at risk of elimination. Elsewhere, the invasion of exotic species has changed the composition, structure, and even the function of whole ecosystems in ways that threaten the continued existence of the historical components of those ecosystems (D'Antonio and Vitousek 1992; Vitousek 1986). Examples include cheat grass in the sagebrush grasslands of the Inland West, which supports fires of unprecedented frequency; and forest diseases, such as chestnut blight in the eastern hardwood forest or white pine blister rust in whitebark pine forests in the western US (Keane and Arno 1993) which have drastically reduced the abundance of formerly dominant trees.

The historical range of variability provides researchers and managers with a reference against which to evaluate present ecosystem change. This is useful both for describing ecosystem dynamics and for measuring the effects of management activities. For example, the historical range of variability is useful as a standard in cumulative effects analyses of environmental impacts resulting from multiple management activities. In the Idaho Cumulative Effects Watershed Analysis Process (Idaho Department of Lands 1993), the amount of sediment likely to be generated through road construction or other activities is compared against the range of variability in reference data for in-stream fine sediments. The reference data have been collected from the 1970s to the present in streams minimally altered by road construction and other human activities and thus likely reflect a portion of the range of conditions inherent in natural systems.

Understanding ecosystem function and the magnitude of current departures from historical conditions can help identify risks to ecosystem sustainability. For example, if a landscape currently dominated by early successional forests had historically supported consistently high amounts of old-growth, this significant departure from historical conditions indicates a potential threat to the long-term sustainability of old-growth dependent species. Identification of such a risk resulted in considerable management attention to the conservation of the ancient forests of western Washington, Oregon, and northern California (FEMAT 1993).

The historical range of variability can also help in the identification of conflicting policies. For example, as has already been discussed, the policy of fire exclusion is inconsistent with the forest fire history of the Inland West. Continued fire exclusion is likely to lead to catastrophic fires, the opposite of the intended effect of the policy. Fire exclusion is also preventing the creation of biologically diverse early successional habitat. In western Oregon, where rare old growth has received most of the attention, some researchers have called post-fire brush "the least common seral stage in the region" (Hansen et al. 1991). Moreover, control of "competing vegetation" and language in the National Forest Management Act encouraging rapid regeneration of harvested sites oppose the conservation of this important habitat type.

Finally, the concept of the historical range of variability is valuable for communicating the dynamics of complex ecosystems. To many, ecosystem processes, such as succession, may seem abstract until illustrated with examples of resultant changes in ecosystem conditions. Historical range of variability provides a tangible illustration of ecosystem dynamics to help managers and the general public to reach a common understanding of the behavior of forest ecosystems, an essential step toward identifying and resolving conflict.

IMPORTANCE OF SCALE

The identification of the spatial and temporal scales relevant to ecosystem patterns and processes are critical to the concept of historical range of variability. The measurement of structure or process (including disturbances) of an ecosystem and the resulting interpretations depend on the scale at which the measurements are made (Turner and Gardner 1991, Urban et al. 1987; Wiens 1985). Scale refers to both spatial (e.g., size of area) and temporal (e.g., length of time) dimensions. The level of biotic organization (e.g., organism to population) should be distinguished from the spatial and temporal scale (Turner and Gardner 1991; Wiens 1985). We endorse the implementation of a national hierarchical ecological classification framework (e.g., ECOMAP 1993) for organizing our spatial information.

Patterns and processes operating at characteristic temporal and spatial scales can be hierarchically arranged (O'Neill et al. 1986; Urban et al. 1987). Nested hierarchies contain finer spatial scales and shorter time scales at lower levels and broader spatial scales and longer time scales at higher levels (Urban et al. 1987). These broader scales provide context and constrain ecosystem structure and dynamics. Explanation of ecosystem properties may be sought in the interactions of the component parts (Allen and Starr 1982; Urban et al. 1987). For instance, the species composition within a forest stand depends upon which species successfully establish within available microsites (a finer scale). At a broader scale, the ecoregion constrains the species which can occur in a stand, and provides a context for evaluating a given stand's unique composition.

The choice of the appropriate temporal and spatial scales should be based upon the objectives of the analysis (Bourgeron and Jensen 1993). Different processes driving ecosystem change operate at different temporal and spatial scales. Site quality develops over geologic time, vegetation reflects successional time, and stream characteristics can change over very brief periods. Broad-scale processes are usually slow (Urban et al. 1987). For example, the uplifting and downcutting of the Idaho Batholith that formed the canyons and river breaklands in the Clearwater and Salmon River drainages occurred over 80,000,000 years (Alt and Hyndman 1989). In contrast, the finer-scale process of landform erosion occurs over mere decades and centuries.

A uniquely appropriate spatial scale can be chosen for most ecosystem patterns and their associated processes based upon the objectives of the analysis. When multiple processes are analyzed, each may occur at a different spatial scale, thus requiring analysis at several scales. Some processes, such as fire, influence ecosystems at multiple temporal and spatial scales. Fire regimes are subject both to regional, long-term climatic influences and local short-term drought cycles (Clark 1988; Swetnam 1993; Swetnam and Betancourt 1990). Furthermore, fire frequency differs with the size of the area described (Arno and Peterson 1983). In such cases, characterizing the historical range of variability across multiple spatial and temporal scales is appropriate.

Relevant Temporal Scales

Temporal scale influences ranges of variability for an area. Over many thousands of years, tectonic plate movements have produced changes in landforms, species assemblages, and isolation (MacArthur 1972). Over 5,000 to 10,000 years, climatic shifts in the Inland West have resulted in areas changing from glaciers to complex vegetation associations (Johnson et al. 1993). In major volcanic eruptions, ash was deposited to considerable depths, changing soil conditions and plant associations. These dramatic fluctuations are important to our understanding of the evolutionary development of current biodiversity, but are of little practical use in predicting the short-term dynamics of contemporary ecosystems in the Inland West.

Historical range of variability should be assessed over a time period characterized by relatively consistent climatic, edaphic, topographic, and biogeographic conditions. For forests, the time should encompass multiple generations of trees, and should be at least a few centuries. In practice, the time period will be constrained by the ability to look backwards through time. Steele (1994) suggested a time frame of 100 to 400 years to be appropriate for interpreting secondary successional dynamics. Hann et al. (1993) suggested at least 100 years for their analysis.

Relevant Spatial Scales

Spatial scale has a major influence on how historical range of variability is used, and on the variables chosen for assessment. Some ecosystem variables have applications for very small areas, such as evaluating a specific stand or site. Variables at this scale might include descriptors of stand composition or structure, soil carbon or nutrient status, standing biomass, or annual productivity (Figure 2).

At a scale encompassing a larger area, such as a watershed, variables such as stream temperature, pool-to-riffle ratios, or in-stream flow might be appropriate (Figure 2). Historical ranges of variability can be used as reference baselines for analyses of the current status of the watershed for

FIGURE 2. The influence of scale on the indentification of appropriate ecosystem variables in relation to historical ranges of variability (HRV).

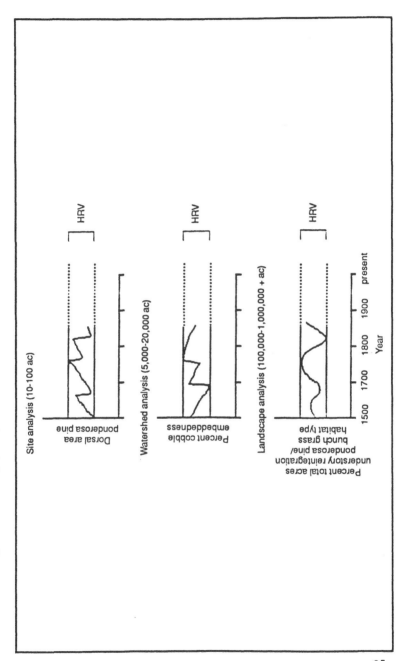

cumulative effects determinations, for fish habitat assessments, or for water quality analyses.

At a landscape scale, relevant variables might include acres of each successional stage (e.g., Oliver and Larson's (1990) understory reinitiation stage) by habitat type present in the landscape (Figure 2). Such analysis would provide insights into the relative abundance of different succession-al stages occurring in the landscape at present and in the past. This in-formation will be useful in determining ranges of desired future conditions needed to maintain biodiversity (Haufler 1994).

It is critical that the historical range of variability be described at the appropriate spatial scales. For example, cobble embeddedness, a measure of the amount of fine sediment in a stream channel and an indicator of fish habitat quality and stream health, was described for stream reaches within the Fish Creek watershed in northern Idaho (Clearwater BioStudies, Inc. 1993). The 68,000 acre watershed was essentially undeveloped. Stream channels were divided into 67 reaches based on Rosgen channel type. Ranges of variability in the percent of cobbles embedded in stream bot-toms were developed by stratifying stream reaches by adjacent landform (Figure 3). Each landform has associated with it particular stream channel characteristics, including gradient, amount and movement of fine sedi-ment, and pool to riffle ratios. If cobble embeddedness were averaged and incorporated into a single historical range of variability for all stream reaches, the resulting range would be of limited utility for understanding stream reaches that occurred adjacent to unusually erosive or stable land-forms.

Thus, while the watershed may be the appropriate scale at which to consider characteristics like stream volume and temperature, landform is the more relevant scale for characterizing other aspects of water quality. Landform also influences site quality, potential vegetation, and fire and other disturbance regimes (Swanson et al. 1988). Because the adjacent landform exerts such influence on site and stream characteristics, land-form provides a link between the aquatic and terrestrial ecosystems.

METHODS FOR DESCRIBING
HISTORICAL RANGE OF VARIABILITY

Methods available for determining the historical range of variability can be applied at a variety of scales and can provide various levels of detail. The objectives of an analysis may require rapid assessment over a large area or may dictate that detailed information be collected at a finer scale of resolution. Variability may be characterized in terms of the states

FIGURE 3. Notched box plots of percent of cobbles embedded in bottoms of stream reaches stratified by adjacent landform for Fish Creek, Idaho watershed. Vertical lines represent 25th percentile, median, and 75th percentile of data, respectively. Notches indicate 95% confidence limits. Horizontal lines show range of data values.

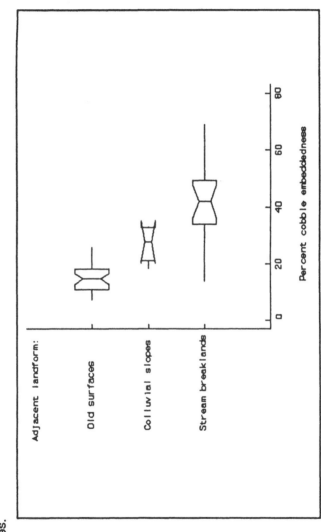

of an ecological system or may be focused on processes (Swanson et al. 1993). Characterization of historical range of variability in terms of both system structures and processes is desirable whenever possible.

Tree Ring Analysis

The annual growth rings of living trees, snags, logs, or stumps can be analyzed to estimate past and present age classes, seral stages, or species composition (Cooper 1960; White 1985). Age structure and fire scar analysis can be used to estimate the pattern of vegetation and disturbance in the past (Arno et al. 1993; Baker and Veblen 1990; Romme 1982).

Dendroecological techniques (Fritts 1971, 1976; Fritts and Swetnam 1989) can be used in forested ecosystems to determine past climatic fluctuations; to describe the past and present structure of forests by age classes, seral stages, or tree density; to characterize patch sizes, frequency, and severity of disturbance regimes, including fire and, to a limited extent, insects and disease. Analysis of the number and width of annual growth rings provides both climatic and ecological information where trees form consistent annual rings. These methods are labor-intensive and time-consuming. The relative effects of climatic fluctuations, competition, insects or disease, and other influences on growth may be difficult to identify and partition. Disturbances such as fires or timber harvest may remove much of the record of past disturbance, and as snags and stumps decompose it becomes more difficult to reconstruct past vegetation structure and disturbance patterns.

Paleoecological Methods

Paleoecological techniques can be used to extend the length of record for characterizing disturbance regimes. Fire frequency over thousands of years can be determined from charcoal layers (which indicate fire occurrence) and pollen content (which indicates vegetation composition). These approaches are limited by lack of detailed information about fire events, spatial extent of inferences, and difficulties in inferring the relationship between pollen abundance and vegetation composition.

Written and Photographic Records

Site-specific data sources and techniques include early land survey records from the 1800s (General Land Office notes) (Bourdo 1956); forest reserve reports from the late 1800s and early 1900s which may contain maps and plot data (e.g., Graves 1899; Leiberg 1900, 1899; Leiberg et al. 1904); early forest inventories (e.g., Lang and Stewart 1910); and comparison of historical and current photographs (e.g., Progulske 1974; Gruell 1983; Gruell et al. 1982; Veblen and Lorenz 1986; Boise National Forest 1993).

Expert opinion (Caraher et al. 1992), comparisons of historical and current photographs, and qualitative descriptions provide general information that can be used to supplement quantitative data or serve as a primary source of information where quantitative data are lacking. Limitations include potential bias and lack of rigor, detail, and length of record (Swanson et al. 1993).

GIS Data Layers

Maps of disturbance events at multiple time intervals and maps of derived disturbance regime classes can be critical for assessment of historical range of variability, if sufficient spatial information exists for their construction. Composite maps created from geographical information systems (GIS) overlays can be used in the various levels of hierarchical ecological classifications (Hann et al. 1993). Overlays of vegetation and topographic variables such as slope, aspect and elevation can be used to indicate fire regime in some forested systems (Barrett and Arno 1991). Other data layers useful in determining ranges of variability include soils, geologic substrate, potential vegetation and climate attributes. Choice of scale and mapping units is dictated by the objectives of the analysis.

Substitutions for Historical Data

Areas believed to have relatively unaltered disturbance regimes can be used to assess ranges of variability, utilizing the full array of available techniques for quantifying pattern and process. Data from large wilderness areas or national parks may be applicable. Unfortunately, the degree to which even large wilderness ecosystems have been altered by human activities may limit their applicability to historical range of variability. Alteration in fire regimes through fire exclusion in western forest systems is pervasive, severely reducing the number of suitable areas for such an analysis. Air pollution, hunting, grazing, mining, fire suppression, and other human activities both within and near wilderness areas have influenced ecosystem structure and function. Also, wilderness areas seldom include a full spectrum of ecological conditions. For instance, there are few low-elevation forest and shrubland types represented within wilderness areas.

When the current range in variability of relatively unaltered areas is used as a surrogate, field sampling, remote sensing and statistical or expert models are useful for characterizing that variability. The range of biophysical environments can be determined using classification techniques or expert models (DeVelice et al. 1993, Engelking et al. 1993). Biotic responses to ranges of environmental variability can be assessed using ordination techniques such as canonical correspondence analysis (ter Braak 1988; Engelking et al. 1993) or generalized linear models (McCullagh and Nelder 1989; Nicholls 1989; Engelking et al. 1993).

Modeling

Simulation models with stochastic properties that incorporate the effects of disturbance on vegetation can be useful analytical tools for the assessment of historical ranges of variability. Individual-based models such as FIRESUM include effects of fire (Keane et al. 1990a, 1990b) and insects and introduced disease (Keane et al. 1990b). Inclusion of spatial relationships in individual-based models, e.g., ZELIG (Smith and Urban 1988) can extend their range of effectiveness. At the stand level, the effect of an insect pest on forests can be assessed using the PROGNOSIS model linked with a spruce budworm model (Wykoff 1985). The impact of climate change on ecosystem structure can be simulated using Forest-BGC (Running and Coughlan 1988). Simulation models can be linked to GIS to provide input and display output. Baker (1992) used a GIS-based spatial simulation model to examine the effects of fire on landscape dynamics in northeastern Minnesota. Well-validated simulation models incorporating disturbance can provide estimates of ranges of variability in ecosystem processes or other components when historical data are incomplete. Sometimes, models can be used to reconstruct past conditions (Hann et al. 1993). The simulation models can also be used to evaluate possible pathways of ecosystem change. Climate simulation models such as MTCLIM (Hungerford et al. 1989) can be used to estimate climate attributes to determine ranges of environmental variability and to support statistical modeling efforts.

IMPLICATIONS FOR MANAGEMENT

Studying the historical range of variability will greatly assist in understanding ecosystem processes and species adaptations to change within ecosystems. This knowledge is helpful in ecosystem management and in sustaining forest health. Indeed, this is the greatest value of assessing the historical range of variability. Increased understanding of the processes that have shaped and will continue to shape ecosystem patterns will aid in the design of management activities that work in concert with, rather than counter to, these processes. Complete understanding is unattainable, but management can be adapted (Walters 1986; Walters and Holling 1990) as new information becomes available.

The risks and probabilities of changes in ecosystems are likely to be related to the magnitude and direction of departures from the historical range of variability. Such risks and probabilities have both ecological (Covington et al. 1994) and societal dimensions (McKetta et al. 1994). When ecosystems are outside the historical range of variability, changes may occur dramatically and rapidly. An investment of money, energy, or human effort may be required to counter processes that would change the desired state of the ecosystem.

The historical range of variability provides managers and researchers with a reference against which to evaluate ecosystem change. It is useful as a monitoring baseline if the goals of management are defined relative to the historical range of variability.

Range of Desired Future Conditions

The historical range of variability in conditions, processes, populations, or structures of variability can be used as a reference in establishing the range of desired future conditions. Societal needs and values should also be considered when identifying the range of desired future conditions.

The historical range of variability can be used to identify the range of future conditions that are sustainable. Such an approach, based upon a strong biological rationale, is scientifically defensible as a strategy for sustaining biodiversity.

The historical range of variability can be used to judge what sort and degree of complexity in a landscape is appropriate as a management goal. Managers may choose to judiciously alter the size or frequency of disturbance in an ecosystem. For example, in ponderosa pine/Douglas-fir forests of the Inland West, Keane et al. (1990a) speculated, based upon simulation modeling results, that partial timber harvest could be combined with periodic prescribed fires at 20-year intervals to maintain the nutrient cycles, stand structures, regeneration, growth and mortality of plants, and the range of stand and landscape compositions experienced when fires historically burned every 10 years. The effects of this alteration on vegetation composition, structure, and other ecological characteristics should be monitored to ensure that biodiversity is maintained.

Use of the historical range of variability approach does not necessarily imply that systems must be maintained within that range (Figure 4), but the risks associated with departure from historical conditions must be acknowledged in the decision-making process. Any departure should be made with caution, and the effects should be rigorously monitored for undesirable results.

Acceptable Bounds for Ecosystem Change

The historical range of variability will be useful in defining the bounds of acceptable ecosystem change, and in determining the minimum acceptable levels for ecosystem elements and processes in managed landscapes (Oliver et al. 1994). The limits of acceptable change approach used in wilderness management embodies one such application (Hendee et al. 1990).

LIMITATIONS OF THE HISTORICAL
RANGE OF VARIABILITY CONCEPT

Although the historical range of variability is a useful concept, its application in particular ecosystems is limited by lack of historical data, difficulties in interpreting the historical record, and societal limitations (Swanson et al. 1993). Additionally, many aspects of the future are likely to be without historical precedent, limiting the usefulness of the historical range of variability as a goal, per se. Lastly, synthesizing historical ranges of variability for multiple characteristics of ecosystems and at multiple spatial scales is difficult.

For many conditions and processes, the historical record reveals little information. For example, very little data exist on historical stream flow; the temperature, sediment content, or nutrient content of water; soil nutrient content or turnover rates; and the influence of many diseases and insects. Records are best where disturbance events were frequent and left an indelible mark, and where ecosystems were little influenced by Euro-American settlement prior to 1900. The historical range of variability concept is most difficult to apply where recent human influence has been great and/or where disturbances were infrequent and catastrophic, as more recent disturbances destroy evidence of previous disturbances (Swanson et al. 1993). Additionally, reconstruction of past vegetation structure and composition is much more difficult where woody plants do not occur or where they do not form distinct annual growth rings. In many cases,

FIGURE 4. Approaches to the use of historical range of variability (HRV) in land management planning for ranges of desired future conditions.

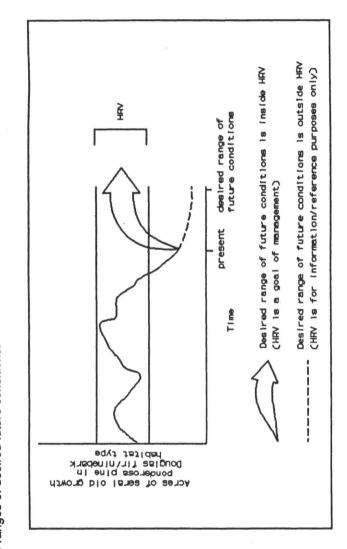

reconstruction of ecosystem structure is limited to the past one to three centuries.

Interpretation of the historical record is limited by incomplete understanding of the causes of observed phenomena. For instance, it is difficult to interpret the relative influence of lightning, Native Americans, and Euro-American settlers on fire frequency and spatial pattern in the Interior West in the mid 1800s (Gruell 1985), and only the results of fires can be observed, not their causes. Similarly, release events observed in tree rings may be the result of fire, wind, insects, or some other disturbance.

Societal values may also limit the designation of historical range of variability as the range of desired future conditions. For example, while managers increasingly understand the importance of large, stand-replacing fires in maintaining many forested ecosystems, the general public is likely to find crown fires undesirable and would prefer to see such fires managed to reduce the risk to human life and property, and to local populations of wildlife and their habitat. More fundamentally, many people do not understand the dynamic nature of ecosystems and may not accept historical range of variability as a valuable concept. In addition, some levels of desired products may only be obtained if ecosystems are managed outside the historical range of variability. In this case, either societal needs and values must change or the costs and consequences to ecosystem sustainability must be understood and accepted.

Finally, humans have changed many ecosystems so dramatically as to limit the usefulness of historical conditions as a guide for the future. Exotic species introductions, native species extinctions, air and water pollution, and climate change result in unprecedented conditions. Where ecosystems are so altered, restoration of historical conditions is impossible or prohibitively expensive. Nevertheless, the historical range of variability is extremely useful in understanding the origins of existing ecosystems, and knowledge of past ecosystem dynamics is invaluable in projecting future ecosystem behavior.

RECOMMENDATIONS AND CONCLUSIONS

We will never completely understand all aspects of the composition, structure, and function of contemporary ecosystems, much less their historical characteristics. This should not discourage investing in additional research into ecosystem structure, composition, and function. Research should be extended to understanding the historical range of variability in these characteristics, as this information is essential to understanding the causes of current conditions and projecting future alternatives.

In summary, we recommend and conclude the following.

1. The concept of historical range of variability in ecosystem structure or process is valuable in helping to understand and illustrate the dynamic nature of ecosystems; the processes that sustain or change ecosystems, especially disturbances; and the current state of the system in relationship to the past. Thus, the historical range of variability is useful in evaluating the sustainability of ecosystem conditions.
2. The historical range of variability can become a reference for determining a range of desired future conditions. Where the range of desired future conditions is defined relative to it, the historical range of variability is useful as a baseline for monitoring.
3. The historical range of variability can be used as a reference in establishing the limits of acceptable change in key ecosystem patterns and processes. Determining the departure of current conditions from the historical range of variability can be helpful in assessing and communicating the risk and probability of change and the inputs required to resist that change. Alternative future conditions can also be evaluated relative to the historical range of variability in ecosystem components and processes.
4. Because ecosystems are structured hierarchically, historical range of variability must be characterized at multiple spatial scales appropriate to ecosystem processes. We endorse the implementation of the hierarchical ecological classification system such as ECOMAP (1993), which will help organize existing knowledge into a useful framework. In addition, we must characterize the historical range of variability at time scales that are both relevant to ecosystem dynamics and management goals.
5. A variety of methods exist to describe the historical range of variability, such as reconstruction of past forest structure and composition from tree rings, pollen in sediment, and historical photographs.
6. Some limitations to the approach include the lack of historical data, difficulties of interpreting the historical record, and societal values. The historical range of variability will be of limited utility as an alternative future model in ecosystems drastically affected by invasion of exotic species, air pollution, climate change, or the extinction or near elimination of some native species. However, even in these situations, the historical range of variability is a useful tool for understanding ecosystem function, which is essential to predicting the behavior of altered as well as natural ecosystems.
7. We must identify and maintain reference areas from which we can draw data to substitute for historical information. Where particular

ecosystems are not well-represented within research natural areas, wilderness, or other appropriate reference areas, additional areas should be identified and managed as reference areas. The few areas that remain relatively unaltered by human activity are of unquestionable value as a window to the past. Even so, many designated reserves have been affected by air pollution, fire exclusion, grazing, mining, and other human activities both within and adjacent to them.

8. Many important sources of information, including old trees, old records, and old people are succumbing to the pressures of time. There is an urgent need to ascertain historical ranges of variability for many ecosystems before these records are irretrievably lost.

REFERENCES

Allen, T. F. H. and T. B. Starr. 1982. Hierarchy: Perspectives for ecological complexity. University of Chicago Press. Chicago.

Alt, D. and D. W. Hyndman. 1989. Roadside geology of Idaho. Mountain Press Publishing Company. Missoula, MT.

Arno, S. F. and T. D. Petersen. 1983. Variation in estimates of fire intervals: A closer look at fire history on the Bitterroot National Forest. Res. Pap. INT-301. U.S. Department of Agriculture, Forest Service, Intermountain Forest and Range Experiment Station. Ogden, UT. 8 p.

Arno, S. F., E. D. Reinhardt and J. H. Scott. 1993. Forest structure and landscape patterns in the subalpine lodgepole type: A procedure for quantifying past and present conditions. Gen. Tech. Rep. INT-294. U.S. Department of Agriculture, Forest Service, Intermountain Research Station. Ogden, UT. 17 p.

Baker, W. L. 1992. Effects of settlement and fire suppression on landscape structure. *Ecology* 73: 1879-1887.

Baker, W. L. and T. T. Veblen. 1990. Spruce beetle and fires in the nineteenth-century subalpine forests of western Colorado, USA. *Arctic and Alpine Research* 22:65-80.

Barrett, S. W. and S. F. Arno. 1991. Classifying fire regimes and defining their topographic controls in the Selway-Bitterroot Wilderness. In: pp. 299-307. Andrews, P. L. and D. F. Potts (Eds.). Proceedings of the 11th conference on fire and forest meteorology, April 16-19, 1991. Missoula, MT. Society of American Foresters. Bethesda, MD.

Boise National Forest. 1993. Snapshot in time. Repeat photography on the Boise National Forest. U.S. Department of Agriculture, Forest Service, Boise National Forest. Boise, ID. 239 p.

Barrows, J. S. 1978. Lightning fires in southwestern forests. Unpublished Report. U.S. Department of Agriculture, Forest Service, Northern Forest Fire Laboratory. Missoula, MT. 154 p.

Botkin, D. B. 1990. Discordant harmonies. A new ecology for the twenty-first century. Oxford University Press. 256 p.

Bourdo, E. A., Jr. 1956. A review of the general land office survey and its use in quantitative studies in former forests. *Ecology* 37: 744-769.

Bourgeron, P. S. and M. E. Jensen. 1993. An overview of ecological principles for ecosystem management. In: pp. 51-64. Jensen, M. E. and P. S. Bourgeron (Eds.). Eastside Forest Ecosystem health Assessment-Volume II: Ecosystem management: principles and applications. U.S. Department of Agriculture, Forest Service, Pacific Northwest Research Station. Portland, OR.

Brooks, D. J. and G. E. Grant. 1992a. New approaches to forest management, part 1. *Journal of Forestry* 90(1):25-28.

Brooks, D. J. and G. E. Grant. 1992b. New approaches to forest management, part 2. *Journal of Forestry* 90(1):21-24.

Caraher, D. L., J. Henshaw, F. Hall et al. 1992. Restoring the health of ecosystems in the Blue Mountains–a report to the Regional Forester and the Forest Supervisors of the Blue Mountain Forests. U.S. Department of Agriculture, Forest Service, Pacific Northwest Region. Portland, OR.

Clark, J. S. 1988. Effect of climate change on fire regimes in northwestern Minnesota. *Nature* 334: 233-235.

Clearwater BioStudies, Inc. 1993. Habitat conditions and salmonid abundance in the Fish Creek drainage, Lochsa Ranger District, Summer 1992. Contract No 53-0276-2-31. Report on file: USDA Forest Service, Clearwater National Forest. Orofino, ID.

Cooper, C. F. 1960. Changes in vegetation, structure, and growth of southwestern pine forests since white settlement. *Ecological Monographs* 30:129-164.

Covington, W. W., R. L. Everett, A. N. D. Auclair, T. A. Daer, L. L. Irwin and R. W. Steele. 1994. Historical and anticipated changes in forest ecosystems of the Inland West of the United States. In: pp. 13-63. Sampson, R.L. and D.L. Adams (Eds.). Assessing Forest Ecosystem Health in the Inland West. Proceedings of the American Forests scientific workshop. The Haworth Press, Inc. New York.

Covington, W. W. and M. M. Moore. 1994a. Southwestern ponderosa forest structure: changes since Euro-American settlement. *Journal of Forestry* 92(1): 39-47.

Covington, W. W., M. M. Moore. 1994b. Postsettlement changes in natural fire regimes and forest structure: Ecological restoration of old-growth ponderosa pine forests. In: pp. 153-181. Sampson, R.L. and D.L. Adams (Eds.). Assessing Forest Ecosystem Health in the Inland West. Proceedings of the American Forests scientific workshop. The Haworth Press Inc. New York.

D'Antonio, C. M. and P. M. Vitousek. 1992. Biological invasions by exotic grasses, the grass/fire cycle and global change. *Annual Review of Ecology and Systematics* 23: 63-87.

DeVelice, R. L., G. J. Daumiller, P. S. Bourgeron and J. O. Jarvie. 1993. Bioenvironmental representativeness of nature preserves: assessment using a combination of a GIS and a rule-based model. In: pp.51-58. Proceedings of the first biennial scientific conference on the Greater Yellowstone ecosystem, September, 1992. National Park Service. Mammoth, WY.

ECOMAP. 1993. National hierarchical framework of ecological units. USDA, Forest Service. Washington, DC.

Engelking, L. E., H. C. Humphries, M. S. Reid, R. L. DeVelice, E. H. Muldavin and P. S. Bourgeron. 1993. Regional conservation strategies: assessing the value of conservation areas at regional scales. In: Jensen, M. E. and P. S. Bourgeron (Eds.). Eastside Forest Ecosystem Health Assessment-Volume II: Ecosystem management: principles and applications. U.S. Department of Agriculture, Forest Service, Pacific Northwest Research Station. Portland, OR.

FEMAT. 1993. Forest ecosystem management: an ecological, economic, and social assessment. Report of the Forest Ecosystem Management Team, July, 1993.

Fritts, H. C. 1971. Dendroclimatology and dendroecology. *Quaternary Research* 1: 419-449

Fritts, H. C. 1976. Tree Rings and Climate. Academic Press. London.

Fritts, H. C. and T. W. Swetnam. 1989. Dendroecology: A tool for evaluating variations in past and present forest environments. *Advances in Ecological Research* 19: 111-188.

Graves, H. S. 1899. The Black Hills Forest Reserve. U.S. Department of Interior, Geological Survey Professional Paper. The nineteenth annual report of the survey, 1897-1898. Part V, Forest Reserves. pp. 67-164.

Gruell, G. E. 1985. Indian fires in the Interior West: A widespread influence. In: pp. 68-74. Lotan, J. E. et al. (Technical Coordinators). Symposium and Workshop on Wilderness Fire. U.S. Department of Agriculture, Forest Service, Gen. Tech. Rep. INT-182.

Gruell, G. E. 1983. Fire and vegetative trends in the Northern Rockies: Interpretations from 1871-1982 photographs. U.S. Department of Agriculture, Forest Service, Gen. Tech. Rep. INT-158.

Gruell, G. E., W. C. Schmidt, S. F. Arno and W. J. Reich. 1982. Seventy years of vegetative changes in a managed ponderosa pine forest in western Montana–implications for resource management. U.S. Department of Agriculture, Forest Service, Gen. Tech. Rep. INT-130. 42 p.

Hann, W., M. E. Jensen, P. S. Bourgeron and M. Prather. 1993. Land management assessment using hierarchical principles of landscape ecology. In: pp. 301-314. Jensen, M. E. and P. S. Bourgeron (Eds.). Eastside Forest Ecosystem health Assessment-Volume II: Ecosystem management: principles and applications. U.S. Department of Agriculture, Forest Service, Pacific Northwest Research Station. Portland, OR.

Hansen, A. J., T. A. Spies, F. J. Swanson and J. L. Ohmann. 1991. Conserving biodiversity in managed forests: lessons from natural forests. *BioScience* 41:382-392.

Haufler, J. B. 1994. An ecological framework for planning for forest health. In: pp. 307-316. Sampson, R.L. and D.L. Adams (Eds.). Assessing Forest Ecosystem Health in the Inland West. Proceedings of the American Forests scientific workshop. The Haworth Press, Inc. New York.

Hendee, J. C., C. H. Stankey and R. C. Lucas. 1990. Wilderness management. 2nd edition, revised. North American Press, Fulcrum Publishing. Golden, CO.

Hungerford, R. D., R. R. Nemani, S. W. Running and J. C. Coughlan. 1989. MTCLIM: a mountain microclimate simulation model. Res. Pap. INT-414. U.S. Department of Agriculture, Intermountain Research Station. Ogden, UT.

Idaho Department of Lands. 1993. A draft cumulative effects analysis process for Idaho. August 13, 1993. State of Idaho, Department of Lands. Boise.

Johnson, C. G., R. Clausnitzer, P. J. Mehringer and C. D. Oliver. 1993. Biotic and abiotic processes of eastside ecosystems: the effects of management on plant and community ecology, and on stand and landscape vegetation dynamics. In: pp. 35-100. Hessburg, P. F. (Ed.) Vol. III Assessment. Ecosystem management: principles and applications. U.S. Department of Agriculture, Forest Service, Pacific Northwest Research Station. Portland, OR.

Keane, R. E. and S. F. Arno. 1993. Rapid decline of whitebark pine in western Montana: Evidence from 20-year remeasurements. *Western Journal of Applied Forestry* 8(2): 44-47.

Keane, R. E., S. F. Arno and J. K. Brown. 1990a. Simulating cumulative fire effects in ponderosa pine/Douglas-fir forests. *Ecology* 71:189.

Keane, R. E., S. F. Arno, J. K. Brown and D. F. Tomback. 1990b. Modelling stand dynamics in whitebark pine (*Pinus albicaulis*) forests. *Ecological Modeling* 51: 73-95.

Keystone Center. 1991. Biological diversity on federal lands. The Keystone Center. Keystone, CO. 97 p.

Kilgore, B. M. 1987. The role of fire in wilderness: a state-of-knowledge review. In: pp. 70-103. Lucas, R. C. (Comp). Proceedings of the national wilderness research conference: issues, state-of-knowledge, future directions. July 23-26, 1985. Fort Collins, CO. Gen. Tech. Rep. INT-220. U.S. Department of Agriculture, Forest Service, Intermountain Research Station. Ogden, UT.

Lang, D. M. and S. S. Stewart. 1910. Reconnaissance of the Kaibab National Forest. U.S. Department of Agriculture, Forest Service Timber Survey (North Kaibab Ranger District), Admin. Report. 35 p. with map.

Leiberg, J. B. 1900. The Bitterroot Forest Reserve (Idaho portion). 20th Annual Report (1898-1899). Washington, DC: U.S. Geological Survey. Part V: 317-410.

Leiberg, J. B. 1899. The Bitterroot Forest Reserve (Montana portion). 19th Annual Report (1897-1898). U.S. Geological Survey. Part V: 253-282. Washington, DC.

Leiberg, J. B., T. F. Rixon and A. Dodwell. 1904. Forest conditions in the San Francisco Mountain Forest Reserve, Arizona. U.S. Department of Interior, Geological Survey Professional Paper No. 22. Series H, Forestry 7. Washington, DC. 95 p.

MacArthur, R. H. 1972. Geographical ecology. Patterns in the distribution of species. Harper and Row, Publ. New York.

McCullagh, P. and J. A. Nelder. 1989. Generalized Linear Models. 2nd edition. Chapman and Hall. New York.

McKetta, C., K.A. Blatner, R.T. Graham, J.R. Erickson and S.S. Hamilton. 1994. Human dimensions of forest health choices. In: pp. 135-149. Sampson, R.L. and D.L. Adams (Eds). Assessing Forest Ecosystem Health in the Inland West. Proceedings of the American Forests scientific workshop. The Haworth Press, Inc. New York.

Mutch, R. W. 1970. Wildland fires and ecosystems–a hypothesis. *Ecology* 51: 1046-1051.

Noble, I. R. and R. O. Slatyer. 1980. The use of vital attributes to predict successional changes in plant communities subject to recurrent disturbances. *Vegetatio* 43: 5-21.

O'Neill, R. V., D. L. DeAngelis, J. B. Waide and T. F. H. Allen. 1986. A hierarchical concept of the ecosystem. Princeton University Press. Princeton, NJ.

Nicholls, A. O. 1989. How to make biological surveys go further with generalized linear models. *Biological Conservation* 50:51-75.

Oliver, C. D., D. Ferguson, A.E. Harvey. H. Malany. J.M. Mandzak and R.W. Mutch. 1994. Managing ecosystems for forest health: An approach and the effects on uses and values. In: pp. 113-133. Sampson, R.L. and D.L. Adams (Eds.). Assessing Forest Ecosystem Health in the Inland West. Proceedings of the American Forests scientific workshop. The Haworth Press, Inc. New York.

Oliver, C. D. and B. C. Larson. 1990. Forest stand dynamics. McGraw-Hill, Inc. New York.

Progulske, D. R. 1974. Yellow hair, yellow ore, yellow pine. A photographic study of a century of forest ecology. Bulletin 616. South Dakota State University, Agricultural Experiment Station. Brookings. 169 p.

Romme, W. H. 1982. Fire history and landscape diversity in Yellowstone National Park. *Ecological Monographs* 52:199-221.

Running, S. W. and J. C. Coughlan. 1988. A general model of forest ecosystem processes for regional applications. I: hydrologic balance, canopy gas exchange and primary production processes. *Ecological Modelling* 42:125-154.

Smith, T. M. and D. L. Urban. 1988. Scale and resolution of forest structural pattern. *Vegetatio* 74:143-150.

Sousa, W. P. 1979. Disturbance in intertidal boulder fields: the nonequilibrium maintenance of species diversity. *Ecology* 60: 1225-1239.

Steele, R. 1994. The role of succession in forest health. In: pp. 183-190. Sampson, R.L. and D.L. Adams (Eds.). Assessing Forest Ecosystem Health in the Inland West. Proceedings of the American Forests scientific workshop. The Haworth Press, Inc. New York.

Swanson, F. J., J. A. Jones, D. O. Wallin, and J. H. Cissel. 1993. Natural variability– Implications for ecosystem management. In: pp. 89-103. M. E. Jensen and P. S. Bourgeron (Eds.). Ecosystem management: principles and applications. Vol. II. Eastside forest ecosystem health assessment. USDA Forest Service, Pacific Northwest Research Station, Forestry Sciences Laboratory. Wenatchee, WA.

Swanon, F. J., T. K. Kratz, N. Caine and R. G. Woodmansee. 1988. Landform effects on ecosystem patterns and processes. *BioScience* 38: 92-98.

Swetnam, T. W. 1993. Fire history and climate change in giant sequoia groves. *Science* 262: 885-889.

Swetnam, T. W., M. A. Thompson and E. K. Sutherland. 1985. Agricultural Handbook 639. U.S. Department of Agriculture, Forest Service. Washington, DC.

Swetnam, T. W. and J. L. Betancourt. 1990. Fire-southern oscillation relations in the southwestern United States. *Science* 249: 1017-1020.

ter Braak, C. J. F. 1988. CANOCO-an extension of DECORANA to analyze species-environmental relationships. *Vegetatio* 75:159-160.

Turner, M. G. and R. H. Gardner. 1991. Quantitative methods in landscape ecology: An introduction. In: pp. 3-14. M. Turner and R. Gardner (Eds.). Quantitative methods in landscape ecology. Springer-Verlag. New York.

Urban, D. L., R. V. O'Neill and H. H. Shugart, Jr. 1987. Landscape ecology. *BioScience* 37(2):119-127.

Veblen, T. T. and D. C. Lorenz. 1986. Anthropogenic disturbance and recovery patterns in montane forests, Colorado Front Range. *Physical Geography* 7: 1-23.

Vitousek, P. M. 1986. Biological invasion and ecosystem properties: how can species make a difference. In: pp. 163-176. Mooney, H. A. and Drake, J. A. (Eds.). Ecology of biological invasions in North America and Hawaii. Springer-Verlag. New York.

Walters, C. J. 1986. Adaptive management of renewable resources. McGraw-Hill. New York.

Walters, C. J. and C. S. Holling. 1990. Large-scale management experiments and learning by doing. *Ecology* 71(6): 2060-2068.

White, A. S. 1985. Presettlement regeneration patterns in a southwestern ponderosa pine stand. *Ecology* 66:589-594.

Wiens, J. A. 1985. Spatial scaling in ecology. *Functional Ecology* 3: 385-397.

Wykoff, W. R. 1985. Introduction to the Prognosis Model-version 5.0. In: pp. 44-52. Van Hooser, E. (Ed.). Growth and yield and other mensurational tricks: Proceedings of a conference, November 6-7, 1984, Logan, UT. General Technical Report INT-193. U. S. Department of Agriculture, Forest Service, Intermountain Research Station. Ogden, UT.

Managing Ecosystems for Forest Health: An Approach and the Effects on Uses and Values

Chadwick D. Oliver
Dennis E. Ferguson
Alan E. Harvey
Herbert S. Malany
John M. Mandzak
Robert W. Mutch

ABSTRACT. Forest health is most appropriately based on the scientific paradigm of dynamic, constantly changing forest ecosystems. Many forests in the Inland West now support high levels of insect infestations, disease epidemics, fire susceptibilities, and imbalances in stand structures and habitats because of natural processes and past management practices. Impending, potentially cata-

---

Chadwick D. Oliver is Professor of Silviculture and Forest Ecology, College of Forest Resources (AR-10), University of Washington, Seattle WA 98195.

Dennis E. Ferguson is Research Silviculturist and Project Leader, and Alan E. Harvey is Principal Plant Pathologist and Project Leader, USDA Forest Service Intermountain Research Station, Forestry Sciences Laboratory, 1221 S. Main, Moscow, ID 83843.

Herbert S. Malany is Forester, Boise Cascade Corporation, Boise, ID 83843.

John M. Mandzak is Forest Research and Management Consultant, 26830 Nine Mile Road, Huson, MT 59846.

Robert W. Mutch is Research Applications Leader, USDA Forest Service, Washington Office/Intermountain Research Station, Intermountain Fire Sciences Laboratory, Missoula, MT 59307.

[Haworth co-indexing entry note]: "Managing Ecosystems for Forest Health: An Approach and the Effects on Uses and Values." Oliver, Chadwick D. et al. Co-published simultaneously in the *Journal of Sustainable Forestry* (The Haworth Press, Inc.) Vol. 2, No. 1/2, 1994, pp. 113-133; and: *Assessing Forest Ecosystem Health in the Inland West* (eds: R. Neil Sampson and David L. Adams) The Haworth Press, Inc., 1994, pp. 113-133. Multiple copies of this article/chapter may be purchased from The Haworth Document Delivery Center [1-800-3-HAWORTH; 9:00 a.m. - 5:00 p.m. (EST)].

113

strophic fires can be avoided or modified through proactive forest health management–protecting, creating, and maintaining stand structures, processes, and species populations at viable levels across substantial landscapes. Proactive management will be less costly than fire fighting and associated rehabilitation, especially if done jointly with environmentally sound production of commodities. Management to achieve a fluctuating balance of patterns, processes, and species can begin while better information-based tools and scientific knowledge are being developed. A wider array of silvicultural and other management tools will be required for forest health management than has traditionally been used for commodity management. Specific changes that will allow forest health management include: recognizing the extent and consequences of present imbalances, decentralizing management decisions, adopting management techniques, coordinating with various landowners through incentives, funding forest health management activities, and supporting appropriate research.

INTRODUCTION

The "constant change" ecological theory (Botkin and Sobel 1975; Stephens 1990; Oliver 1992a,b) fits forest ecosystem processes in the Inland West better than the "steady state" theory. Disturbances of all types, climate changes, and species migrations have strongly influenced the constantly changing structure, composition, and function of these ecosystems (Johnson et al. 1993).

Many Inland West forests presently contain or soon will have high levels of biomass as well as large insect and disease infestations (Hessburg 1993). The result will be catastrophic fires and–perhaps even more destructive–reburns of the dead trees created by the first fire (Wellner 1973). These fires will be expensive to fight and recover from and will result in erosion, loss of property (and probably human life), and imbalances of habitats.

These catastrophic fires can be avoided, or their adverse impacts reduced, by using management practices to mimic, avoid, and regulate natural disturbances. For example, thinning an overly dense ponderosa pine forest would begin to return a "park-like" condition, common before fire control, by removing excess stems. The thinning would also reduce the remaining trees' susceptibility to insects, diseases, and fires, thus reducing the potential for catastrophic fires (Oliver et al. 1994). If desired, controlled surface fires could then be reintroduced into these thinned stands. This management would approximate the historical range of disturbances but would apply the disturbances at scales, intensities, and frequencies which would maintain all habitats and species within a landscape area (Everett et al. 1993b).

National Parks and Wildernesses are presently being managed with prescribed natural fires or manager-ignited fires to perpetuate fire-dependent ecosystems on a landscape scale within the historical range of variability. Outside of Parks and Wildernesses, forests can be managed both to sustain healthy ecosystems and to produce such other amenities as wildlife habitat, recreation, timber, fish, and water.

This paper will describe how forests can be managed to restore and maintain forest health. Conditions creating the problems and what specific targets can be managed for will be addressed first, followed by a description of the available tools of management. A discussion of ways of reducing the cost of management and of dealing with uncertainty and a description of present barriers to managing for forest health and ways to overcome them are also included.

THE PROBLEM

Forests fluctuate in their amount of biomass and their structure. Extreme conditions of biomass–too much or too little–or extreme amounts of a single structure can create problems in forest health, as will be described below.

Biomass–"Carbon"–Fluctuations

Forest biomass (organic matter) is primarily composed of carbon, hydrogen, oxygen and (to a lesser extent) other elements. For this paper, fixation of carbon will be used to refer to accumulations of all elements in biomass. Carbon accumulation, movement, and loss are unique to and indicative of the general condition (health) of forest ecosystems and of the balance between fire and biological decomposition. This balance is an especially important characteristic of forests in the Inland West (Harvey 1994; Harvey et al. 1979, 1987, 1994). Forests exist as a balance between the fixation of carbon through photosynthesis and the release of carbon through respiration (biological decomposition) and oxidation (fire) (Harvey 1994; Harvey et al. 1979, 1994) (Figures 1a,b).

Forests in humid areas such as the tropics often maintain a balance of carbon by rapidly releasing it through rotting of dead trees (biological decomposition). In the Inland West, however, dead material does not decompose rapidly (Edmonds 1991) because moisture and warm temperatures are not simultaneously available to promote rotting. Consequently,

FIGURE 1a. Schematic depiction of carbon stored and lost through combustion and biological decay in forest development in a frequent fire environment (left) and in an environment of naturally or artificially infrequent fires (right). Successive bars to the right represent 10-year time intervals for ponderosa pine, 25-year intervals for Douglas-fir, and 100-year intervals for cedar/hemlock forests.

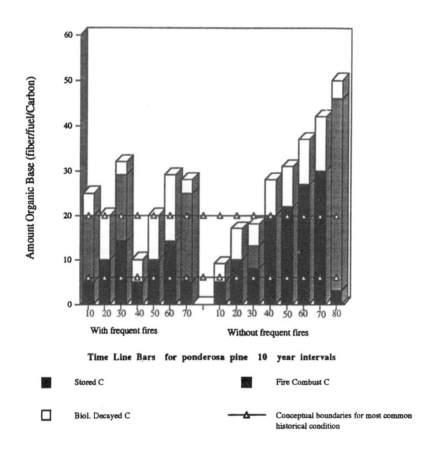

carbon fixation exceeds decomposition in virtually all ecosystems in the Inland West (Olsen 1963). Carbon is a useful indicator of forest health since forests with high carbon accumulations are at risk from insect and disease activities and from high intensity fires.

Accumulations of carbon in excess of biological decomposition either are stored as peat, coal, or similar products or are released by fire. Forests

FIGURE 1b. Timely removal of "potentially available" carbon from these forests preserves both the cycle and the fire regime, including soil organic reserves. Removal also allows sequestration of carbon in long-lived timber products that help offset the cost of removal. Bars represent 10-year time intervals for ponderosa pine. For different species the time interval would vary.

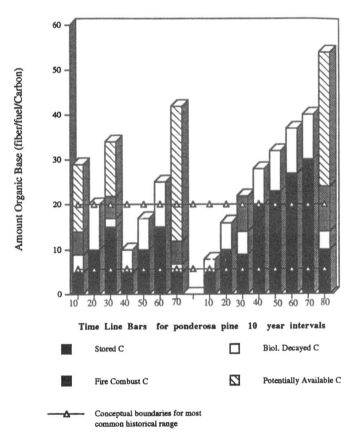

in the Inland West generally release the carbon through fires. The fires can be frequent and of low intensity if little flammable biomass is allowed to build up, but they can be infrequent and catastrophic if much biomass has accumulated (Habeck and Mutch 1973). Extreme wildfire events can produce excessive loss of carbon from the site and create excessive erosion and slow regrowth of vegetation (Harvey et al. 1989; Jurgensen et al. 1990).

Stand Structures

Accumulation of carbon is a characteristic of stand structures, depending on local conditions of soil, climate, disturbance,and other factors (Oliver and Larson 1990) (Figure 2).

These structures change with time and disturbances and eventually become susceptible to insects, diseases, fires, and other agents (Harvey 1994). Following disturbances and the associated release of carbon, the forests regrow through similar or different structures, again depending on local conditions.

Historically, these disturbances often covered large areas, but left "refuges" of "old growth" and other closed forest habitats where plants and animals requiring these structures survived. Plants and animals requiring open areas increased in numbers and avoided genetic inbreeding by out-

FIGURE 2. Stands in the Inland Western United States develop after both stand-replacing and partial disturbances. They change through a variety of structures as they grow. Each structure is suitable for some species and not others. White tree crowns represent shade intolerant species (e.g., pines and larch); dark crowns represent shade tolerant species (e.g., Douglas-fir and true firs). From Johnsen et al. 1993.

A. DEVELOPMENT AFTER STAND-REPLACING DISTURBANCE

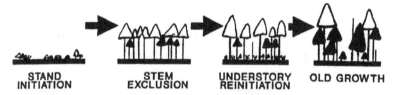

STAND INITIATION   STEM EXCLUSION   UNDERSTORY REINITIATION   OLD GROWTH

B. DEVELOPMENT AFTER PARTIAL DISTURBANCE

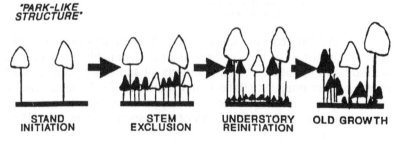

"PARK-LIKE STRUCTURE"

STAND INITIATION   STEM EXCLUSION   UNDERSTORY REINITIATION   OLD GROWTH

breeding during times of abundant habitat immediately after disturbances. With forest regrowth, "open" (stand initiation) areas became more isolated and eventually became refuges where small populations of "open" species survived. Concomitantly, the "closed" forest increased, as did the ability of closed forest species to outbreed. This fluctuation may have been beneficial both to avoid genetic inbreeding of open and closed species and to allow genetic drift during times of isolation.

PRESENT CONDITIONS

Most forests in the Inland West now have large excesses of carbon and an excess of closed forests (especially the "stem exclusion" stage), thus setting the stage for widespread, catastrophic fires. They may be lacking "open" structures–creating shortages of species which utilize open areas (e.g., butterflies [Young 1992], goshawks [Reynolds 1992], bighorn sheep, lynx, *Penstemon lemhiensis, Penstemon attenuatus* var. militaris, *Cirsium subniveum*, and *Juncus hallii* [USDA Forest Service 1990]). The presently volatile condition is the result of natural processes in some places and past management in others.

Many forest ecosystems in the northern Rocky Mountains and Cascade Range developed naturally to high levels of insect infestation and then burned through catastrophic fires (Wellner 1973) at about 100 year intervals (Johnsen et al. 1993; Oliver et al. 1993). Many of these forests burned about 100 years ago; their present extreme insect and fire susceptibilities are probably within the natural range, although not necessarily desired by people.

Other forest ecosystems–such as the ponderosa pine forests of the southern Inland West, the Sierra Nevada Range, and low elevations of the Cascades and northern Rockies–historically had relatively benign, low intensity surface fires which kept the stands in open, park-like structures (Figure 2) (Wellner 1973; Covington and Moore 1994; Johnsen et al. 1993). Many decades of fire exclusion have allowed these fire-adapted ecosystems to grow from open, park-like structures to overly crowded stands with high levels of biomass and insect infestations. These conditions are not within the historical range of variability of these stands and will probably result in catastrophic, stand-replacing fires and reburns–which are also not part of the historical condition of these stands (Mutch et al. 1993).

If the fires occur naturally, they will probably be on such extreme scales that they will reduce the amount of closed forest habitat enough to threaten or extirpate species which need these conditions, as well as endanger people, property, and the future productivity of affected forests.

OPPORTUNITIES TO IMPROVE FOREST HEALTH THROUGH MANAGEMENT

Forest health management can prevent extremely severe fires over large areas by removing excess carbon. The carbon which is removed can be utilized as timber products, thus both substituting for more polluting products and keeping the carbon from being released to the atmosphere as carbon dioxide. This approach can incorporate a wide range of tradeoffs (Figure 1b).

To manage for forest health, the various stand structures, levels of carbon accumulation, and populations of plants and animals would be allowed to fluctuate across each landscape, but minimum amounts of each structure and process would be maintained within a target range. Even with unlimited funds, it will not be possible to engineer forest landscape patterns in great detail because of natural and management-induced variations in stand structures, disturbances, soils, and climate. Across a broad area, management of different landscapes could be roughly coordinated to maintain the fluctuations of stand structures out of synchrony. For example, when one landscape area had large amounts of "old growth" and minimum amounts of "open" areas, a contiguous landscape area could have large amounts of "open" areas and minimum amounts of "old growth."

The appropriate range of variability for structures, processes, and populations can be described for each ecosystem. The historical range of variability would be used as a reference to judge appropriate ranges; however, it may sometimes be appropriate to manage outside the historical range of variability to protect endangered species or valuable properties, to emphasize other values, or to account for changes in climate or species. When outside this historical range of variability, scientific recognition of the departure and recognition of the possible consequences are important. Research can help refine reference ranges of variability.

Exact target amounts of each structure, process, and population will not be necessary to begin management. The information which can be gathered about forests is infinite, and delaying management until more information is gathered can create long delays, during which time catastrophic wildfires of hundreds of thousands of acres will probably have occurred. Delaying operations can also cause equipment owners and skilled labor and professionals to leave the region, so that operations can no longer be performed even if a decision is made to do them. Much management can be done with existing knowledge, since there has been a legacy of over 50 years of intensive research on forests (Baker 1934; Toumey and Korstian 1937; Lutz and Chandler 1946; Smith 1962 and 1986; Daniel et al. 1979; Spurr and Barnes 1980; Kimmins 1987; Hunter 1990; Oliver and Larson 1990).

Management of ecosystems for forest health, like other biological processes, never did or will achieve precise conditions because of the variability of biological and logistic factors. Neither human nor natural disturbance (or other processes) ever provided perfectly predictable structures or patterns in ecosystems; consequently, all forest species are adaptive in their requirements. It will probably be most effective for managers to proceed with trying to maintain or create conditions where all species, patterns, and processes are maintained across the landscape while improving these targets through interacting with the scientific community and through adaptive management (Walters 1986; Everett et al. 1993a).

TOOLS FOR MANAGING FOR FOREST HEALTH

Many tools exist or are rapidly being developed to allow management to create and maintain a targeted variety of forest conditions across the landscape and over time. Some tools are new, and others are modifications of well known practices. Analytical tools are used to determine what silvicultural operations to apply in each stand within a landscape to achieve and maintain the desired balance of conditions. Specific silvicultural operations are needed to achieve these desired structures.

Landscape Management Tools

Analytic tools are rapidly being developed which allow the projection of different stand structures and landscape patterns through time across landscape areas. These tools integrate traditional inventory systems, satellite imagery, geographic information systems, stand projection models, and silvicultural decision keys (Boyce 1985; Oliver et al. 1992; O'Hara and Oliver 1991). They allow rapid visualization of consequences of various management alternatives. They will be very efficient both at determining costs and tradeoffs of various alternatives and implementing desired management directions once the costs and tradeoff decisions have been made.

These tools are not yet readily accessible; however, management can proceed while these tools are being developed. The local manager can subdivide forests into landscape management units along topographic boundaries. On each landscape management unit, the manager can then begin doing operations in those structures which are in extreme excess to create more of those structures which are in extreme shortage. This initial management is highly unlikely to interfere with the more detailed plans developed through the landscape tools, since both objectives will be to reduce the extremes of structures. The initial management will also give local managers understanding and ability to contribute to developing the landscape tools. As the tools become more refined, the ability to determine appropriate ranges of structures can be refined to account for differences in geomorphologic origin, climate, species ranges, and other factors.

Silvicultural Operations

Specific silvicultural operations can be used to create the desired stand structures and biomass accumulations within each stand. These operations include:

- control of tree density and species composition;
- salvage of dead or dying trees to reduce the amount of carbon on the site–and so reduce the potential for unplanned fires or reburns;
- regeneration to achieve the targeted mixture of tree, shrub, and herb species and genotypes at the desired time and place;
- site preparation to reduce undesired fuel, soil, or vegetation conditions;–competition control to encourage targeted species and avoid excesses or non-targeted species (e.g., introduced weeds);
- grazing with domestic or wild animals to obtain and maintain appropriate tree, shrub, and herb mixtures. Silvicultural regimes can also be designed either to encourage or discourage animal grazing;
- fires to reduce carbon buildup in dry fuels or living trees, to control species compositions, and to recycle nutrients;
- productivity enhancement through fertilization, which may also increase tree resistances to insects and diseases;
- gene management for trees, shrubs, and herbs to develop races which are resistant to introduced pests;
- evolving operations to achieve new targets; such operations include creating snags for cavity nesting birds, and creating down logs in appropriate conditions for mushroom production.

These and other silvicultural operations can be used to alter stand structures and landscape patterns. They can be used to enhance wildlife habitat, fish and riparian habitat, forest health, recreation, timber production, and other values as needed.

Integrated Management of Wildlife, Fish, and Pests

Wildlife, fish, and pest management can utilize both silvicultural operations and other techniques to control habitats and population sizes of target, predator, and prey species. This integrated management is most advanced in pest control through "integrated pest management." Through this technique, insect and disease populations are controlled, but not eliminated, using an integration of pest management tactics in a systems approach based on anticipated consequences of the pest behavior.

Maintaining forest health will often require more intensive management than is needed for single commodity production. It is pro-active management. Various portions of the ecosystem will need to be accessed to thin stands. Where roads are undesirable, too expensive, or liable to cause too much stream sedimentation, other forms of transport could be used. Helicopters or balloons may be an alternative. Like other management operations, the cost of a transport system may need to be justified through the joint production of commodity and non-commodity values, including forest health.

OFFSETTING THE COST
OF FOREST HEALTH MANAGEMENT

The various operations needed to maintain forest health will create employment but may represent a cost, although probably not as high a cost as fighting and recovering from uncontrolled fires and pest epidemics. The removed trees can be manufactured into timber products to help offset the cost of removing the trees. Timber products are also an environmentally sound substitute for energy-intensive steel, aluminum, plastic, and concrete (Kershaw et al. 1993), or for timber removed from other places (and other countries) with fewer environmental safeguards.

Managing for forest health while jointly producing other values such as timber can be more cost-efficient than managing for forest health alone (Lippke and Oliver 1993a,b; Oliver et al. 1994). The cost of joint production, however, will often be higher than the cost of managing for a single product such as timber. Forest health management would sometimes mean that a forest area would forego maximum timber production to maintain certain uneconomical (for timber production) structures so the landscape maintains all desired patterns and processes.

Where forest health, wildlife habitat, recreation, or other values are high, these values can help pay the cost of creating and maintaining the structures which are uneconomical solely from a timber production perspective. The joint production of two or more values reduces the cost of producing either value individually. Such joint production practices have been considered as "below-cost" timber sales when values other than timber were not considered to benefit from the operation.

OVERCOMING BARRIERS
TO FOREST HEALTH MANAGEMENT

Various institutional barriers have inadvertently developed which impede forest health management. These barriers include organizational structures, conflicting laws, funding, and research needs.

Organizational Structure

Organizational Structure and Management–Forest management can be more responsive to local or impending forest health conditions if local managers have the authority and responsibility to plan and implement management over local areas–a bottom-up management approach. This management approach allows improved knowledge to be incorporated efficiently in the diverse forest areas through a "flexible systems" approach (Reich 1983). Upper levels of management are useful in a bottom-up approach to coordinate information, direction, resources, and effort to allow local managers to do their job more efficiently .

This flexible systems (bottom-up) approach allows rapid incorporation of new knowledge into management; it allows specific prescriptions to be applied to local circumstances; it allows rapid response to changes induced by unexpected perturbations; and it allows rapid implementation so that skilled labor is not lost and biological opportunities are not foregone during a planning period.

Although the U.S. Forest Service was initially organized to allow local flexibility, many existing laws and practices discourage this "bottom-up" approach. For example, the long time required to change a forest plan means local managers are either operating under outdated plans or are doing no operations until the plans are completed–both of which can be counterproductive. NEPA (National Environmental Protection Act) guides and sets standards, rather than allowing local managers to respond to local conditions. State Forest Practice laws controlling private lands also generally set uniform regulations which do not allow adjustment to local conditions.

Centralized planning and management directions have been shown to be extremely inefficient by the collapse of the Soviet Union. "Top-down" planning relies on the creativity of a few planners, who simply cannot anticipate all conditions. Consequently, others in the system use much of their creativity to circumvent inefficiencies created by central planning.

A bottom-up management style could be effective in forest management through utilizing elements of the U.S. income tax system and medical system. Similar to the payment of income taxes, each local manager would be responsible for managing for the common goal of forest health, but could solicit advice from technical and scientific experts. Line officers would be evaluated and rewarded on their contributions to managing for forest health. Resource specialists in recreation, wildlife, timber production, and other functions would be evaluated and rewarded on how well they used their special knowledge to contribute to the overall objective of forest health, rather than on how well they defended their functional specialty.

A group of "auditors" would occasionally inspect various local management systems, both to offer constructive criticism and to stop any willful deviations from primary objectives. The U.S. medical doctors similarly prescribe treatments on a case-by-case basis, rather than through uniform standards. The doctor is, however, responsible and accountable for using up-to-date scientific and technical knowledge.

Organizational Structure and the Public–Part of the public has lost confidence in the technical ability of foresters. Their credibility needs to be regained through an organizational system which encourages management decisions supported on a scientific basis, a commitment to managing forests using techniques that achieve historical ranges of variability, and a commitment to full public participation in forest management decision making.

Problems in management organization and training of public and private forest management entities have led to failure to communicate with the public and between hierarchial levels of organizations. This "dysfunction" has not allowed different levels of management or the public to realize or understand either the change in ecological theory from the "steady state" to "constant change" or the change in management paradigm from primarily commodity production to primarily forest health maintenance.

An organizational approach which reduces dysfunction was first developed in the United States in the 1930s (Feigenbaum 1951) and expanded in Japan under such names as "statistical process control" and "total quality control" (Deming 1982). It is recently being adopted by many organizations in the United States.

As the concept of forest health management is being developed, these organization and management techniques can help with public coordination, with empowerment of local managers through bottom-up management processes, and with applying adaptive management.

Conflicting Laws

A strong role for politicians and upper level managers will be to ensure that local public and private managers are not encumbered by directives and federal, state, and local government rules and laws which work at cross-purposes. It may require a public decision to identify forest values as a "target" segment of the U.S. economy, followed by detailed examination and rewriting of pertinent legislation, to create the most favorable climate for management of forests for biodiversity. Such targeting and rewriting could produce extremely favorable results for maintenance of biodiversity, other commodity and non-commodity values, and employment under economically efficient conditions.

Specific laws could include incentives for managing for forest health on private lands, air quality regulations relative to controlled burning, and anti-trust laws requiring private landowners to coordinate harvest activities in order to maintain stand structure balances within a landscape of multiple ownerships.

Incentives for Managing for Forest Health on Private Lands–Forest health management on private lands will require more intensive silviculture (and more costs) than traditional commodity management. Except for prevention of pest outbreaks and fires, the public–not the landowner–derives most of the other forest health benefits such as species diversity. The private landowner, therefore, is asked to provide a public value. Mandating the providing of this value through regulations forces the landowner to spend more money on management than is otherwise efficient, which means the cost of timber increases. This raised cost shifts consumption to wood extracted from areas of fewer environmental safeguards or to non-wood substitutes (Perez-Garcia 1993) which are less environmentally sound (Kershaw et al. 1993).

Incentives would reward landowners for providing the public values, rather than punish them as regulations do (Lippke and Oliver 1993a,b). Incentives could be efficiently handled through a free market bidding of the contract to produce certain forest health values (e.g., a needed stand structure) within a landscape area. The landowner who could produce the value at the lowest cost through joint production with timber management and other values would be rewarded by receiving the bid.

Air Quality Regulations and Prescribed Burning–Scientists will need to describe–and regulatory agencies and the public will need to adjust to–the tradeoffs between smoke from increased prescribed fires to avoid wildfires, smoke from the otherwise inevitable wildfires, and ecosystem health requirements which include the presence of fires. There may need to be changes in regulations or basic laws to allow these adjustments.

Anti-Trust Laws and Managing Landscapes with Multiple Owners–Forest health management will require coordination of timber harvesting among private landowners in some areas. The landowners become liable for violation of the Sherman Anti-Trust Act if such practices are done in a cooperative manner. The competitive bidding described above can avoid this liability. Another approach is to require each landowner to maintain certain proportions of each structure and/or species populations on their lands, but with the ability to trade (buy/sell) acres of structure and numbers of animals with other landowners within a landscape unit (Oliver 1992b). This system is similar to the "tradeable pollution quotas" of the 1990 Clean Air Act. Such systems would be less subject to anti-trust violations if state laws mandate them.

Funding Allocations

Forest health management in the Inland West must also focus on the monetary means to achieve it. Special appropriations will probably be needed to do the extra forest operations necessary to maintain forest health. The degree to which the balance can be maintained will be directly related to both the efficiency of management and the funding available.

Management will be most efficient if funds are directed toward ecosystem health, rather than toward traditional functional targets such as timber, wildlife, recreation, grazing, or similar single values. The funds will also be most efficiently used if forest health management is done jointly with production of commodities, as discussed earlier.

Several sources of public funds have been described by Lippke and Oliver (1993b). Since forest health management would reduce the costs of fighting otherwise inevitable wildfires, it may be appropriate to fund forest health management from the same source that pays for very large forest fires.

Research Needs

It is important that some research and development be targeted to ongoing specific problem areas related to ecosystems and forest health. For example:

- Specific areas where basic information is needed to improve management can be identified by local managers;

- Incorporating research scientists into the adaptive management process will then increase the scientific accuracy and statistical soundness of monitoring and analysis;
- Continued research on forest patterns, disturbance processes, and populations will help maintain critical features, provided the research concentrates equally on all seral stages, species, and habitats;
- Research and development on information systems, management tools, and equipment will increase the efficiency of planning and implementing management at the landscape and stand levels;
- Research into commodities such as wood products that increase their ease of management or usefulness (and value) to society will reduce the cost (and increase the incentives) for a landowner to grow forests for forest health through the joint production process, as discussed earlier.

A SYSTEMATIC APPROACH
TO ACHIEVING FOREST HEALTH

The above principles can be incorporated into a systematic process to manage for forest health. For example, an eight-step process is described below. All steps can begin simultaneously and be improved reiteratively in an "adaptive management" process to avoid further delays in improving forest health.

Step 1. Describe a historical range of variability of such elements as disturbance regimes, stand densities, species, and vegetation structures using all available information. Knowledge of this variability provides a reference for assessing the current condition and the range of desired conditions of the landscape–although historical ranges are references, not recipes (Everett et al. 1993).

Step 2. Assess possible current departures from historical ranges of variability to determine whether present trends are within or outside of these ranges.

Step 3. Involve scientists, the public, managers, and all landowners in determining an acceptable range of variability using this historical range as a reference.

Step 4. Involve all partners in developing strategies on a landscape scale to return trajectories within bounds of acceptable variation. To involve the public meaningfully, the process must do more than ask the public to ratify foregone conclusions and must be more than the compromises among well-represented interest groups (Hanrahan 1993).

Step 5. Use large areas to feature alternative forest health management strategies. The large areas and alternative strategies ensure that opportunities to promote forest health are not lost because they do not follow the prevalent theory (Walters 1986).

Step 6. Practice adaptive management techniques to be in a position to modify actions and prescriptions as new feedback information becomes available.

Step 7. Highlight consequences of alternative management actions so that various publics appreciate the effects of different alternative courses of action.

Step 8. Conduct education and awareness programs at all levels, both internally and externally, so that people understand the application of forest health management principles.

CONCLUSIONS

The serious decline of forest health in the Inland West can be attributed both to natural conditions and to management actions that did not fully recognize the role of disturbances in sustaining forest health. Diseases, insects, and fires have periodically occurred within the Inland West with different ranges of variability for different areas. The effective exclusion of fire over the past six to seven decades has changed the structure, species composition, and diversity of many forests. Other forests are naturally returning to conditions where they are susceptible to insects, diseases, and catastrophic fires. People, their property, and natural resources are at risk.

The solution to this forest health problem seems straightforward, but it has some significant barriers. The techniques, costs, and values of forest health management are different from timber extraction, fire fighting, or individual endangered species preservation. The difference is a shift toward proactive management to prevent excessive fires and insect epidemics and species declines, and away from reactive management to recover from them. The solution could start with thinning and harvesting to reduce the density of overcrowded stands. Where large quantities of standing dead trees are present, salvage logging could be encouraged to remove accumulations of fuels and to obtain wood products. Diverse silvicultural operations applied systematically within a landscape area can maintain biodiversity and reduce insect and fire hazards by creating a diversity of stand structures and levels of biomass. The silvicultural operations can include thinning, regenerating, selection and even-age harvesting, using prescribed fire, and creating snags, among others. Integrated management of pests, wildlife, and fish will help ensure their balance as well.

Successful management for forest health in the Inland West will depend on how well internal and external barriers can be overcome to allow the management to be done on a large enough scale to make a difference. Embarking on a major paradigm shift toward forest health management will also require a significant paradigm shift toward the public perceiving ecosystems as dynamic and changing. Placing the priority on valuing the health of entire ecosystems will require increased scientific understanding on the part of natural resource specialists and managers, as well as on the part of the general public, politicians, and regulatory agencies.

REFERENCES

Baker, F.S. 1934. Theory and practice of silviculture. McGraw-Hill New York. 502 p.

Botkin, D.B. and M.T. Sobel. 1975. Stability in time-varying ecosystems. *The American Naturalist* 109 (970): 625-646.

Boyce, S.G. 1985. Forestry decisions. USDA Forest Service General Technical Report SE-35. 318 p.

Covington, W.W. and M.M. Moore. 1994. Postsettlement changes in natural fire regimes and forest structure: Ecological restoration of old-growth ponderosa pine forests. In: pp. 153-181. Sampson, R.L. and D.L. Adams (Eds.). Assessing Forest Ecosystem Health in the Inland West. Proceedings of the American Forests scientific workshop, November 14-20, 1993, Sun Valley, ID. The Haworth Press, Inc. New York.

Daniel, T.W., J. Helms and F.S. Baker. 1979. Principles of silviculture. 2nd edition. McGraw-Hill. New York. 500 p.

Deming, W.E. 1982. Out of the crisis. Massachusetts Institute of Technology, Center for Advanced Engineering Study. Cambridge. 507 p.

Edmonds, R.L. 1991. Organic matter decomposition in Western United States forests. In: pp. 90-194. Harvey, A.E. and L.F. Neuenschwander (Comps.) Proceedings-Management and productivity of western montane forest soils. USDA Forest Service, Intermountain Research Station. Ogden, UT.

Everett, R., C. Oliver, J. Saveland, P. Hessburg, N.Diaz, and L. Irwin. 1993. Adaptive ecosystem management. In: pp. 361-375. M.E. Jensen and P.S. Bourgeron (Eds.). Eastside forest ecosystem health assessment, Volume II. Ecosystem Management: Principles and Applications. USDA Forest Service, National Forest System, Forest Service Research. Wenatchee, WA.

Everett, R., P.F. Hessburg, M. Jensen, and B. Bormann. 1993. Eastside forest ecosystem health assessment, Volume I. Executive Summary. USDA Forest Service, National Forest System, Forest Service Research. Wenatchee, WA. 57 p.

Feigenbaum, A.V. 1951. Total quality control. McGraw-Hill, Inc. New York. 851 p.

Habeck, J.R. and R.W. Mutch. 1973. Fire-dependent forests in the northern Rocky Mountains. *Quaternary Research* 3: 408-424.

Hanrahan, T.P. 1993. Soft systems analysis and public lands planning. Unpublished Masters of Science thesis. Washington State University, Department of Natural Resource Sciences. Pullman. 98 p.

Harvey, A.E. 1994. Integrated roles for insects, diseases and decomposers in fire dominated forests of the Inland Western United States: Past, present, and future forest health. In: pp. 211-220. Sampson, R.L. and D.L. Adams (Eds.). Assessing Forest Ecosystem Health in the Inland West. Proceedings of the American Forests scientific workshop, November 14-20, 1993, Sun Valley, ID. The Haworth Press, Inc. New York.

Harvey, A.E., M.F. Jurgensen and R.T. Graham. 1989. Fire-soil interactions governing site productivity in the Northern Rocky Mountains. In: pp. 9-18. Baumgartner, D.M. et al. (Comps.). Prescribed fire in the Intermountain Region, Symposium Proceedings, Washington State University, Cooperative Extension. Pullman, WA.

Harvey, A.E., M.F. Jurgensen, M.J. Larsen and R.T. Graham. 1987. Decaying organic materials and soil quality in the Inland Northwest: a management opportunity. General Technical Report INT-225. USDA Forest Service, Intermountain Research Station. Ogden, UT. 15 p.

Harvey, A.E., M.J. Larsen and M.F. Jurgensen. 1979. Fire-decay: interactive roles regulating wood accumulation and soil development in the Northern Rocky Mountains. Research Note INT-263. USDA Forest Service, Intermountain Forest and Range Experiment Station. Ogden, UT. 4 p.

Harvey, A.E., G.I. McDonald, M.F. Jurgensen and M.J. Larsen. 1994. Microbes: drivers of long-term ecological processes in fire-influenced cedar-hemlock-white pine forests in the Inland Northwest. In: Baumgartner, D.A. (Comp.). Interior cedar-hemlock-white pine forests: Ecology and Management. March 2-4, 1993. Spokane, WA. Washington State University, Cooperative Extension. Pullman, WA. In press.

Hessburg, P.F. 1993. Eastside Forest Ecosystem Health Assessment. Volume III. Assessment. USDA Forest Service, National Forest System, Forest Service Research. Wenatchee, WA. 750 p.

Hunter, M.L., Jr. 1990. Wildlife, forests, and forestry. Regents/Prentice Hall. Englewood Cliffs, NJ. 370 p.

Johnsen, C.G., R.R. Clausnitzer, P.J. Mehringer, and C.D. Oliver. 1993. Biotic and abiotic processes of eastside ecosystems: the effects of management on plant and community ecology, and on stand and landscape vegetation dynamics. In:

pp. 36-100. Hessburg, P.F. (Ed.). Eastside Forest Ecosystem Health Assessment. Volume III. Assessment. USDA Forest Service, National Forest System, Forest Service Research. Wenatchee, WA.

Jurgensen, M.F., A.E. Harvey, R.T. Graham, M.J. Larsen, J.R. Tonn and D.S. Dumroesse. 1990. Soil organic matter, timber harvesting, and forest productivity in the Inland Northwest. In: pp. 392-415. Gessel, S.P. et al. (Eds.). Sustained productivity of forest soils. Proceedings of the 7th North American Forest Soils Conference. University of British Columbia, Faculty of Forestry Publication. Vancouver.

Kershaw, J.A. Jr., C.D. Oliver and T.M. Hinckley. 1993. Effect of harvest of old growth Douglas-fir stands and subsequent management on carbon dioxide levels in the atmosphere. *Journal of Sustainable Forestry* 1(1): 61-67.

Kimmins, J.P. 1987. Forest ecology. Macmillan. New York. 531 p.

Lippke, B. and C.D. Oliver. 1993a. An economic tradeoff system for ecosystem management. In: pp. 337-345. Jensen, M.E. and P.S. Bourgeron (Eds.). Eastside forest ecosystem health assessment, Volume II. Ecosystem Management: Principles and Applications. USDA Forest Service, National Forest System, Forest Service Research. Wenatchee, WA.

Lippke, B. and C.D. Oliver. 1993b. Managing for multiple values. *Journal of Forestry* 91: 14-18.

Lutz, H.J. and R.F. Chandler, Jr. 1946. Forest soils. John Wiley & Sons. New York. 514 pp.

Mutch, R.W., S.F. Arno, J.K. Brown, C.E. Carlson, R.D. Ottmar and J.L. Peterson. 1993. Forest health in the Blue Mountains: a management strategy for fire-adapted ecosystems. USDA Forest Service, Pacific Northwest Research Station. General Technical Report PNW-GTR-310. 14 p.

O'Hara, K.L. and C.D. Oliver. 1991. Silviculture: achieving new objectives through stand and landscape management. *Western wildlands* 17: 28-33.

Oliver, C.D. 1992a. Enhancing biodiversity and economic productivity through a systems approach to silviculture. In: pp. 287-293. The Silviculture Conference. Forestry Canada. Ottawa, Ontario.

Oliver, C.D. 1992b. A landscape approach: achieving biodiversity and economic productivity. *Journal of Forestry* 90: 20-25.

Oliver, C.D., D.R. Berg, D.R. Larsen, and K.L. O'Hara. 1992. Integrating management tools, ecological knowledge, and silviculture. In: Naiman, R. (Ed.). New perspectives for watershed management. Springer-Verlag. New York.

Oliver, C.D. and B.C. Larson. 1990. Forest Stand Dynamics. McGraw-Hill. New York. 467 p.

Oliver, C.D., L.L. Irwin and W.H. Knapp. 1993. Eastside forest management practices: Historical overview, extent of their application, and their effects on sustainability of ecosystems. In: pp. 216-288. Hessburg, P.F. (Ed.). Eastside Forest Ecosystem Health Assessment. Volume III. Assessment. USDA Forest Service, National Forest System, Forest Service Research. Wenatchee, WA.

Oliver, C.D., C. Harrington, M. Bickford, R. Gara, W. Knapp and G.Lightner. 1994. Maintaining and creating old growth structural features in previously disturbed stands typical of the eastern Washington Cascades. In: pp. 353-387. Sampson, R.L. and D.L. Adams (Eds.). Assessing forest ecosystem health in the Inland West. Proceedings of the American Forests scientific workshop, November 14-20, 1993, Sun Valley, ID. The Haworth Press, Inc. New York.

Olsen, J.S. 1963. Energy storage and balance of producers and decomposers in ecological systems. *Ecology* 44: 322-331.

Perez-Garcia, J.M. 1993. Global forestry impacts of reducing softwood supplies from North America. CINTRAFOR Working Paper 43. University of Washington College of Forest Resources Center for International Trade in Forest Products. 35 p.

Reich, R.B. 1983. The next American frontier. The New York Times Book Co. New York. 324 p.

Reynolds, R.T. 1992. Quoted from Seattle Post-Intelligencer Newspaper, p. A-1 & A-6. November 9, 1992. Mr. Reynolds is a research wildlife biologist, USDA Forest Service, Rocky Mountain Station. Fort Collins, CO.

Smith, D.M. 1962. The practice of silviculture. 7th edition. John Wiley & Sons. New York. 578 p.

Smith, D.M. 1986. The practice of silviculture. 8th edition. John Wiley & Sons. New York. 527 p.

Spurr, S.H. and B.V. Barnes. 1980. Forest Ecology. 3rd Edition. John Wiley & Sons. New York. 687 p.

Stevens, W.K. 1990. New eye on nature: the real constant is eternal turmoil. New York Times. Science article. Tuesday, July 31, 1990. B5-B6.

Toumey, J.W, and C.F. Korstian. 1937. Foundations of silviculture upon an ecological basis. John Wiley & Sons. New York. 456 p.

USDA Forest Service. 1990. Final environmental impact statement. Trail Creek Timber Sale, Beaverhead National Forest. Dillon, MT. 122 p.

Walters, C. 1986. Adaptive management of renewable resources. Macmillan Publishing Company. New York. 374 p.

Wellner, C.A. 1973. Fire history in the northern Rocky Mountains. In: pp. 42-64. The role of fire in the Intermountain West: Proceedings of the Intermountain Fire Research Council. October 27-29, 1970. University of Montana. Missoula.

Young, M.R. 1992. Conserving insect communities in mixed woodlands. In: pp. 277-296. Cannell, M.G.R., D.C. Malcolm, and P.A. Robertson (Eds.). The Ecology of Mixed-Species Stands of Trees. Blackwell Scientific Publications. Boston, MA.

Human Dimensions
of Forest Health Choices

Charley McKetta
Keith A. Blatner
Russell T. Graham
John R. Erickson
Stanley S. Hamilton

ABSTRACT. Forests in the Inland West have been shaped by over a century of human migration and forest use. Relative forest health is still defined by human objectives in forest management. Economic analyses can help people understand the forest health implications of management choices between forest preservation, commodity production, or ecosystem management. Neo-classical microeconomics predictably casts the optimal forest health question as a cost-benefit trade-off even though valuation of forest health benefits must be inferred. Alternative schools of thought, such as institutional economics, may reach different conclusions. Lack of public consensus on forest health objectives and acceptable remedies conflict with traditional deterministic analytic tools. An alternative planning ap-

Charley McKetta is Forest Economist, Department of Forest Resources, University of Idaho.

Keith A. Blatner is Associate Professor of Forest Economics, Department of Natural Resource Sciences, Washington State University, Pullman, WA 99164-6410.

Russell T. Graham is Research Silviculturist, USDA Forest Service, Intermountain Research Station, Moscow, ID.

John Erickson is Wildlife Manager, USDA Forest Service, Boise National Forest.

Stanley S. Hamilton is Director, Idaho Department of Lands.

[Haworth co-indexing entry note]: "Human Dimensions of Forest Health Choices." McKetta, Charley et al. Co-published simultaneously in the *Journal of Sustainable Forestry* (The Haworth Press, Inc.) Vol. 2, No. 1/2, 1994, pp. 135-149; and: *Assessing Forest Ecosystem Health in the Inland West* (eds: R. Neil Sampson and David L. Adams) The Haworth Press, Inc., 1994, pp. 135-149. Multiple copies of this article/chapter may be purchased from The Haworth Document Delivery Center [1-800-3-HAWORTH; 9:00 a.m. - 5:00 p.m. (EST)].

135

proach that makes problem definition the focus of analysis may facilitate public forest health decisions.

INTRODUCTION

Forest health is a condition of forests reflecting the complexity of their ecosystems while providing for human needs (O'Laughlin et al. 1993). Health is both a relative and a normative concept depending on management objectives and the standard of reference. Among the varied forests of the Inland West, managerial and societal perceptions of optimal forest health can vary significantly.

Inland forests evolved on extremes of land forms and climatic conditions which produced diverse forest ecosystems. Their successional patterns are closely linked with historic disturbance events, especially insect, disease and fire cycles. Forest types associated with frequent natural disturbances are experiencing dramatic changes that appear to be outside the historic range of variability (HRV). Some observers have labeled changes in species composition, stand structure and disturbance patterns as poor forest health and called for a new forest health initiative to heal sick Inland forests (Everett et al. 1993).

We address socio-economic forest health considerations relevant to different forest management strategies. We consider:

1. What kinds of socio-economic pressures shaped these forests and might affect future forest management strategies?
2. Are techniques available to help people better understand the variability of possible futures and options?
3. How can managers and society consider forest health trade-offs in managing Inland Forests?

THE INLAND FOREST SETTING

Inland forests grow in the valleys and on the slopes of the numerous mountain ranges. Private forest lands occupy the more accessible lower terrain and patchwork railroad grants. They are surrounded by extensive areas of usually higher elevation federal forest lands. Across the region both industrial and non-industrial private forests usually have slightly lower than average site class lands (Green and Van Hooser 1983). These forests have younger age class distributions, higher stocking and higher ratios of harvest to inventory.

State lands are widely distributed due to the original land office allocation patterns and subsequent land trades. Most western states have land organizations which operate forests to maximize long-term financial returns to school endowment funds.

Federal forests occupy the largest area, but due to settlement patterns and late forest establishment, they have many less-accessible sites with higher elevations. Large roadless, wilderness, and reserved areas occupy the most inaccessible and frequently higher terrain. Across the Inland West, public forest area which is open to commodity production is less heavily stocked with trees than other ownerships, has a greater variety of ecosystems represented, and a broader distribution of age classes, but not size classes (Green and Van Hooser 1983). Regional averages may not hold locally (e.g., north Idaho–size classes are skewed higher than elsewhere).

Resource interests have described many forest health problems on these public forests. During recent years, because of dry weather and forest conditions, many forests have experienced severe mortality and subsequent record forest fire years (O'Laughlin et al. 1993).

Over the last century and a half, rural sawmill and ranch service communities were established to utilize commodities produced from adjacent, usually public, forests. While some communities exhibit stabilizing economic diversity, most smaller communities are commodity resource based. They are populated by third and fourth generation families who are closely tied to specific resource oriented lifestyles (Carlson 1994). Forest health influences the flow of commodities which directly affects the socioeconomic health of local human communities.

In recent decades, urban professionals began relocating to rural communities in the Inland West, especially near resorts and wilderness areas for forest amenity values (Rudzitis and Johanson 1989). The ability of communities to respond to changes in forest policy, away from traditional commodity production objectives, depends on the extent of socio-economic change already taking place. Specifically, in some communities, high rates of transferred income and other business lifestyles have already replaced higher salaried commodity based lifestyles (Carlson 1994).

MANAGERIAL ANTECEDENTS
FOR FOREST HEALTH DECISIONS

Private Industrial and Non-Industrial Forests–Most private forests are in their second or third cycle of timber production with little consideration of non-timber uses. Early entries extracted valuable species such as old-growth western white pine (Pinus monticola) and ponderosa pine (Pinus ponderosa). "High grading" left large bare areas or stands of mixed conifers disrupting successional patterns. Regeneration and the resulting forest structure was an afterthought.

Twentieth century re-entries focused on each remaining species as their market values increased. Logging favored extensive roading with tractor and ground cable yarding systems, disregarding effects on soil, water and future stand structure, until state forest practice acts and private management focused on the sustainability of these increasingly valuable forests. Most private (industrial and non-industrial) forests typically maintain low growing stock densities and narrow young age class distributions to maximize the efficiency of forest capital. Forest health concerns on private land have traditionally focused on tree/stand health.

National Forest Lands–Legislative mandates direct public forest management. The Organic Administration Act of 1897, the Multiple-Use Sustained Yield Act of 1960 (MUSYA), the National Environmental Protection Act of 1969 (NEPA), and the National Forest Management Act of 1976 (NFMA), regulate national forest decisions. Commodities and values emphasized on Inland forests were highly variable, but resource value extraction potential and resiliency to disturbance meant that dedication of lands to specific uses was not uncommon (Overbay 1992).

Early forest management was custodial, primarily fire protection. Fire exclusion and suppression transcended wilderness and National Park lands. The value of forests for timber, scenery, watershed protection, wildlife habitat, and forage for domestic animals encouraged aggressive fire suppression policies, especially after large fires of the early 1900s. Other scientists in this volume argue that such fire policies may be a major cause of contemporary forest health problems.

In the 1930s, the National Forests became significant producers of timber. After World War II, intensive "timber" management became emphasized (Kennedy and Quigley 1993). Forest management for timber production often focused analyses and treatments at the stand level without considering landscape interactions. Domestic livestock grazing on forest lands also had a profound impact on forest ecosystems.

Non-market commodity production, including watershed protection and recreation, was mandated by MUSYA. Big game hunting and herd enhancement influenced management and habitat treatments on many areas. Activities oriented to direct human uses of forest resources, primarily timber along with forage, big game, and recreation commodities have dominated the management strategies and budgets of most public forests.

Commodity and small spatial orientations eventually generated conflicts with non-commodity forest users. Pressure for non-commodity issues to direct forest management has been increasing. The Wilderness Act of 1964 and state wilderness bills specifically dedicated areas for intangible benefits of untrammeled space. Under the Endangered Species Act (ESA), large forested tracts are being reserved as habitat to protect individual species such as the spotted owl, and the Snake River anadromous species. Unfortunately, resource and species protection zones are often protected from fire, but receive little other biological management designed to ensure long-term forest health.

STRATEGIC OPTIONS
FOR FUTURE FOREST MANAGEMENT

For purposes of discussion consider the following three abstract strategic philosophies (referred to herein as "strategies") for managing Inland forest resources: (1) preservation, (2) mixed commodity production, and (3) ecosystem management. Each strategy implies different definitions of optimal forest health and each implies different risks in their applications. Although simplistic, these strategies reflect styles of management currently under debate in the Inland West.

Preservation

This strategy zones forests for specific non-commodity objectives such as endangered species protection and wilderness areas. Spatial reserve options frequently presume that ecosystems do not change, or change at very slow rates, and ignore forest health.

A more sophisticated and sustainable variation of this strategy is constant site maintenance to hold succession in abeyance so that a finely tuned ecosystem matches static preservation needs. While this variation seems to grade into ecosystem management, the difference is that "ecosystem management" recognizes the landscape scale of a dynamic system as well as some continued provision of commodities.

Mixed Commodity Production

This strategy sees forests as mechanisms for directly producing human needs in the broadest sense. Forest ecosystems utilized for commodity production (e.g. timber) are often pushed outside natural HRVs to enhance output characteristics. Sustainability within this strategy is premised on artificially maintaining stand characteristics through regular silvicultural treatment. Manicured "natural" park vistas, tree farms with enhanced game habitat, or silvo-pastoral systems are examples. A forest's health is directly linked to its technical efficiency in commodity production.

Ecosystem Management

Ecosystem management is new and being implemented on federal forest lands. This approach focuses on the interactive functions of all ecosystem components at multiple spatial and temporal scales. Ecosystem management precludes neither commodity production nor maintaining specific components. Its purpose is to sustain ecosystems using HRV as a standard of reference. Human expectations are tempered by the limitations of a naturally sustainable ecosystem that retains its natural functions and processes. Ecosystem management is often considered as a mechanism for achieving forest health. Under this strategy a "healthy ecosystem" may be consistent with high levels of tree mortality.

CONTRIBUTIONS AND LIMITATIONS OF ECONOMIC THEORY

There is persistent disagreement among resource economists about the appropriate use and limitations of analytical tools and economic theory in public resource policy applications. Although disconcerting to non-economists, this disagreement reflects fundamentally different philosophies and methodologies common to resource economics (Randall 1985). The application of economic theory to the problems of forest health is no different.

Forest economics has been dominated by neoclassical economists who are oriented to rational comprehensive planning. Neoclassical thought cannot be separated from the evolution of benefit-cost analysis procedures (Schulze and Howe 1985). Neoclassical economists strongly believe in resource allocation efficiency to maximize either financial returns or net social benefits depending on forest ownership criteria. In the case of forest resources this implies evaluation of forest capital efficiency.

Traditional Capital Theory Considerations

Forest health is a normative concept that will vary depending on the forest management strategy chosen. "Good" forest health implies a forest ecosystem with a structure that is complementary to its set of long-run management objectives. "Poor" forest health exists when a forest's ecosystem condition detracts from the potential productivity of that forest, as defined by one of the strategic philosophies. Costs of forest health decisions can be measured either as tangible cash flows or economically as trade-offs imposed, regardless of how measured or who bears them.

Economic Costs of Forest Management–The largest cost of forest management never appears on the books–the opportunity cost of forest capital that has accumulated in growing stock. It is predicated on other forest uses foregone once a particular forest allocation is made.

In the commodity strategy, capital costs cause private forests to lower stand densities and narrow the age class distribution and reduce species complexity. This lowers risks and costs of forest protection and maximizes net sustainable forest cash income. Public forests ignore the opportunity value of accumulated forest capital so the application of the same strategy maintains relatively more biomass and recognizes more commodities. In both cases, these simplified ecosystems are artificially sustained, although not at natural levels of complexity. Both jointly produce many non-commodity benefits, but at lower levels than may be possible from more complex ecosystems. Financial and biological risks are a function of the magnitude of forest capital, so there is a close association between capital parsimony and forest health. Where forest capital is thought to have no opportunity value, its loss through mortality or catastrophic destruction is often ignored. Preserved ecosystems may not be sustainable in the long-run, as significant protection efforts may actually destabilize these ecosystems. Instability imposes costs in the additional risks that forest assets might be totally dissipated by catastrophic change.

Both preservation and ecosystem management strategies imply that complex ecosystem structures are maintained. These conditions tend to emphasize and protect different sets of non-commodity benefits. Ecosystem management has other correlated management and protection costs. First, forest revenues decline by the extent that commodity targets are lowered to achieve naturally sustainable ecosystems. Second, more costly intervention may be necessary to maintain and protect these complex systems.

Measuring Financial Costs of Ecosystem Management–Currently, both private and public forests use timber-based single resource accounting and budgeting approaches even though timber activities jointly produce non-commodities. Cash accounting obviously biases against discretionary non-timber objectives. The bias in public timber accounting causes many timber programs to appear financially below cost even though they may generate significant complimentary social benefits.

Financial costs of achieving public timber and non-timber objectives such as forest health appear in two forms: direct and indirect. For all three strategies, the direct costs are silvicultural practices to establish, maintain, and enhance forest ecosystems towards a desired state. Manipulation to achieve forest structural objectives also causes indirect costs: revenue losses from attaching enhancement practices to timber sales (e.g., expanded road networks, improved wildlife habitat, forest health maintenance, etc.) Timber sale losses financially mitigate non-timber activities. These costs are real and may never appear on timber program balance sheets, but can be estimated (Schuster and Niccolucci 1989) and are increasing (McKetta 1994).

The Costs and "Value" of Forest Health–To the extent that forest health is to be achieved through management, its cost is like any other non-commodity objective, being the sum of actual achievement expenditures, plus the opportunity cost of any activities forgone (e.g., commodity sales).

As long as decision makers are explicitly willing to incur costs to achieve a forest health state, it is unnecessary to value forest health itself. This logic comes from fire protection analyses that show that minimizing the least cost plus net value change in the protected resources is equivalent to maximizing protection values (Gorte and Gorte 1979). The net value changes of intangible protected resources (particularly under preservation and ecosystem management) continue to present difficult valuation problems.

The value of "good" forest health becomes the reduction of costs imposed by "poor" forest health. The cost of "poor" forest health can be estimated by the reduced value of the productive forest asset, plus the costs of mitigating the bad health (e.g., protecting against fire). Where undesirable fire risks are a consequence of a management strategy, this component of "poor" health cost is the difference of effective fire protection costs between the two health states.

The Logic of Investing in Forest Health–Managers of all three strategies would invest in forest health improvement only when the discounted and risk adjusted gains from healthy forests, plus the deferred costs of protecting a sick forest, outweigh the treatments necessary to change its health. Where forests are sustainably productive (relative to specific strategy objectives) in a subperfect health condition, improvements may not be rational. However, investment to prohibit forest health from slipping below that level may be rational. There are two considerations: first, productive losses and protection costs rise at increasing rates, so their combined discounted impacts rise; second, costs of restoring deteriorated forest health are probably larger than the costs of maintaining it.

Four costs are relevant. Productivity of commodity and non-commodity objectives drops (although timber salvage may temporarily rise), the risk of forest asset dissipation rises, annual protection costs rise, and deferred restoration increases its costs. The sum of these may exceed the costs of restoration implied under ecosystem management.

Managerial Behavior in Risk Decisions–There is a risk/cost trade-off and people will hedge their bets by operating within the healthier standard deviation of any forest health reference point. How far they operate from the mean depends on subjective evaluation of the marginal risk versus the marginal costs of reducing risk further. People who face the costs of temporally expanding risk, or operate under uncertainty, will time-weight their decisions. It is common to prefer known current states even if they are inferior to uncertain future states, and foresters are probably more risk averse than most decision makers. Decision maker risk adversity also leads to fewer decisions where controversial trade-offs must be faced. Public foresters, especially, bear the cost of controversy or error in risk taking, but rarely the costs of delay, failure to innovate, or inefficiency. Accumulated bet hedging from one authority level to the next compounds this behavior. If forest health initiatives serve as political as well as resource risk insurance policies, and budgets are not constrained, investment in them could exceed socially optimal levels.

ALTERNATIVE VIEWS

Even though the neoclassical view has dominated resource decisions in recent decades, at least three other schools of economic thought are influential in resource decisions (Randall 1985). Alternative economic views could result in largely different forest health conclusions. For example, institutional economists utilize a holistic methodology where neoclassical economists focus on the aggregate behavior of individuals. The institutional methodology reflects subjective thoughts, feelings and values differently. Institutionalists tend to see social values beyond the revealed individual preferences measured by neoclassicists. "The institutionalist perspective . . . does not so much refute the neoclassical/efficiency perspective as incorporate and supersede it" (Wandschneider 1986:95). Thus, an analysis of the forest health problem from an institutionalist perspective could result in substantially different conclusions and higher investments for forest health.

Increasingly negative public opinion of traditional resource management decisions suggests that relative public values of different resources appear to be sharply changing. Forest efficiency remains important, but is measured in the achievement of a different set of forest outputs. The growing support for preservation is a direct outgrowth of changing public values. Iverson and Alston (1993) argue for broader economic focus in resource decisions beyond the strict limitations of the neoclassical viewpoint. Ecosystem management is an effort by resource managers to incorporate the public's growing concern for environmental quality in resource management, while recognizing the dynamic nature of ecosystems.

POTENTIAL VECTORS OF CHANGE

Resource managers will be confronted with a wide array of new predictable pressures and stochastic events. These include both ecological and social pressures.

Bio-Technical Pressures

Over the next century, radical changes in climate may cause forests to undergo further changes in species makeup, appearance (structure), and vitality (function). The Inland region is vulnerable to global warming and to extreme moisture stress (Auclair 1994). As CO_2 can be sequestered by forests, there may be pressures to maintain biomass, and growth rates could accelerate.

Science and technology will further change human expectations from forests. New products constantly appear (e.g., the use of Pacific yew for taxol production). Milling and harvest efficiencies continue to use less wood and labor to produce the same volume of products, albeit frequently of lower product quality.

Socio-Economic Pressures

As societies develop they spend proportionally less on basic resources. U.S. wood consumption is rising more slowly than population. Almost all primary resources, including wood, are becoming economically less scarce (Barnett 1979). Reduced societal commodity dependence frees forests for non-wood purposes, although temporary local scarcity may occur from policy-induced forest reallocations.

Changing demographics such as increasing age, socio-economic class, education and cultural influences also change non-market forest demands.

Intangible amenity values have risen relative to commodities and will continue, but forms will change. With aging, dispersed trail use may give way to developed and road recreation. The rate of rising environmentalism could abate. The second wave of the environmental movement has ethics that are more proportioned to other high priority issues, such as national defense and deficit spending–or "peace" and "pocket book"–that have emerged over the last two decades.

Public perceptions of forests are symbolic and they often view the forest resource in relatively simple terms. They are less aware of the ecological or production dynamics of the forest system than they are of human motives for forest control. "Last-one-in" and "not-in-my-backyard" protectionism are examples of inconsistent compartmentalized thinking that need to be addressed. Otherwise, forest health programs will be adversely affected by public distaste for the tools essential to achieve forest health, such as prescribed fire and harvesting, which will conflict with public unwillingness to accept the consequences of catastrophic fire.

An increasingly polarized public will continue to debate issues regarding the utilitarian versus preservation role of forests. There are extreme personal differences in their interest or understanding of ecological concepts, their symbolic valuations of forests, and their degree of dependence upon forest resources. Value polarization will continue the trend of resource management by litigation. The definition of optimal forest health will remain equally contentious.

BUILDING FOREST HEALTH
INTO PUBLIC FOREST PLANNING

Natural resource planning on public land is complicated by the ever growing number of legal interpretations of a large body of laws and regulations. NFMA required the U.S. Forest Service to develop integrated forest planning with less emphasis on timber. FORPLAN was adopted as the analysis tool that could accommodate integrated planning. FORPLAN provides good answers to logical resource allocation problems, suitability, and cost/benefit analysis where objectives are clear. However, it is not well-suited to either stochastic processes or spatial interactions necessary to simulate ecosystem management. Its incorrect application led to widespread disillusionment and its rejection as an effective planning tool (Barber and Rodman 1990).

Systems analysis is evolving to match the increased complexity and comprehensiveness of forest planning. As Iverson and Alston (1986) note,

> This movement has been away from independent and functional analysis toward multidisciplinary, then integrated evaluation of all functions and alternative actions in the context of ecosystem management. (p. i)

Planning in the context of ecosystem management introduced a plethora of complex questions regarding interactions among and between the physical/biological resources, economics and sociology. The Office of Technology Assessment (OTA) argued that the U.S. Forest Service needed a different, more collaborative relationship with the public.

> Problems in public management of natural resources and environmental quality necessarily involve technical, biophysical questions, e.g., what is feasible, what results from specific practices, what various practices cost. They also involve human, socioeconomic questions as well as what should be the goals, what values are important, and what practices are acceptable. (U.S. Congress, Office of Technology Assessment 1992:89)

This requires that planning methodologies and tools be flexible and further evolve.

One Alternative Approach

Soft systems methodology (SSM) is one planning approach which incorporates all of these needs. In contrast to FORPLAN and other computer algorithms (hard systems), premised largely on assumed social objectives and criteria, SSM focuses on fundamental issues such as the nature of the problems, what constitutes improvements, and appropriate problem solving techniques (Wilson and Morren 1990).

SSM begins by recognizing that public resource problems are typically type B problems, which are characterized by the lack of clear objectives or any consensus on appropriate methods to achieve them (Davis and Johnson 1987). Thus, the first task is to define objectives to bring technical and value worlds together.

The first three steps in SSM define the nature and extent of the problem, including: perceiving and describing the situation, fact finding, mutual learning, and the application of systems thinking to understand the situation (Wilson and Morren 1990). Forest health expectations would be addressed within this phase. Implementing the SSM approach to planning does not ignore the strength of traditional analytical models, but uses them within a broader context. In the case of resource management decisions, mathematical modeling, e.g., linear programming, geographical information systems, ecosystem simulation models and decision support models, are applied simultaneously and iteratively. In the later stages of SSM there is a comparison of different models, modifications, implementation and iteration, in an effort to move towards an improved forest condition.

SSM is no panacea. It is just one of a variety of potentially useful tools to aid in reducing environmental conflict and in dealing with forest health considerations on public forests. Further integration of public participation with scientific and technical analyses and a broader conception of economics into more flexible decision processes is needed.

SUMMARY OF SIGNIFICANT POINTS

- Major forest management trends are currently centered around commodity production (market and non-market) and single-use non-commodity production (preservation).
- The relevant definition of optimal forest health is inseparable from management objectives.
- Decision maker risk adversity will lead to fewer decisions where controversial trade-offs must be faced and more use of initiatives such as "forest health" designed to reduce public controversy.
- Resource managers will be confronted with a wide array of added/new biological/physical and socio-economic pressures which will tax their ability to respond.
- Each of the three abstract future forest management strategies, protection, commodity production, and ecosystem management has its own forest health risk profiles.
- Accounting systems don't reflect all the costs of producing both commodity and non-commodity goods from the forest, nor will they capture the costs of achieving forest health.
- Broader economic philosophies may be required to understand and resolve more complex public forest resource management questions.
- Forest health declines impose four costs: commodity and non-commodity production losses (although timber salvage may temporarily rise), dissipation of the forest asset itself, increased annual protection costs, and rapidly escalating restoration costs.
- Planning in the context of forest health must recognize interactions among and between the physical/biological resources, economics and sociology requiring methodologies and tools to be flexible and to further evolve.

* Resource planning should seek to develop public agreement on such fundamental issues as the nature of the problems, what would constitute an improvement and the appropriate methodology for problem solving.

In summary, the authors felt the following quote from Hutchinson and Winters (1942:57) was particularly timely.

There is no shortage of solutions. The problem is to select the one which least disrupts the existing scheme of things and which, therefore, invites the public support necessary to transform it into an action program. Also it is necessary to recognize that the course which is best from a purely local standpoint may not serve the national interest. Therefore, any federal participation in the program should be conditioned by the national viewpoint.

Maybe we haven't advanced our thinking very far over the past few decades!

REFERENCES

Auclair, A.N.D. and J.A. Bedford. 1994. Conceptual Origins of Catatrophic Forest Mortality in the Western United States. In: pp. 249-265. Sampson, R.N. and D.L. Adams (Eds.). Assessing Forest Ecosystem Health in the Inland West. Proceedings of the American Forests scientific workshop, November 14-20, 1993, Sun Valley, ID. The Haworth Press, Inc. New York.

Barber, K. H. and S. A. Rodman. 1990. FORPLAN: The Marvelous Toy. *Journal of Forestry* 88(5):26-30

Barnett, H. J. 1979. Scarcity and Growth Revisited. In: pp. 163-217. Smith, V.K. (Ed.). Scarcity and Growth Reconsidered. Johns Hopkins University Press. Baltimore.

Carlson, J. E. 1994. Social Impacts of a Reduction in the Timber Industry in Three Eastern Oregon Counties. Draft report to county commissioners of Baker, Union and Wallowa counties.

Davis, L. S. and K. N. Johnson. 1987. Forest management 3rd edition. McGraw-Hill Publishing Company. New York. 790p.

Everett, R., P. Hessburg, M. Jensen and B. Bormann. 1993. Eastside forest ecosystem health assessment. Volume I, Executive Summary. USDA, National Forest System. 57p.

Gorte, J. K. and R. W. Gorte. 1979. Application of Economic Techniques to Fire Management–A Status Review and Evaluation. USDA-Forest Service. Intermountain Forest and Range Experiment Station. Ogden, UT. General Technical Report INT-53.

Green, A. W. and D. D. Van Hooser. 1983. Forest Resources of the Rocky Mountain States. USDA-Forest Service. Intermountain Forest and Range Experiment Station. Ogden, UT. Resource Bulletin INT-33.

Hutchinson, S. B. and R. Winters. 1942. Northern Idaho forest resources and industries. USDA Misc. Pub. No. 508. Washington, DC. 75p.

Iverson, D. C. and R. M. Alston. 1986. The genesis of FORPLAN: A historical and analytical review of forest service planning models. USDA Forest Service, General Technical Report INT-214. 31p.

Iverson, D. C. and R. M. Alston. 1993. Ecosystem-based forestry requires a broader economic focus. *J. Sustain. For.* 1(2):97-106.

Kennedy, J. J. and T. M. Quigley. 1993. Evolution of forest service organizational culture and adaptation issues in embracing ecosystem management. In: pp. 19-29. Jensen, M. E. and P. S. Bourgeron (Eds.). Eastside forest ecosystem health assessment. Vol. II Ecosystem management: principles and applications. USDA Forest Service.

McKetta, C. W. 1994. Wallowa-Whitman National Forest Below Cost Timber Program: Are the Costs at Fault? Draft report to county commissioners of Baker, Union and Wallowa counties.

O'Laughlin, J., J. G. MacCracken, D. L. Adams, S. C. Bunting, K. A. Blatner and C. E. Keegan. 1993. Forest Health Conditions in Idaho. Report No. 11, Idaho Forest, Wildlife and Range Policy Analysis Group, University of Idaho. Moscow.

O'Laughlin, J. 1994. Assessing forest health conditions in Idaho with forest inventory data. In: pp. 221-247. Sampson, R.N. and D.L. Adams (Eds.). Assessing Forest Ecosystem Health in the Inland West. Proceedings of the American Forests scientific workshop, November 14-20, 1993, Sun Valley, ID. The Haworth Press, Inc. New York.

Overbay, J. C. 1992. Ecosystem management. In: pp. 3-15. Proceedings of the national workshop: taking an ecological approach to management. April 27-30. Salt Lake City, UT. WO-WSA-3. USDA Forest Service, Watershed and Air Management. Washington, DC.

Randall, A. 1985. Methodology, ideology and the economics of policy: why resource economists disagree. *Am. J. Agr. Econ.* 67:1022-1028.

Rudzitis, G. and H. E. Johanson. 1989. Migration into Western Wilderness Counties: Cases and Consequences. *Western Wildlands* 15(1):19-23

Schulze, W. D. and C. W. Howe. 1985. Observations on the frontiers and fringes of the neoclassical paradigm. *Am. J. Agr. Econ.* 67:1035-1038.

Schuster, E. G. and M. J. Niccolucci. 1989. Separable Costs of Provisions for Nontimber Resources in Forest Service Timber Sales. *Western Journal of Applied Forestry* 4(4):119-124

U.S. Congress, Office of Technology Assessment. 1992. Forest Service planning: accommodating uses, producing outputs and sustaining ecosystems. OTA-F-505. Government Printing Office. Washington, DC.

Wandschneider, P. R. Neoclassical and institutionalist explanations of changes in Northwest water institutions. *J. Econ. Issues* 20(1):87-107.

Wilson, K. and G. E. B. Morren, Jr. 1990. Systems approaches for improvement in agriculture and resource management. Macmillan Publishing Company. New York. 361p.

SECTION III

This section is divided into three parts: Ecological and Historical Perspectives; Processes, Models, and Tools; and Management and Policy. These papers contributed in-depth analyses of some of the factors involved in the forest health situation in the Inland West.

These papers were submitted by individual participants at the beginning of the workshop and each was reviewed by three qualified scientists during the course of the workshop. The comments from reviewers were incorporated by the authors in the weeks following the workshop.

ECOLOGICAL AND HISTORICAL PERSPECTIVES

Postsettlement Changes in Natural Fire Regimes and Forest Structure: Ecological Restoration of Old-Growth Ponderosa Pine Forests

W. W. Covington
M. M. Moore

W. W. Covington is Professor, and M. M. Moore is Associate Professor, School of Forestry, Northern Arizona University, Flagstaff, AZ 86011.

The authors thank Robert W. Mutch, Lyn Morelan, Penny Morgan, and Maia Enzer for many useful comments and suggestions for improving this manuscript.

This research was supported by research agreements including Northern Arizona University (Organized Research Program and Bureau of Forestry Research), the McIntire-Stennis Cooperative State Research Service, and the Salt River Project. In kind support was provided by Kaibab Renewable Resources.

An early unrefereed version (Covington and Moore 1992) of this paper was published in the proceedings of a workshop on old-growth forests held in Portal, Arizona, March 9-13, 1992.

[Haworth co-indexing entry note]: "Postsettlement Changes in Natural Fire Regimes and Forest Structure: Ecological Restoration of Old-Growth Ponderosa Pine Forests." Covington, W. W. and M. M. Moore. Co-published simultaneously in the *Journal of Sustainable Forestry* (The Haworth Press, Inc.) Vol. 2, No. 1/2, 1994, pp. 153-181; and: *Assessing Forest Ecosystem Health in the Inland West* (eds: R. Neil Sampson and David L. Adams) The Haworth Press, Inc., 1994, pp. 153-181. Multiple copies of this article/chapter may be purchased from The Haworth Document Delivery Center [1-800-3-HAWORTH; 9:00 a.m. - 5:00 p.m. (EST)].

ABSTRACT. Heavy livestock grazing, logging, and fire exclusion associated with Euro-American settlement has brought about substantial changes in forest conditions in western forests. Thus, old-growth definitions based on current forest conditions may not be compatible with the natural conditions prevalent throughout the evolutionary history of western forest types. Detailed analysis of data from two study areas in the southwestern ponderosa pine type suggests that average tree densities have increased from as few as 23 trees per acre in presettlement times to as many as 851 trees per acre today. Associated with these increases in tree density are increases in canopy closure, vertical fuel continuity, and surface fuel loadings resulting in fire hazards over large areas never reached before settlement. In addition, fire exclusion and increased tree density has likely decreased tree vigor (increasing mortality from disease, insect, drought, etc.), herbaceous and shrub production, aesthetic values, water availability and runoff, and nutrient availability, and also changed soil characteristics and altered wildlife habitat. To remedy these problems and restore these forest ecosystems to more nearly natural conditions, and maintain a viable cohort of old age-class trees, it will be necessary to thin out most of the postsettlement trees, manually remove heavy fuels from the base of large, old trees, and reintroduce periodic burning.

Between the two extremes of passively following nature on the one hand, and open revolt against her on the other, is a wide area for applying the basic philosophy of working in harmony with natural tendencies. (H. J. Lutz 1959)

INTRODUCTION

Understanding natural ecological conditions and processes, before significant impact by Euro-American settlement (Kilgore 1985), is central to developing ecologically coherent forest management programs (e.g., Vogl 1974; Franklin 1978; Harris 1984; Kilgore 1985; Forman and Godron 1986). This is particularly true for management of landscape diversity with remnant natural patches of old-growth forests (Bonnicksen and Stone 1985; Parsons et al. 1986; Moir and Dieterich 1988; Forman and Godron 1986; Booth 1991) or managing for the development of old-growth and other stages of forest development (Thomas 1979; Hoover and Wills 1984; Moir and Dieterich 1988). However, heavy livestock grazing, logging, and fire suppression associated with Euro-American settlement have brought about substantial changes in forest conditions in western forests, so much so that current conditions may be decidedly "unnatural." In particular, the exclusion of natural fires has led to increased tree densities and associated shifts in ecosystem structure, fire hazard, disturbance regimes, and wildlife habitat in some western forest types. Thus, old-growth definitions based on current forest structure may not be compatible with the natural conditions prevalent throughout the evolutionary history of the organisms living in western forests. For these reasons, we believe that planning and management for old-growth within ponderosa pine and other forest types must include an understanding of past (presettlement), present, and future conditions. Failing to do so may lead to less than desirable forest conditions in the future.

In this paper we present a general discussion of changes in natural fire regimes, then examine in more detail the evidence for such changes in the southwestern ponderosa pine type and how these changes have affected overall ecological conditions. Finally, we close with a brief discussion of possible methods for remedying some of the problems associated with postsettlement changes in western forests.

NATURAL FIRE REGIMES

Understanding the natural disturbance regimes under which a particular species evolved is central to predicting the ecological consequences of management activities (e.g., Bormann 1981; White 1979). Periodic wildland fire has played a central role in the evolution of forest and woodland ecosystems throughout the western United States (Parsons 1981; Kilgore 1981; Covington et al. 1994b). In fact, many species and forest types worldwide appear to be dependent upon a particular frequency and intensity of fire for their survival (Mooney 1981; Parsons 1981).

Fire regimes have been classified according to frequency, intensity, size, and type (Heinselman 1981; Kilgore 1981; Sando 1978). Frequency, or burning interval, has been defined as the average return period for fire burning through a particular vegetation type. Sando defined frequent fires as fires which occur at intervals of 1-10 years; infrequent fires by his definition are fires occurring at intervals greater than 10 years, often as infrequently as once every 20-300 years. Kilgore (1981) separated frequent from infrequent fires at 25 years.

Although fire intensity has been used as a qualitative term (e.g., light surface fire vs. severe crown fire), some authors argued for a quantitative definition such as fire line intensity (Sando 1978; Kilgore 1981). Specifically, they recommended Byram's fire line intensity, a product of heat yield per unit area (BTU per square foot) and the rate of fire spread (feet per second). The resulting units are BTUs per foot per second. To avoid confusing qualitative and quantitative definitions, we recommend that qualitative differences be referred to as fire severity, and that the term fire intensity be reserved for a more quantitative measure.

Fire regimes have also been characterized according to size (Table 1). However, no generally agreed upon classification exists; certainly other size classes may be more appropriate for specific applications.

Fire type has been classified into as few as two (surface vs. crown) to six or more categories. For example, Heinselman (1981) differentiated fire type into light surface fires, severe surface fires, crown fires, and combinations of the three. Kilgore (1981) used the terms low intensity surface fire, high intensity surface fire, stand replacement fire, and combinations of these three categories. Several authors have related fire severity to fire intensity (Byram 1959; Van Wagner 1973; Albini 1976; Sando 1978). Integrating these views on intensity: severity relationships, Sando (1978) concluded that at low to moderate (0-1200 BTU/ft/sec) fire line intensity, complete mortality of overstory vegetation would not occur. At levels above 1200 BTU/ft/sec, nearly complete overstory mortality would be expected.

For this discussion we will use Sando's (1978) classification of natural fire regimes. However, more comprehensive classification strategies should be developed for designing adaptive management experiments.

TABLE 1. Examples of fire size classification.

Fire size (acres)	USDA Forest Service Wildfire Classes	Heinselman (1981) Classes
<0.25	A	small
0.26-9	B	small
10-99	C	small
100-299	D	medium
300-999	E	medium
1,000-4,999	F	large
5,000-9,999	G	large
>10,000	G	very large

Nonetheless, for brevity's sake we will use the following four types in this paper:

Type one: frequent fires (1-25 yr) of low to moderate intensity (< 1200 BTU/ft/sec) (e.g., ponderosa pine, lower elevation [warmer and drier sites] mixed conifer, giant sequoia, southern pine forests, short grass and mixed grass prairies, savannahs).

Type two: infrequent fires (> 25 yr) of high intensity (> 1200 BTU/ft/sec)(e.g., boreal and subalpine spruce-fir, higher elevation [cooler and wetter] mixed conifer forests, temperate rain forests).

Type three: frequent fires of high intensity (e.g., tall grass prairie).

Type four: infrequent fires of low to moderate intensity (e.g., deserts, tundra, mesic deciduous forests).

Fire exclusion affects each of these fire regimes differently. Although it typically causes major shifts in ecosystem structure and function (see below), from a fire control perspective the central concern is when shifts occur between fire regimes. Only small remnants of type three exist, virtually all of it having been converted to agriculture. However, under natural conditions, fire suppression would not have been practical in this type because of its extreme fire behavior. Fire exclusion in type four has little effect on the fire regime because excess flammable organic matter rarely accumulates.

Over time, fire exclusion in type two will result in rescaling of fire size as more patches reach a condition which will support crown fire. For example, if we assume that 100 years are necessary for a particular vegetation type to accumulate sufficient fuel to support a crown fire, then after 100 years of successful fire exclusion, all of the area in that type would be capable of supporting crown fire (Figure 1). Thus, in type two, fires become larger over time.

The greatest change in fire regimes following fire exclusion is in type one. Here the natural fire regime was frequent enough to keep surface fuel loads low and to thin out trees so that canopy fuels were separated both vertically and horizontally. With fire exclusion in this type, surface fuels accumulate and trees become established gradually, providing a fuel ladder and increasing canopy closure. These changes in fuel structure lead to a shift from light surface fires to intense stand replacement crown fires characteristic of the type two fire regime. Continued fire suppression in this type might well lead to the same changes seen in type two fire regimes, i.e., the coalescing of patches into larger and larger areas capable of supporting crown fire (Figure 1). The classic example of a type one natural

FIGURE 1. Effects of fire exclusion in landscapes with an infrequent, high intensity natural fire regime (Type two).

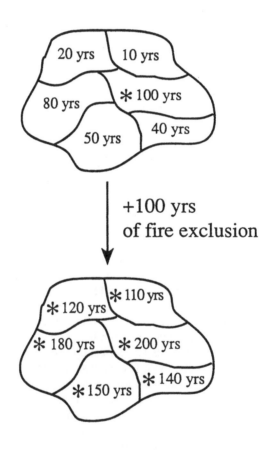

Assume 100 years to accumulate fuel
loads which support crown fires
(✱indicates crown fire potential)

fire regime is ponderosa pine ecosystems (Biswell 1972; Kilgore 1981; Weaver 1974).

Now we will turn to a more detailed discussion of ecological consequences of shifts in fire regimes by examining postsettlement changes in southwestern ponderosa pine ecosystems.

POSTSETTLEMENT CHANGES IN SOUTHWESTERN PONDEROSA PINE FORESTS

It is widely acknowledged that fire exclusion and other factors associated with European settlement have greatly altered forest conditions in southwestern ponderosa pine (Cooper 1960; Weaver 1951; Covington and Sackett 1984; White 1985; Covington and Sackett 1986; Covington and Moore 1994a). However, postsettlement changes in the ponderosa pine type are not unique to the Southwest. In fact, studies in Utah (Madany and West 1983; Stein 1987), Montana (Gruell et al. 1982), Idaho (Barrett 1988; Steele et al. 1986), Washington (Weaver 1959), and California (Laudenslayer et al. 1989) suggested that increased tree density, fuel loading, and crown fire occurrence are common consequences of fire exclusion throughout the ponderosa pine type (Kilgore 1981). Simulation studies (van Wagtendonk 1985; Keane et al. 1990) indicated that this phenomenon occurs not only in the pure ponderosa pine type, but also in ponderosa pine/Douglas-fir and mixed conifer forests as well. Thus, although many questions remain regarding the ecological and multiresource implications of postsettlement changes in ponderosa pine forests, there is a wide consensus that today's forests are radically different from those present before European settlement.

Reports from early travelers illustrate the changes in appearance of the ponderosa pine forest since settlement. E. F. Beale's 1858 report is quoted by C. F. Cooper (1960) as follows:

> We came to a glorious forest of lofty pines, through which we have travelled ten miles. The country was beautifully undulating, and although we usually associate the idea of barrenness with the pine regions, it was not so in this instance; every foot being covered with the finest grass, and beautiful broad grassy vales extending in every direction. The forest was perfectly open and unencumbered with brush wood, so that the travelling was excellent. (Beale 1858)

Cooper (1960) stated "The overwhelming impression one gets from the older Indians and white pioneers of the Arizona pine forest is that the entire forest was once much more open and park-like than it is today."

Before European settlement of northern Arizona in the 1860s and 70s, periodic natural surface fires occurred in ponderosa pine forests at frequent intervals, perhaps every 2-12 years (Weaver 1951; Cooper 1960; Dieterich 1980). Extensive study of fire scars suggests that the natural fire size was approximately 3,000 acres (Swetnam and Dieterich 1985; Swetnam 1990).

Several factors associated with European settlement caused a reduction in natural fire frequency and size. Roads and trails broke up fuel continuity. Domestic livestock grazing, especially overgrazing and trampling by cattle and sheep in the 1880s and 1890s, greatly reduced herbaceous fuels. Active fire suppression, as early as 1908 in the Flagstaff area, was a principal duty of early foresters in the Southwest. A direct result of interrupting and suppressing these naturally occurring, periodic fires has been the development of overstocked forests.

Changes in the forest structure (e.g., tree density, cover, and age distributions) in southwestern ponderosa pine forests since European settlement have been blamed for many forest management problems (Biswell 1972; Cooper 1960; Weaver 1974; Covington and Sackett 1990; Covington and Moore 1994a). Forest management problems attributed to fire exclusion and resulting increased tree density in southwestern ponderosa pine include:

1. overstocked sapling patches;
2. reduced tree growth;
3. stagnated nutrient cycles;
4. increased disease, insect infestation, and parasites (e.g., root rot, bark beetle, dwarf mistletoe);
5. decreased forage quality and quantity;
6. increased fuel loading;
7. increased vertical fuel continuity due to dense sapling patches;
8. increased severity and destructive potential of wildfires;
9. increased tree canopy closure;
10. decreased on-site water availability;
11. decreased stream-flow and ground water recharge;
12. shifts in habitat quality for biota; and
13. decreased diversity of native flora and associated food webs.

Evidence for the shift from a type one to a type two fire regime in ponderosa pine since settlement comes from a study by Barrows (1978), later updated by Swetnam (1990). Using USDA Forest Service wildfire statistics they determined that lightning-caused crown fires had increased from 10,127 acres per year in the 1940s to 15,117 acres per year in the 1980s. In the 1970s an average of 33,801 acres per year were burned by wildfire in the Southwest. Both Barrows (1978) and Swetnam (1990) observed that lightning-caused wildfires in the Southwest are getting larger and larger over time, with some fires reaching 10,000-20,000 acres, in contrast to the 3,000 acre surface fires of presettlement times (Swetnam and Dieterich 1985; Swetnam 1990). This represents a three- to six-fold increase in average fire size. Thus, we may be witnessing in the ponderosa pine type the kind of shift in size observed in the type two fire regimes, i.e., the coalescing of patches into larger and larger areas capable of supporting very large (> 10,000 acres) crown fires.

There is little quantitative information on conditions of presettlement forests and woodlands of the Inland West (Covington et al. 1994b). The major writings and research in the southwestern ponderosa pine type deal only with tree densities. As mentioned earlier, Cooper (1960) cited the writings of early expedition leaders, Whipple (1856) and Beale (1858). They reported that the condition of the southwestern ponderosa pine forest " . . . was open and park-like with a dense grass cover." These early descriptions of the open nature of presettlement ponderosa pine forests are in agreement with results of recent research which found that canopy coverage by trees of presettlement origin range from 17% (Covington and Sackett 1986) to 22% (White 1985) of the surface area for unharvested sites near Flagstaff, AZ. In addition, Pearson (1923) noted that "rarely does [ponderosa pine] crown cover reach more than 30% and usually not over 25%."

Cooper (1960) stated that the structure of the southwestern ponderosa pine type in the White Mountains of east-central Arizona is actually that of an all-aged forest composed of even-aged groups. He noted great variation in diameter within a single age class. Using contiguous quadrat analysis (Grieg-Smith 1952) in two stands, Cooper determined that the presettlement trees aggregated into areas ranging from 0.16 to 0.32 acres. White (1985), in a study conducted on the Pearson Natural Area near Flagstaff, noted that successful establishment of ponderosa pine in presettlement times was infrequent (as much as four decades between regeneration events). White also quantified the strong aggregation of ponderosa pine. Using the nearest neighbor method (Clark and Evans 1954), White demonstrated that the aggregation ranged from 3 to 44 stems within a group, with a group occupying an area that ranged from 0.05-0.70 acres. "Ages of stems within a group were also variable with the most homogeneous group having a range of 33 years and the least having a range of 268 years (White 1985)." White's findings of a pattern of uneven-aged groups near Flagstaff are in contrast to the results of Cooper (1960) for the White Mountains. However, Cooper's subsequent observations have lead him to conclude that even-agedness in presettlement ponderosa forests was rare (C.F. Cooper, San Diego State University, San Diego, CA, personal communication 1992). For the southwestern ponderosa pine type, therefore, the data suggest that, at the group-level, the trees are basically all-aged and have sporadic regeneration events (White 1985; Covington and Moore 1994a). At the landscape and regional levels (square miles and larger in size), however, several studies have shown simultaneous regeneration events that were correlated with simultaneous surface fires, and favorable climatic oscillations (e.g., La Niña [opposite pattern of El Niño], Kerr 1988; Swetnam 1990; Savage 1989).

Madany and West (1983) discussed the effects that many years of heavy grazing and fire suppression have had on ponderosa pine regeneration in southern Utah (Zion National Park). They suggested that ponderosa pine seedling survival was probably greater in the early 1900s than in the presettlement days due to reduced competition of grasses (through grazing) with pine seedlings, and the reduced thinning effect that fires once had on seedlings in presettlement times.

Moir and Dieterich (1988) pointed out the importance of understanding the role of the natural, presettlement fire regime in directing successional processes toward ponderosa pine old-growth development and in keeping fuel loading low enough for large trees to survive wildfires. They stated that most of the old-growth in southwestern ponderosa pine forests has deteriorated because recurrent natural fires have been suppressed. Finally, they present an eleven stage model of succession from open meadow through sapling, pole, yellow pine, and dead snag dominated landscape units. Although they do not describe the scale of these units, Cooper's (1960) and White's (1985) results indicate that these units were on the order of a few tenths of an acre.

DETAILED ANALYSIS OF TWO STUDY AREAS IN ARIZONA PONDEROSA PINE

To better understand postsettlement changes in southwestern ponderosa pine, we studied changes in forest conditions for two areas in northern Arizona. One, the Bar-M study area, has soils of volcanic origin and is on the Mormon Lake Ranger District of the Coconino National Forest. The other has soils of limestone origin and is on the North Kaibab Ranger District of the Kaibab National Forest. More detailed descriptions of the study area and methods used are available in Covington and Moore (1994a) and Covington and Moore (in preparation).

Study Areas

The Bar-M Canyon study area is located approximately 25 miles south of Flagstaff. The Bar-M watershed is part of the Mogollon Rim Plateau. It is a gently rolling landscape, dissected by many steep canyons. Elevations range from 6360-7710 feet. Our plots were located between 6800 and 7200 feet. The bedrock underlying the area consists of igneous rocks of volcanic origin. The soils, developed on basalt and cinders, are mostly silty clays and silty clay loams less than 2.6 feet deep (Brown et al. 1974).

The average annual precipitation for the area is 25.0 inches. There are two major precipitation seasons. Sixty-four percent of the precipitation falls during the winter–October through April. Thirty-two percent falls during the summer–particularly July and August (Brown et al. 1974).

The North Kaibab study area is located on the Kaibab Plateau of north central Arizona, approximately 100-120 miles north of Flagstaff. Like Bar-M canyon it is a gently rolling landscape, dissected by steep canyons. The elevation of our plots ranged from 6800-7800 feet. The bedrock underlying the area consists primarily of Kaibab limestone. The soils, developed from limestone, are mostly sandy and gravelly loams and loams. Average annual precipitation for the North Kaibab ponderosa type is 22 inches (Brewer et al. 1991). Predominant vegetation composition of the study areas is described in Table 2.

Field Procedures

On both sites we used a stratified systematic sampling procedure. The areas were stratified by soil type and topography, using the U.S. Forest Service Terrestrial Ecosystem (TE) Survey (USDA Forest Service 1987; Brewer et al. 1991). At Bar-M, map unit #582 was the most common soil-slope-vegetation unit (Typic Argiboroll and Mollic Eutroboralf; Low Sun Cold, with ponderosa pine and Gambel oak as the dominant trees, 0-15% slope). Within the North Kaibab study area, map units #293 and #294 were the most common (Mollic Eutroboralf; Low Sun Cold, with ponderosa pine and Gambel oak as the dominant trees, 0-15% slope (#293) the most common and 14-40% slope (#294) also represented).

At the Bar-M study area, seventy 0.62 acre (one-quarter hectare) plots were systematically located within map unit #582. The large plot size was chosen to incorporate the patchy nature of ponderosa pine old-growth and for spatial analysis in the future. Sixty-two of these plots were labeled extensive; only the presettlement trees were sampled on these plots. All presettlement trees were stem-mapped (exact x,y location recorded on the plot). In addition to location, presettlement tree species, dbh, condition

TABLE 2. Predominant vegetation composition of the ponderosa pine/ bunchgrass ecosystems of the Kaibab Plateau and Bar-M study areas.

Common Name	Scientific Name	Growth Form
Ponderosa pine	*Pinus ponderosa*	Tree
Gambel oak	*Quercus gambelii*	Tree
Juniper	*Juniperus* spp.	Tree
Quaking aspen	*Populus treniuloides*	Tree
Douglas-fir	*Pseudotsuga menziesii*	Tree
Spruce	*Picea* spp.	Tree
Fir	*Abies* spp.	Tree
New Mexican locust	*Robinia neomexicana*	Shrub
Gambel oak sprouts	*Quercus gambelii*	Shrub
Buckbrush	*Ceonothus Jendlerii*	Shrub
Oregon grape	*Berberis repens*	Shrub
Showy aster	*Aster commutatus*	Forb
Spreading fleabane	*Erigeron divergens*	Forb
Showy goldeneye	*Viguiera multiflora*	Forb
Western ragweed	*Ambrosia psilostachya*	Forb
Snakeweed	*Gutierrezia* spp.	Forb
Lupine	*Lupinus* spp.	Forb
Mutton bluegrass	*Poa fendleriana*	Grass
Pine dropseed	*Blepharoneuron tricholepis*	Grass
Black dropseed	*Sporobolus interruptus*	Grass
Blue grama	*Bouteloua gracilis*	Grass
Bottlebrvsh squirreltail	*Sitanion hysterix*	Grass
Long-tongue mutton bluegrass	*Poa longiligula*	Grass
Arizona fescue	*Festuca arizonica*	Grass

(live, snags, stumps, and down), and density (number per acre and basal area) were recorded.

Eight of these seventy 0.62 acre plots were sampled more intensively at Bar-M. Information on all trees was gathered on these plots. In addition to location, the species (live, snags, stumps, and down), size class (e.g., seedling, sapling, etc.), dbh, and density were also recorded. A ten percent sample of all trees less than 14.6 and greater than 3.9 inches dbh was selected to determine an age distribution and approximate date of post-settlement tree establishment. All trees greater than or equal to 14.6 inches were aged as was any pine tree with yellow bark. Our logic for these dbh and bark criteria was similar to that of White (1985) who determined statistically that the majority of presettlement ponderosa pine at a similar site would be > 14.6 in., and that those which were not would have yellow bark. This latter criterion is based on the observation that, in the Southwest ponderosa pine, bark changes color from "black" to "yellow" as the tree ages (Pearson 1950).

At the North Kaibab study site we made our plot locations compatible with an earlier inventory (Lang and Stewart 1910). Our sample plots were located in the center of systematically selected quarter sections. Forty-six 0.62 acre (0.25 ha) plots were located in TE map units #293 and #294. Presettlement trees were sampled on all plots, while additional information on postsettlement trees was gathered on 36 of these plots. Data collected and sampling techniques used were the same as described above for the Bar-M study site. The results for only 16 intensive plots (data on all trees) from the North Kaibab are presented in this paper because we are still analyzing the remainder.

The tree rings of all presettlement trees from both study sites were counted and measured to determine total tree age and to determine the diameter of each tree at a point before Euro-American settlement (1867-Bar M; 1881-North Kaibab), and to determine average annual growth since settlement. If the tree was a stump, snag, or down material then the year of death was estimated (Thomas 1979; Maser et al. 1979; Cunningham et al. 1980; Rogers et al. 1984), and the presettlement diameter calculated by regression.

Simulation Models

To understand how forest structure and patterns and resource conditions changed over time we linked the spatial data from the stem-mapped intensive plots (8 from Bar-M and 16 from North Kaibab) to the ECOSIM multiresource forest growth and yield simulation model (Rogers et al. 1984). The tree growth and yield model used in ECOSIM is based on the FREP/STEMS model (Belcher et al. 1985; Brand 1981) calibrated with continuous forest inventory data from Arizona and New Mexico. The water yield model is based on the "Baker-Kovner" model (Brown et al. 1974) in which streamflow is a function of winter precipitation, aspect, slope, and tree density. For herbage production, a modification of Clary's (1978) model was used, where herbage is a function of annual precipitation, tree density, and range site class. Forest floor accumulation is estimated as the difference between litterfall (calculated from tree density and canopy biomass) and decomposition, using Fogel and Cromack's (1977) estimates for ponderosa pine decomposition rates. Near-view scenic beauty is estimated using equations developed by Daniel and Boster (1976) and Schroeder and Daniel (1981). This index of scenic beauty is calculated as a function of number of large (> 16 inch dbh) trees, number of mid-sized trees (5-16 inches dbh), amount of logging slash, amount of herbage, and amount of shrubs. For more detail on the simulation techniques the reader is referred to Rogers et al. (1984).

Each simulation run consisted of entering the dbh of all presettlement trees (live trees, snags, downed trees and stumps) by diameter class as stand conditions in 1867 for Bar-M and 1881 for the North Kaibab. Settlement of the Flagstaff area preceded that of the North Kaibab area, thus grazing and hence fire exclusion began earlier at the Bar-M study area. This paper is designed to examine the inherent properties of ponderosa pine population irruptions when fire is excluded; therefore we did not simulate historical tree harvesting in this analysis. This stands in contrast to our analytical procedure which included historical harvest at Bar-M published elsewhere (Covington and Moore 1994).

After initialization to presettlement conditions, we entered trees into the stand at appropriate intervals in simulated time, based on the regeneration events inferred from the age distribution of the postsettlement trees. The output from these computer runs was a series of tables (from 1867 or 1881 through 2021 or 2027) which quantitatively estimate the changes in forest density in southwestern ponderosa pine since Euro-American settlement. This information on forest density was used to run fuel loading, herbage, water yield, and esthetics models (Rogers et al. 1984). We used the model results to draw inferences about temporal changes in multiresource conditions since European settlement and to forecast future trends.

Results and Discussion

Tree density at both study areas has increased greatly since the late 1800s (Table 3). Presettlement tree density was higher by a factor of two on the North Kaibab study area than at the Bar-M study area. Postsettlement tree density was much greater at Bar-M than at North Kaibab. However, because of the small sample size (8), the postsettlement tree densities at Bar-M should be viewed with caution. Our estimates of presettlement tree densities are consistent with estimates from other sources (Table 4).

The results from the simulation model analysis are presented in Figures 2 through 5. We initialized the simulation in 1867 for Bar-M (although we begin plots at 1887 for chronological comparability with NKRD) and in 1881 for North Kaibab as follows:

1. Tree density–we entered the trees by dbh and species present in 1867 at Bar-M and 1881 at North Kaibab.
2. Site index = 75, the site index for Terrestrial Ecosystem Survey map units #582, #293, and #294.
3. Soil rating factor = 9 for Bar-M and 12 for North Kaibab, based on the forage production potential for map units #582, #293, and #294. Soil rating factor is an index varying from 0 for poor range sites to 12 for the best.
4. Fuel loading = 0.1 t/ac of fermentation + humus layers and 0.1 t/ac of litter layer of the forest floor. This is based on the fuel loading data from Covington and Sackett's (1986) 2-year interval prescribed burning plots in ponderosa pine with the assumption that tree canopy covered approximately 20% of the surface of the land (Covington and Sackett 1986; White 1985).
5. For Bar-M, average annual precipitation = 25 inches; average annual winter precipitation = 16.9 inches. This is the 22 year average for the ponderosa pine watersheds in the Beaver Creek drainage (Campbell and Ryan 1982). Precipitation was held constant throughout the simulation. For North Kaibab, average annual precipitation = 22 inches; average annual winter precipitation = 12.1 inches, based on climatic data reported in the Terrestrial Ecosystem Survey (Brewer et al. 1991) for north Kaibab map units #293 and #294.

To simulate the establishment of postsettlement trees we entered seedling density corresponding with the number of postsettlement trees for each species. Seedlings were entered beginning at the point in simulated time which represented the earliest establishment data observed on the plot. Similarly, the last date of seedling establishment was used as the last seedling establishment date in simulated time.

The results of the simulation model provide an estimate of how forest conditions have changed since Euro-American settlement in Arizona ponderosa pine type and how these trends might continue into the future (Figures 2-5). Increases in tree density through 1990 are estimated to have caused substantial declines in average herbage production (decreases of over 1,000 lbs/ac at Bar-M and 350 lbs/ac on the North Kaibab). This decreased herbage in conjunction with increased small diameter tree density has caused a striking decline in near-view scenic beauty (Covington and Moore 1994a). At the same time, forest floor and fuel loading is estimated to have increased from less than 1 t/ac before settlement to an average of

TABLE 3. Changes in the tree density since Euro-American settlement for two ponderosa pine study sites. Data are means ($\bar{x}$) based on varying sample sizes (n).

Study Area	Presettlement[1]		Current[1]	
	- - - - - - trees/acre - - - - - -			
	$\bar{x}$	(n)	$\bar{x}$	(n)
North Kaibab[2] (limestone)	55.9	(36)	276.3	(30)
Bar-M Canyon[2] (volcanic)	22.8	(70)	851.0	(8)

[1]includes stumps, snags, and down trees
[2]TE map units #293 and #294 NKRD; TE map unit #582 Bar-M; North Kaibab sampling and data analysis are still in progress

TABLE 4. Density of southwestern ponderosa pine presettlement or yellow pine trees reported in the literature and in this study.

Location	Trees/acre
Specific studies in the Southwest:	
Ft. Valley, Coconino N.F.[1]	15
Bar-M, Coconino N.F.[2]	23
North Kaibab R.D., Kaibab N.F.[3]	56
North Kaibab R.D., Kaibab N.F.[4]	40-45
White Mountains, Apache-Sitgreaves N.F.[5]	35-45
Southern Utah, Zion N.P.[6]	22.7
USFS Bulletin 101:[7]	
(some of the heavily stocked yellow pine stands)	
Coconino N.F.	27
Kaibab (Tusayan) N.F.	35
Carson N.F.	26-47

[1]White, 1985; [2]Covington and Moore, 1994a; [3]Covington and Moore, 1994b; [4]Rasmussen, 1941; [5]Cooper, 1960; [6]Madany and West, 1983; [7]Woolsey, 1911

FIGURE 2. Tree basal area (ft²/ac) since European settlement in the Arizona ponderosa pine type at Bar-M and at the North Kaibab Ranger District (NKRD).

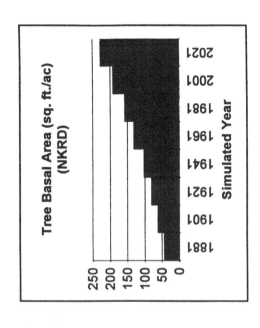

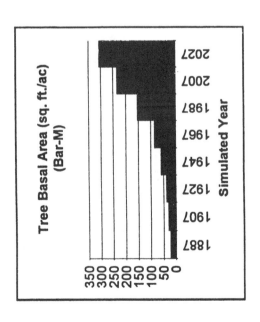

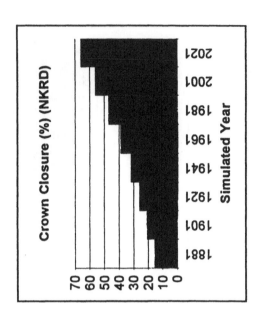

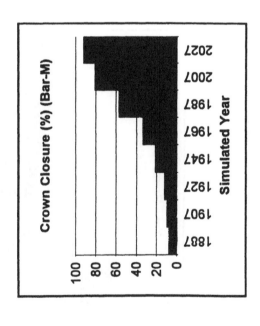

FIGURE 3. Crown closure (%) since European settlement in the Arizona ponderosa pine type at Bar-M and at the North Kaibab Ranger District (NKRD).

FIGURE 4. Fuel loading (t/ac) since European settlement in the Arizona ponderosa pine type at Bar-M and at the North Kaibab Ranger District (NKRD).

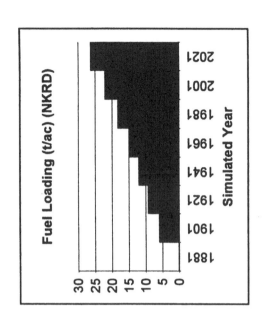

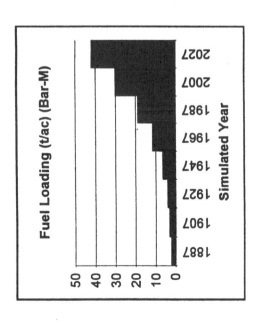

172

FIGURE 5. Herbage production (lbs/ac) since European settlement in the Arizona ponderosa pine type at Bar-M and at the North Kaibab Ranger District (NKRD).

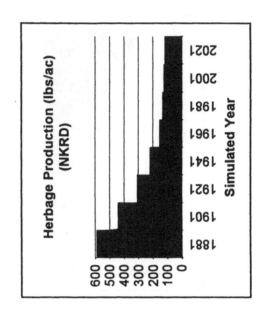

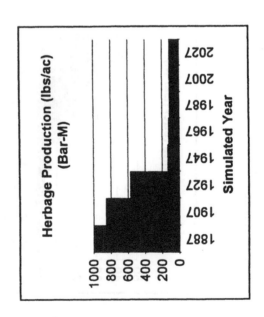

over 20 t/ac at both study areas. Vertical diversity, fuel ladder continuity (as estimated by diameter distribution of trees), and crown closure have also increased substantially (Figures 2-5).

To estimate changes in wildlife habitat characteristics since 1867, we used the simulation output for tree density by diameter class to classify the simulated stand into vegetation structural stages (Thomas 1979) and then used the Forest Service Southwestern Region's forest planning wildlife report (Byford et al. 1984) to determine the change in relative habitat value from 1867 to the present. This analysis indicates that there has been a shift away from the grass-forb structural stage common in the late 1800's to a seedling dominated structural stage after the turn of the century, and then to mature timber and finally old-growth trees growing over dense sapling and poles from the 1960s on.

These changes in vegetation since settlement indicate a shift in foraging habitat from one favoring grassland/savannah species (e.g., pronghorn antelope, grasshoppers, bluebirds, and turkeys) to one favoring species feeding in dense forests (Abert squirrel, porcupine, bark beetle, pygmy nuthatches, and perhaps Mexican spotted owl). In sum, the shift in tree density seems to have favored species dependent on closed forest conditions at the expense of those which require some portion of their habitat in grass/forb or savannah. Gruell et al. (1982) noted similar changes in wildlife habitats for ponderosa pine/Douglas-fir forests in western Montana.

Under the assumption that no substantial tree mortality occurs between 1987 and 2027, many of these trends are predicted to continue (Figures 2-5). Undoubtedly, the increased tree density is a key factor contributing to the increased occurrence of large crown fires in southwestern ponderosa pine (Barrows 1978; Swetnam 1990). Fire simulation studies of ponderosa pine and related forest types (van Wagtendonk 1985; Keane et al. 1990) are consistent with these conclusions.

Furthermore, increased density is responsible for decreased growth rates of presettlement trees (Sutherland 1983), and hence decreased vigor (Waring 1983), which increases susceptibility to bark beetle attack (Sartwell 1971; Sartwell and Stevens 1975) and other agents of mortality. Numerous studies have demonstrated that the mortality of ponderosa pine increases with both diameter and stand density (McTague 1990). A trend toward increasing rates of mortality, especially of the largest and oldest presettlement trees, is supported by an analysis of the Pearson Natural Area data (Covington and Moore 1994a).

Thus, the increase in tree density following Euro-American settlement has resulted, on the one hand, in a major increase in forest canopy cover and vertical diversity within the tree canopies, and on the other hand striking decreases in herbage production. Our results are consistent with studies in Interior (Gruell et al. 1982; Keane et al. 1990) and California (van Wagtendonk 1985; Laudenslayer et al. 1989) ponderosa pine forests which describe changes in tree density, fuel loads, wildlife habitat, and esthetics after fire exclusion. Of particular concern is the increased risk of mortality, especially for the oldest age classes of presettlement trees, from crown fire, bark beetles, and other agents (such as root rot), and the implications for old-growth forest management in the ponderosa pine type.

IMPLICATIONS FOR OLD-GROWTH ECOLOGY AND MANAGEMENT

Disruptions of natural disturbance regimes coupled with postsettlement anthropogenic disturbances have led to forest and woodland conditions which may bear little resemblance to natural conditions. Thus, old-growth definitions and management objectives based only on current stand structure may not be compatible with the conservation biology goal of preserving species diversity by providing for the habitats in which species have evolved (Hunter 1991). In fact, detailed analysis of tree density, size, crown closures, and down logs indicates that, before settlement, none of the plots sampled at Bar-M nor at the North Kaibab would meet all of the current minimum criteria for old-growth used in the Southwest[1] (USDA Forest Service 1992). In addition, when we compared the current conditions of the 36 intensive plots on the North Kaibab (all were site index 70-80) to these same criteria for old-growth, only three plots met the old-growth trees/acre requirements; however, in all three cases only half of the dominant trees met the 180 year requirement, and only one of these plots met the snag, while none met the down log, requirement.

Setting aside old-growth ponderosa pine stands which most closely meet current old-growth definitions may have unexpected consequences. These stands that have higher than normal canopy closures when compared to presettlement times are likely to be the most susceptible to crown fire, low tree vigor, and mortality from drought, insects, and diseases. Further deterioration from natural conditions is inevitable if restorative actions are not taken immediately.

---

1. Minimum structural attributes for old-growth inventory in the USDA Forest Service's Southwestern Region–for high sites (Minor site index > 55): large dominant/codominant trees–20 trees/acre ≥ 18 inches dbh; average 180 years; total tree basal area of 90 ft²/acre; total canopy cover of 50%; dead tree component of 1 snag/acre of 14 inches dbh and 25 feet in length; 2 down logs/acre of 12 inches dbh and 15 feet in length; canopies single or multiple storied.

Thus, definitions of old-growth should take into account natural conditions before Euro-American settlement, particularly the natural fire regime and patchy nature of the forest. In managing toward natural (using presettlement conditions as the "yardstick") old-growth, it will be necessary to design and apply treatments for restoring candidate stands (e.g., thinning from below, manual fuel treatments, prescribed burning).

RESTORATION
OF SOUTHWESTERN PONDEROSA PINE

To summarize, numerous lines of evidence point to striking postsettlement changes in southwestern ponderosa pine forests. A combination of livestock grazing, fire exclusion, and logging disturbances has resulted in increases in tree density, canopy closure, vertical diversity, aerial fuel continuity, and surface fuel loads. At the same time, herbaceous and shrub production have likely declined. As a consequence, the forests of today differ substantially from the natural conditions before Euro-American settlement. These changes in ecosystem structure imply changes in wildlife habitat, water relations, nutrient cycling, soils, species diversity, ecosystem health, and other resource characteristics.

Furthermore, these changes in ponderosa pine structure have led to a shift away from the natural fire regime of frequent, low intensity surface fires to high intensity crown fires. Most recently, the occurrence of larger and larger crown fires in the ponderosa pine type may indicate a further shift to a regime characterized by very large (> 10,000 acre) crown fires.

To remedy these problems and restore these forests to more natural conditions immediate action is essential. Although it is beyond the scope of this paper to recommend specific ecological restoration prescriptions, sufficient understanding (both scientific and expert knowledge) exists for developing and testing site specific ecological restoration hypotheses at levels in the landscape hierarchy from small plots to entire landscapes. For example, in southwestern ponderosa pine/bunchgrass ecosystems such as those reported in this paper, existing knowledge indicates that in dense sites with heavy forest floor accumulations, heavy fuels must be removed from the base of large, old trees (or larger replacement postsettlement trees in the absence of old-growth), periodic burning must be reintroduced, and native understory species must be reestablished.

However, simultaneous with widescale restoration treatments, it is necessary to determine postsettlement changes in ecological conditions (Bonnicksen and Stone 1985) for a broader range (e.g., a variety of soil types, topographies, and vegetation types) of forest and woodland types in the

Rocky Mountain and Southwest regions. Then a combination of an adaptive resource management approach (Walters 1986) and process-oriented simulation modeling (e.g., van Wagtendonk 1985; Keane et al. 1990; Covington and Moore 1994a) could be used to design restoration management regimes appropriate for each set of conditions. Simultaneously, small (10-20 acre) plot studies could be established to examine the ecological effects and practicality of various restoration treatment scenarios. Using such an integration of historical studies, simulation modeling, and management experiment approaches should ensure that we address our wildland management issues in a more coherent and ecologically sound manner.

REFERENCES

Albini, F.A. 1976. Estimating wildfire behavior and effects. USDA Forest Service General Technical Report INT-30. 92 p.

Barrett, S.W. 1988. Fire suppression's effects on forest succession within a central Idaho wilderness. *Western Journal of Applied Forestry* 3(3):76-80.

Barrows, J.S. 1978. Lightning fires in southwestern forests. Unpublished Report. Northern Forest Fire Laboratory, USDA Forest Service. Missoula, MT.

Beale, E.F. 1858. Wagon road from Fort Defiance to the Colorado River. 35 Cong. 1 Sess., Sen. Exec. Doc. 124.

Belcher, D.M., M.R. Holdaway and G.J. Brand. 1985. A description of STEMS, the stand and tree evaluation and modeling system. USDA Forest Service General Technical Report NC-79.

Biswell, H.H. 1972. Fire ecology in ponderosa pine grassland. Proceedings Tall Timbers Fire Ecology Conference 12:69-97.

Bonnicksen, T.M. and E.C. Stone. 1985. Restoring naturalness to national parks. *Environmental Management* 9:479-486.

Booth, D.E. 1991. Estimating prelogging old-growth in the Pacific Northwest. *Journal of Forestry* 89(10):25-29.

Bormann, F.H. 1981. Introduction. In: pp. 1-3. Fire regimes and ecosystem properties. Proceedings of the conference held in Honolulu, HI, December 11-15, 1978. USDA Forest Service General Technical Report WO-26. 594 p.

Brand, G.J. 1981. GROW–A computer subroutine that projects the growth of trees in Lake States' forests. USDA Forest Service Research Paper NC-207. 11 p.

Brewer, D.G., R.K. Jorgensen, L.P. Munk, W.A. Robbie and J.L. Travis. 1991. Terrestrial Ecosystems Survey of the Kaibab National Forest. USDA Forest Service Administrative Report. 319 p.

Brown, H.E., M.B. Baker, Jr., J.J. Rogers, W.P. Clary, J.L. Kovner, F.R. Larson, C.C. Avery and R.E. Campbell. 1974. Opportunities for increasing water yields and other multiple use values on ponderosa pine forest lands. USDA Forest Service Research Paper RM-129. 36 p.

Byford, K., L. Fager, G. Goodwin, J. MacIvor and R. Wadleigh. 1984. Wildlife coefficients technical report. USDA Forest Service, Southwestern Region. Albuquerque, NM.

Byram, G.M. 1959. Combustion of forest fuels. In: K. P. Davis (Ed.). Forest fire control and use. McGraw-Hill Book Co. New York.

Campbell, R.E. and M.G. Ryan. 1982. Precipitation and temperature characteristics of forested watersheds in central Arizona. USDA Forest Service General Technical Report RM-93. 12 p.

Clark, P.J. and F.C. Evans. 1954. Distance to nearest neighbor as a measure of spatial relationships in populations. *Ecology* 35:445-453.

Clary, W.P. 1978. Producer-consumer biomass in Arizona ponderosa pine. USDA Forest Service General Technical Report RM-56. 4 p.

Cooper, C.F. 1960. Changes in vegetation, structure, and growth of southwestern pine forest since white settlement. *Ecological Monographs* 30:129-164.

Covington, W.W. and M.M. Moore. 1992. Postsettlement changes in natural fire regimes: implications for restoration of old-growth ponderosa pine forests. Paper presented at the Old-growth Forests in the Southwest and Rocky Mountain Regions, Portal, AZ, March 9-13, 1992. In: pp. 81-99. USDA Forest Service General Technical Report RM-213. 201 p.

Covington, W. W. and M. M. Moore. (1994). Southwestern ponderosa forest structure: changes since Euro-American settlement. *Journal of Forestry.* 92(1):39-47.

Covington, W.W. and M.M. Moore. [in preparation]. Historical and anticipated changes in ecological conditions of ponderosa pine/bunchgrass ecosystems, Kaibab Plateau, AZ.

Covington, W.W. and S.S. Sackett. 1984. The effect of a prescribed burn in southwestern ponderosa pine on organic matter and nutrients in woody debris and forest floor. *Forest Science* 30:183-192.

Covington, W.W. and S.S. Sackett. 1986. Effect of periodic burning on soil nitrogen concentrations in ponderosa pine. *Soil Science Society America Journal* 50:452-457.

Covington, W.W. and S.S. Sackett. 1990. Fire effects on ponderosa pine soils and their management implications. In: pp. 105-111. Effects of fire management of southwestern natural resources. USDA Forest Service General Technical Report RM-191. 293 p.

Cunningham, J.B., R.P. Balda and W.S. Gaud. 1980. Selection and use of snags by secondary cavity-nesting birds of the ponderosa pine forest. USDA Forest Service Research Paper RM-222. 15 p.

Daniel, T.C. and R.S. Boster. 1976. Measuring landscape esthetics: the scenic beauty estimation method. USDA Forest Service Research paper RM-167. 66 p.

Dieterich, J.M. 1980. Chimney Spring forest fire history. USDA Forest Service Research Paper RM-220. 8 p.

Fogel, R. and K. Cromack, Jr. 1977. Effect of habitat and substrate quality on Douglas-fir litter decomposition in western Oregon. *Canadian Journal of Botany* 55:1632-1640.

Forman, R.T.T. and Michel Godron. 1986. Landscape ecology. John Wiley and Sons. New York. 619 p.

Franklin, J.F. 1978. Wilderness ecosystems. In: pp. 191-122. Hendee, J.C. et al. (Eds.). USDA Forest Service, Washington Office. Miscellaneous Publication 1365.

Grieg-Smith, P. 1952. The use of random and contiguous quadrats in the study of the structure of plant communities. *Annals of Botany* 16:293-316.

Gruell, G.E., W.C. Schmidt, S.F. Arno and W.J. Reich. 1982. Seventy years of vegetative changes in a managed ponderosa pine forest in western Montana–implications for resource management. USDA Forest Service General Technical Report INT-130. 42 p.

Harris, L. D. 1984. The fragmented forest. University of Chicago Press. Chicago. 211 p.

Heinselman, M.L. 1981. Fire intensity and frequency as factors in the distribution and structure of northern ecosystems. In: pp. 7-57. Fire regimes and ecosystem properties. Proceedings of the conference held in Honolulu, HI, December 11-15, 1978. USDA Forest Service General Technical Report WO-26. 594 p.

Hoover, R.L. and D.L. Wills (Eds.). 1984. Managing forested lands for wildlife. Colorado Division of Wildlife, Denver. 459 p.

Hunter, M. 1991. Coping with ignorance: the coarse filter strategy for maintaining biodiversity. In: pp. 256-281. Kohn, K.A. (Ed.). Balancing on the brink of extinction–the Endangered Species Act and lessons for the future. Island Press. Washington, DC.

Keane, R.E., S.F. Arno and J.K. Brown. 1990. Simulating cumulative fire effects in ponderosa pine/Douglas-fir forests. *Ecology* 71:189-203.

Kerr, R.A. 1988. La Niña's big chill replaces El Niño. *Science* 241:1037-1038.

Kilgore, B.M. 1985. What is "natural" in wilderness fire management? In: pp. 57-66. Lotan, J.E., B.M. Kilgore, W.C. Fischer and R.M. Mutch (Technical Coordinators). Proceedings–Symposium and workshop on wilderness fire. USDA Forest Service General Technical Report INT 182. 434 p.

Kilgore, B.M. 1981. Fire in ecosystem distribution and structure: western forests and scrublands. In: pp. 58-89. Fire regimes and ecosystem properties. Proceedings of the conference held in Honolulu, HI, December 11-15, 1978. USDA Forest Service General Technical Report WO-26. 594 p.

Lang, D.M. and S.S. Stewart. 1910. Reconnaissance of the Kaibab National Forest. USDA Forest Service Timber Survey (North Kaibab Ranger District), Admin. Report. 35 p. with map.

Laudenslayer, W.F., Jr., H.H. Darr and S. Smith. 1989. Historical effects of forest management practices on Eastside pine communities of northeastern California. In: pp. 26-34. Tecle, A., W.W. Covington, and R.H. Hamre (Technical Coordinators). Multiresource management of ponderosa pine forests. USDA Forest Service General Technical Report RM-185. 282 p.

Lutz, H.J. 1959. Forest ecology, the biological basis of silviculture. The H.R. MacMillan Lectureship Address. Publications of the University of British Columbia. Vancouver. 8 p.

Madany, M.H. and N.E. West. 1983. Livestock grazing–fire regime interactions within montane forests of Zion National Park, Utah. *Ecology* 64:661-667.

Maser, C., R.G. Anderson, K. Cromack, Jr., J.T. Williams and R.E. Martin. 1979. Ch. 6: Dead and down woody material, In: pp. 78-95. Thomas, J.W. (Ed.). Wildlife habitats in managed forests–the Blue Mountains of Oregon and Washington. USDA Agricultural Handbook 553. 512 p.

McTague, J.P. 1990. Tree growth and yield in southwestern ponderosa pine forests. In: pp. 24-120. Tecle, A. and W.W. Covington (Technical Coordinators). Multiresource management of southwestern ponderosa pine forests: the status of knowledge. USDA Forest Service Southwestern Region. 410 p.

Moir, W.H. and J.H. Dieterich. 1988. Old-growth ponderosa pine from succession in pine-bunchgrass forests in Arizona and New Mexico. *Natural Areas Journal* 8(1):17-24.

Mooney, H.A. 1981. Adaptations of plant to fire regimes: integrating summary. In: pp. 322-323. Fire regimes and ecosystem properties. Proceedings of the conference held in Honolulu, HI, December 11-15, 1978. USDA Forest Service General Technical Report WO-26. 594 p.

Parsons, D.J. 1981. The role of fire management in maintaining natural ecosystems. In: pp. 469-488. Fire regimes and ecosystem properties. Proceedings of the conference held in Honolulu, HI, December 11-15, 1978. USDA Forest Service General Technical Report WO-26. 594 p.

Parsons, D.J., D.M. Graber, J.K. Agee and J.W. van Wagtendonk. 1986. Natural fire management in national parks. *Environmental Management* 10:21-24.

Pearson, G.A. 1923. Natural reproduction of western yellow pine in the Southwest. USDA Bulletin No. 1105. 142 p.

Pearson, G.A. 1950. Management of ponderosa pine in the Southwest. USDA Agriculture Monograph 6. 218 p. Forest Service. Washington, DC.

Rasmussen, D.I. 1941. Biotic communities of Kaibab Plateau, Arizona. *Ecological Monographs* 11:229-276.

Rogers, J.J., J.M. Prosser, L.D. Garrett and M.G. Ryan. 1984. ECOSIM: A system for projecting multiresource outputs under alternative forest management regimes. USDA Forest Service, Admin. Report. Rocky Mountain Forest and Range Experiment Station. Fort Collins, CO. 167 p.

Sando, R.W. 1978. Natural fire regimes and fire management–foundations for direction. *Western Wildlands* 4(4):34-44.

Sartwell, C. and R.E. Stevens. 1975. Mountain pine beetles in ponderosa pine: prospects for silvicultural control in second growth stands. *Journal of Forestry* 73:136-140.

Sartwell, C. 1971. Thinning ponderosa pine to prevent outbreaks of mountain pine beetle. In: pp. 41-52. D. Baumgartner (Ed.). Proc. Precommercial thinning of coastal and intermountain forests in the Pacific Northwest. Washington State University, Coop. Ext. Serv. Pullman.

Savage, M. 1989. Structural dynamics of a pine forest in the American Southwest under chronic human disturbance. Ph.D. dissertation. University of Colorado. Boulder. 185 p.

Schroeder, H. and T.C. Daniel. 1981. Progress in predicting the perceived scenic beauty of forest landscapes. *Forest Science* 27:71-80.

Steele, R., S.F. Arno and K. Geier-Hayes. 1986. Wildfire patterns change in central Idaho's ponderosa pine/Douglas-fir forest. *Western Journal of Applied Forestry* 1(1):16-18.

Stein, S.J. 1987. Fire history of the Paunsaugunt Plateau in southern Utah. *Great Basin Naturalist* 48:58-63.

Sutherland, E.K. 1983. Fire exclusion effects on ponderosa pine growth. M.S. Professional Paper. University of Arizona. Tucson. 21 p. + figures.

Swetnam, T.W. and J.H. Dieterich. 1985. Fire history of ponderosa pine forests in the Gila Wilderness, New Mexico. In: pp. 390-397. Lotan, J.E., B.M. Kilgore, W.C. Fischer and R.M. Mutch (Technical Coordinators). Proceedings–Symposium and workshop on wilderness fire. USDA Forest Service General Technical Report INT 182. 434 p.

Swetnam, T.W. 1990. Fire history and climate in the southwestern United States. In: pp. 6-17. Effects of fire management of southwestern natural resources. USDA Forest Service General Technical Report RM-191. 293 p.

Thomas, J.W. (Ed.). 1979. Wildlife habitats in managed forests–the Blue Mountains of Oregon and Washington. USDA Forest Service Agricultural Handbook No. 553. 512 p.

USDA Forest Service. 1987. Terrestrial Ecosystem Survey Handbook. USDA Forest Service Southwestern Region. Albuquerque, NM.

USDA Forest Service. 1992. Recommended old-growth definitions and descriptions and old-growth allocation procedure. USDA Forest Service, Southwestern Region. Albuquerque, NM. 53 p.

Van Wagner, C.E. 1973. Height of crown scorch in forest fires. *Canadian Journal of Forest Research* 3:373-378.

van Wagtendonk, J. W. 1985. Fire suppression effects on fuels and succession in short-fire-interval wilderness ecosystems. In: pp. 119-126. Lotan, J.E., B.M. Kilgore, W.C. Fischer, and R.M. Mutch (Technical Coordinators). Proceedings–Symposium and workshop on wilderness fire. USDA Forest Service General Technical Report INT 182. 434 p.

Vogl, R.J. 1974. Ecologically sound management: modern man's road to survival. *Western Wildlands* 1(3):7-10.

Walters, C. 1986. Adaptive management of renewable resources. MacMillan Publ. Co. New York. 374 p.

Waring, R.H. 1983. Estimating forest growth efficiency in relation to canopy leaf area. *Advances in Ecological Research* 13:327-354.

Weaver, Harold. 1959. Ecological changes in the ponderosa pine forest of Cedar Valley in southern Washington. *Journal of Forestry* 57:12-20.

Weaver, Harold. 1951. Observed effects of prescribed burning on perennial grasses in the ponderosa pine forest. *Journal of Forestry* 49:267-271.

Weaver, Harold. 1974. Effects of fire on temperate forests: Western United States. In: pp. 279-319. Kozlowski, T.T. and C.E. Ahlgren (Eds.). Fire and Ecosystems. Academic Press. New York.

Whipple, A.W. 1856. Report of explorations for a railway route near the thirty-fifth parallel of North latitude, from the Mississippi River to the Pacific Ocean. Pacific Railroad Repts., Vol. 3. 33 Cong. 2 Sess., House Exec. Doc. 91.

White, P.S. 1979. Pattern, process, and natural disturbance in vegetation. *Botanical Review* 45:229-297.

White, A.S. 1985. Presettlement regeneration patterns in a southwestern ponderosa pine stand. *Ecology* 66:589-594.

Woolsey, T.S, Jr. 1911. Western yellow pine in Arizona and New Mexico. USDA Bulletin 101. 64 p.

The Role of Succession in Forest Health

Robert Steele

ABSTRACT. Forest health has become a major problem in much of the west and is closely linked to forest succession. Disturbances such as fire, flood, or windstorm periodically interrupt forest succession and recreate earlier seral conditions. The frequency and intensity of disturbance controls the extent of succession or the number of seral stages that may occur on a site. When the frequency of these disturbances is interrupted such as with fire control, succession can carry species composition and density beyond the historical range of seral stages. Historically, sites which were frequently disturbed are the first to get beyond their historical range and experience forest health problems when these disturbances are interrupted. Whereas, sites which, historically were less frequently disturbed may experience more complex successions that include outbreaks of insects and disease as part of the historical range of conditions.

INTRODUCTION

Forest health has become a major land management issue in much of the Inland West. Although attention has focused on the Blue Mountains of eastern Oregon (McLean 1992, Mutch et al. 1993), forest health problems have occurred during the past few years from Idaho to New Mexico (American Forests 1992). The intent of this paper is to briefly emphasize the relationship of this forest health problem with time, succession, and disturbance patterns.

Forest health is defined here as the capacity for self-renewal, the ability

Robert Steele is Research Forester, Boise Forestry Sciences Laboratory, Intermountain Research Station, 324 25th Street, Ogden, UT 84401.

[Haworth co-indexing entry note]: "The Role of Succession in Forest Health." Steele, Robert. Co-published simultaneously in the *Journal of Sustainable Forestry* (The Haworth Press, Inc.) Vol. 2, No. 1/2, 1994, pp. 183-190; and: *Assessing Forest Ecosystem Health in the Inland West* (eds: R. Neil Sampson and David L. Adams) The Haworth Press, Inc., 1994, pp. 183-190.

to recover from natural and human-caused stress and disturbance (O'-Laughlin 1993; USDA Forest Service 1993). This concept implies health of the ecosystem rather than the forest stand yet it is "stand health" that is usually diagnosed. Assessing the stand as part of the ecosystem and its normal processes allows us to evaluate forest health over time. Time is an important element when considering forest health and succession. Yet, the slow successional changes that occur over decades are so difficult for most people to recognize and interpret that they have referred to them as "the invisible present" (Magnuson 1990).

Forest health problems can become rapidly evident when wildfire and insect epidemics occur but the conditions that give rise to these episodes take decades or even centuries to develop. Plant succession coupled with human influence and extremes in weather are the primary ingredients of forest health problems; insects and disease are often the symptoms (Vasechko 1983). Of these ingredients, succession is the most difficult for people to recognize due to the long time perspective that is required.

SUCCESSION

The dictionary definition of succession is the coming of one thing after another in some kind of order, a replacement process (The English Language Institute of America, Inc. 1974). In relation to forests, succession is the orderly change in species composition, a predictable trend in plant communities (Oosting 1956). The relative competitive ability of each species maintains order and predictability of successional trends (Steele 1984). The potential end-point of forest succession is termed the climax community (Daubenmire 1968), a condition seldom achieved in the Inland West due to periodic burning.

The absence of disturbance allows forest succession to occur even though historically disturbance was initiated by natural events such as fire, flood, or windstorm. These disturbances occur periodically, creating successional cycles or a historical range of plant communities for each site. These episodes have a renewal effect on the ecosystem because they recycle nutrients and provide regeneration opportunities for many plant species (Van Cleave and Viereck 1981). Fire is the predominate recycling force in Rocky Mountain forests (Wellner 1970). The burning of organic matter releases nutrients into the soil (Wells et al. 1979). For example, in stands of ponderosa pine greater than 10.8 inches in diameter at 4.5 feet, prescribed burning increased the ammonium-nitrogen content by as much as 80 times that of similar unburned stands (Ryan and Covington 1986). Prescribed burning may also help reduce losses of trees to certain root rots

(Froelich et al. 1978; Reaves et al. 1990) thereby increasing resilience of the stand.

Clearly, sites that experience frequent burning every 10 to 20 years would support few plant communities because there is little opportunity for successional development. Some cold sites such as boreal forests also have few plant communities in spite of infrequent disturbance (Heinselman 1981); apparently the low temperatures allow few species to participate in the succession. When burning frequencies are interrupted, such as with fire control, succession can carry species composition and densities beyond their historical norm for the site. In the Inland West, the result is overly dense stands composed of late seral or climax tree species. The increased density results in greater flammability of the forest and suppression of wildlife forage (Hall 1981). It also creates severe demands among the plants for moisture and nutrients, resulting in physiological stress that predisposes the plant community to attack by various insects and diseases (Waring and Pitman 1985; Wargo and Harrington 1991).

Sites which burned most frequently are the first to get beyond their historical range and experience forest health problems. For example, the low elevation ponderosa pine communities in southwest Idaho historically experienced a 10 to 20 year fire-return interval (Steele et al. 1986). The frequent fires maintained open stands of seral tree and browse species. Low fuel levels were the norm. Ponderosa pine remained dominant on the site and could regenerate whenever fire prepared a seedbed. The open stand condition and frequent burning enabled the pine to resist attack by bark beetles (Sartwell 1971) and dwarf mistletoe (Hawksworth 1961).

With the exclusion of fire since the early 1900s, many of these stands have progressed to a near climax condition of dense Douglas-fir or grand fir. Recently, they have experienced epidemics of western spruce budworm, Douglas-fir tussock moth, western pine beetle, and Douglas-fir beetle as well as stand-destroying wildfire which killed even the large pines (Arno 1988, Wickman 1992). These devastating insect, disease, and wildfire occurrences are simply nature's response to abnormal conditions brought about by forest succession that has progressed beyond its historical range of plant communities. By ignoring efforts to exclude fire, environmental assessments refer erroneously to this scenario as the "no action alternative" giving the impression that if left alone, a natural balance will somehow prevail. But continued attempts to exclude fire from these warm, dry ecosystems without compensating the system by thinning the stands and reducing fuels can only lead to stressed plant communities and repeated devastation by insects, disease, and wildfire (Mutch et al. 1993).

These abnormal disturbances have a dysfunctional effect on the ecosystem because the regenerative power of the plant community, particularly the pine, is reduced through loss of site protection, seed sources, and, in extreme cases, soil.

At the other extreme of burning frequency are the cool high-elevation forests of lodgepole pine, Engelmann spruce, and subalpine fir. Here, fire-return intervals may take centuries, and succession may progress to a near climax condition as part of the historical range of plant communities (Romme and Knight 1981; Romme and Despain 1989). In these conditions, native insect and disease outbreaks and stand-destroying wildfire are part of the normal process (Haack and Byler 1993; van der Kamp 1991). For example, a fire-originated stand of lodgepole pine may shade the site and enable Engelmann spruce and subalpine fir seedlings to establish. As the lodgepole pine reaches maturity, mountain pine beetle kills the pine and the spruce and fir gain more growing space (Amman 1977). As the spruce reach maturity, spruce beetle may kill these trees (Schmid and Hinds 1974). Ultimately, the fir dominates until killed by fire which in turn prepares a seedbed for lodgepole pine as well as the spruce and fir. In some cases, fire revisits the site sooner and recycles dead and dying lodgepole pine into a new crop of seedlings, or it may return every 20 to 50 years and maintain a mosaic of multi-aged lodgepole pine with scattered spruce, and fir (Monnig and Byler 1992). The important feature of these forests is that the regenerative power or integrity of the ecosystem is maintained in spite of large insect outbreaks and stand-destroying wildfire. Yet, the same level of insect activity and stand-destroying wildfire would have a debilitating effect on the ponderosa pine ecosystem.

Clearly, forest managers must understand the successional processes of the individual ecosystems. Forest health is not simply a function of insect epidemics, disease, or destructive wildfire. These episodes may be symptoms of poor forest health in certain ecosystems but in other ecosystems they are a normal part of the forest cycle (Monnig and Byler 1992). Forest health is primarily a function of the normal range of cycling processes within a particular ecosystem, often referred to as the historical range of variability. Forest managers should use the normal historical range of these processes as a management guide. Managers may not always want to duplicate historical stand conditions, but the historical range of conditions provides a basis for assessing resiliency of existing stands. This is the focus of ecosystem management and a logical approach to managing our forest health problems.

TIME INTERVALS AND HISTORICAL RANGE
OF VARIABILITY

In order to characterize the historical range of succession for management purposes, one must first define a time interval of appropriate scale (Clark 1985; Ritchie 1986). The time interval should coincide with an environment that is relatively uniform and the successional direction of the historical vegetation should have the same potential end-point (climax) as the present vegetation. The upper parameter of this time interval is strongly influenced by such forces as climate and geology. These forces change continually but at a much slower rate than successional change. They create intermittent thresholds of change in vegetation and ecosystem processes (Holling 1973; May 1977). These kinds of changes may occur within a 500 to 1,000 year interval (Webb 1986; Tausch et al. 1993) which means that we should not use an interval longer than 500 years from the present unless we can be certain that no change in potential climax has occurred. The lower parameter is the time required for the full range of normal successional variability to take place. In the Inland West, this parameter is generally determined by the fire return-interval. The very cool forests that are infrequently burned may take 200 to 400 years to express their normal range of successional variability (Romme 1981). The more frequently burned warm, dry forests may only need 10 to 20 years or less but a 100 year record would yield a stronger data base and is usually obtainable from fire scar analysis.

A time interval of 100 to 400 years seems appropriate for defining the historical range of variability of most forests in the Inland West. This interval must be applied prior to the time of European settlement so as to avoid the interruption of fire and possibly other events that were part of the historical disturbance pattern. European settlement generally occurred from the mid to late 1800s in the northern Rocky Mountains. Therefore, the period from 1850 to 1750 could be used to determine the historical range of variability for the warm, dry, frequently disturbed ecosystems. The cool ecosystems with infrequent disturbance could be based on the period from 1850 to 1450. In some cases, objective data such as fire scars and tree ages may be difficult to obtain for the longer time intervals but the range of successional stages should be evident from field observation.

The time interval of 1850 to 1450 coincides with a cooler period known as the Little Ice Age (Webb 1986). Although some plant species distributions shifted near their environmental limits during this period (Webb 1986; Tausch et al 1993), there is little evidence that major changes in forest composition occurred (Davis 1986; Ritchie 1986). Apparently, presettlement forests of the Inland West were influenced more strongly by natural and anthropogenic wildfire than by climatic change during this period. Similar effects of fire and other anthropogenic factors overriding minor climatic change have been noted in other parts of the world (Ritchie 1986). It appears that data collected during this time interval would reflect the historical range of variability provided sample sites avoid the environmental extremes of plant communities.

CONCLUSION

Forest health should be evaluated in terms of the historical range of plant succession as it relates to insect, disease, and wildfire occurrence. The length of various successions vary between ecosystems due to the frequency of recurring disturbance. Likewise, historical levels of insects, disease, and wildfire vary between ecosystems so that severe episodes may be normal or healthy in one ecosystem and abnormal or unhealthy in another. Understanding how the ecosystem works, the normal range of processes, is the key to maintaining healthy forests and minimizing loss of resources to forest health problems.

REFERENCES

American Forests 1992. Health emergency imperils western forests. In: Gray, G.J. (Ed.). Resource Hotline *American Forests.* 98(9).

Amman, G.D. 1977. The role of mountain pine beetle in lodgepole pine ecosystems: Impact on succession. In: pp. 3-18. Mattson, W.J. (Ed.). The Role of Arthropods in Forest Ecosystems. Springer-Verlag. New York.

Arno, S.F. 1988. Fire ecology and its management implications in ponderosa pine forests. In: pp. 133-139. Baumgartner, D.M. and J.E. Lotan (Eds.). Ponderosa Pine the Species and Its Management. Symposium proceedings. Sept 29-Oct 1, 1987. Spokane, WA. Washington State University. Pullman.

Clark, W.E. 1985. Scales of climate impacts. *Climatic Change* 7:5-27.

Daubenmire, R. 1968. Plant communities, a textbook of synecology. Harper and Row. New York. 300p.

Davis, M.B. 1986. Climatic instability, time lags, and community disequilibrium. In: pp. 269-284. Diamond, J. and T.J. Case (Eds.). Community Ecology. Harper and Row. New York.

Froelich, R.C., C.S. Hodges, Jr. and S.S. Sackett. 1978. Prescribed burning reduces severity of *Annosus* root rot in the south. *Forest Science* 24(1):93-100.

Haack, R.A. and J.W. Byler. 1993. Insects and pathogens, regulators of forest ecosystems. *J. of Forestry* 91(9):32-37.

Hall, F.C. 1981. Fire history-Blue Mountains, Oregon. In: pp. 75-81. Stokes, M.A.

and J.H. Dieterich (Technical Coordinators). Proceedings of the Fire History Workshop. Oct. 20-24. Tucson, AZ. USDA Forest Service. Rocky Mountain Forest and Range Experiment Station. General Technical Report RM-81.

Hawksworth, F.G. 1961. Dwarf mistletoe of ponderosa pine in the southwest. USDA Forest Service, Washington, DC. Technical Bulletin 1246. 112p.

Heinselman, M.L. 1981. Fire and succession in conifer forests of northern North America. In: pp. 374-405. West, D.C., H.H. Shugart and D.B. Botkin (Eds.). Forest Succession, concepts and application. Springer-Verlag. New York.

Holling, C.S. 1973. Resilience and stability of ecological systems. *Annu. Reo. Ecol. Systematics* 4:1-23.

Magnuson, I.I. 1990. Long-term ecological research and the invisible present. *Bioscience* 40: 495-501.

May, R.M. 1977. Thresholds and breakpoints in ecosystems with a multiplicity of stablestates. *Nature* 269:471-477.

McLean, H.E. 1992. The Blue Mountains, Forest out of control. *American Forests* 98:32-35, 58, 61.

Monnig, E. and J. Byler. 1992. Forest health and ecological integrity in the northern Rockies. USDA Forest Service. Northern Region FPM Report 92-7. unpaginated.

Mutch, R.W., S.F. Arno, I.K. Brown, C.E. Carlson, R.D. Ottmar and J.L. Peterson. 1993. Forest health in the Blue Mountains: a management strategy for fire-adapted ecosystems. USDA Forest Service, Pacific Northwest Research Station. General Technical Report PNW-GTR-310. 14p.

O'Laughlin, J.O. 1993. Exploring the definition of forest health. In: pp. 9-14. Adams, D.L. and L. Morelan (Co-Chairs). Forest Health in the Inland West. Symposium abstracts. June 1-3, 1993. Boise, ID. University of Idaho, Moscow, ID.

Oosting, H.J. 1956. The study of plant communities. W.H. Freeman and Co. San Francisco. 440p.

Reaves, J.L., C.G. Shaw III and J.E. Mayfield. 1990. The effects of *Trichoderma* spp. isolated from burned and non-burned forest soils on the growth and development of *Armillaria ostoyae* in culture. *Northwest Science* 64(1):39-44.

Ritchie, J.C. 1986. Climate change and vegetation response. *Vegetatio* 67:65-74.

Romme, W.H. 1981. Fire frequency in subalpine forests of Yellowstone National Park. In: pp. 27-30. Stokes, M.A. and J.H. Dieterich (Technical Coordinators). Proceedings of the Fire History Workshop. Oct 20-24, 1980. Tucson, AZ. USDA Forest Service, Rocky Mountain Forest and Range Experiment Station. General Technical Report RM-81.

Romme, W.H. and D.H. Knight. 1981. Fire frequency and subalpine forest succession along a topographic gradient in Wyoming. *Ecology* 62(2):319-326.

Romme, W.H. and D.G. Despain 1989. Historical perspective on the Yellowstone fires of 1988. *Bioscience* 39(10):695-699.

Ryan, M.G. and W.W. Covington. 1986. Effect of a prescribed burn in ponderosa pine on inorganic nitrogen concentrations of mineral soil. USDA Forest Service, Rocky Mountain Forest and Range Experiment Station. Research Note RM-464. 5p.

Sartwell, C. 1971. Thinning ponderosa pine to prevent outbreaks of mountain pine beetle. In: pp. 41-52. Precommercial Thinning of Coastal and Intermountain Forests in the Pacific Northwest. Proceedings of short course. Feb 3-4, 1971. Washington State University. Pullman, WA.

Schmid, J.M. and T.E. Hinds 1974. Development of spruce-fir stands following spruce beetle outbreaks. USDA Forest Service, Rocky Mountain Forest and Range Experiment Station. Research Paper RM-131. 16p.

Steele, R. 1984. An approach to classifying seral vegetation within habitat types. *Northwest Science* 58:29-39.

Steele, R., S.F. Arno and K. Geier-Hayes. 1986. Wildfire patterns change in central Idaho's ponderosa pine-Douglas-fir forest. *Western Journal of Applied Forestry* 1(1):16-18.

Tausch, R.J., P.E. Wigand and J.W. Burkhardt. 1993. Viewpoint: Plant community thresholds, multiple steady states, and multiple successional pathways: legacy of the Quaternary? *J. Range Management* 46(5):439-447.

The English Language Institute of America, Inc. 1974. The living Webster Encyclopedic Dictionary of the English language. The English Language Institute of America, Inc. Chicago. 1158p.

USDA Forest Service 1993. Northeastern area forest health report. NA-TP-03-93, Northeastern State and Private Forestry. Durham, NH. 57p.

Van Cleave, K. and L.A. Viereck. 1981. Forest succession in relation to nutrient cycling in Boreal Forest in Alaska. In: pp. 185-211. West, D.C., H.H. Shugart and D.B. Botkin (Eds.). Forest Succession, concepts and application. Springer-Verlag. New York.

van der Kamp, B.J. 1991. Pathogens as agents of diversity in forested landscapes. *The Forestry Chronicle* 67(4):353-354.

Vasechko, G.I. 1983. An ecological approach to forest protection. *Forest Ecology and Management* 5:133-168.

Wargo, P.H. and T.C. Harrington. 1991. Host stress and susceptibility. In: pp. 88-101. Shaw, C.G. and G.A. Kile (Eds.). Armillaria Root Disease. USDA Forest Service. Washington, DC. Agriculture Handbook No. 691.

Waring, R.H. and G.B. Pitman. 1985. Modifying lodgepole pine stands to change susceptibility to mountain pine beetle attack. *Ecology* 66(3):889-897.

Webb, T. III. 1986. Is vegetation in equilibrium with climate? How to interpret late-quaternary pollen data. *Vegetatio* 67:75-91.

Wells, C.G., R.E. Campbell, L.F. DeBano, C.E. Lewis, R.L. Fredriksen, E.C. Franklin, R.C. Froelich and P.H. Dunn. 1979. Effects of fire on soil. USDA Forest Service. Washington, DC. General Technical Report WO-7. 34p.

Wellner, C.A. 1970. Fire history in the northern Rocky Mountains. In: pp. 42-64. The Role of Fire in the Intermountain West. Symp. Proc. Oct 27-29, 1970. Missoula, MT. University of Montana Forestry School. Missoula.

Wickman, B.E. 1992. Forest health in the Blue Mountains: the influence of insects and diseases. USDA Forest Service. Pacific Northwest Research Station. General Technical Report PNW-GTR-295. 15p.

The Role of Nutrition in the Health of Inland Western Forests

John M. Mandzak
James A. Moore

ABSTRACT. In this paper we present evidence supporting the premise that forest nutrition, nutrient cycling and nutrient management are critical factors for the health of Inland Western forests and for management of the ecosystems. The current state of knowledge suggests that inadequate tree nutrition, particularly potassium shortage, influences tree chemistry such that inadequate plant defensive compounds are produced. Pathogens and insects are unusually successful in attacking such trees. Forest managers have substantial influence on the forest nutritional environment through nearly all types of silvicultural treatments.

INTRODUCTION

In this paper we summarize evidence supporting the premise that nutrition, nutrient cycling and nutrient management are critical factors to consider when assessing Inland Western forest health and ecosystem management.

John M. Mandzak is Senior Forester, Land and Water Consulting, Missoula MT.

James A. Moore is Director, Intermountain Forestry Nutrition Cooperative and Professor at the University of Idaho College of Forestry, Wildlife and Range Sciences.

[Haworth co-indexing entry note]: "The Role of Nutrition in the Health of Inland Western Forests." Mandzak, John M. and James A. Moore. Co-published simultaneously in the *Journal of Sustainable Forestry* (The Haworth Press, Inc.) Vol. 2, No. 1/2, 1994, pp. 191-210; and: *Assessing Forest Ecosystem Health in the Inland West* (eds: R. Neil Sampson and David L. Adams) The Haworth Press, Inc., 1994, pp. 191-210. Multiple copies of this article/chapter may be purchased from The Haworth Document Delivery Center [1-800-3-HAWORTH; 9:00 a.m. - 5:00 p.m. (EST)].

191

The Nature of Inland Western Forests and Current Forest Health Issues

The Inland West is roughly defined for our purposes as an area generally east of the Cascade crest in Washington and Oregon, extending into British Columbia and including the drier portions of the Klamath and Sierra Ranges in Oregon and California east to the Great Plains of the U.S. and Canada. A defining feature of these forests is a midsummer drought period typically most intense in late July through August.

Forest health is currently entwined with the developing concepts of ecosystem management and is thus difficult to separately define. However, there is considerable interest in doing so to determine the future of these forests. In turn, potential future and desired future conditions will shape forest management planning goals. Riiters et al. (1990), stated that no widely accepted definition of forest health exists. Leonard, O'Laughlin, and Marshall presented papers with various definitions and concepts of forest health during a recent symposium on Forest Health in the Inland West (Adams and Morelan 1993). The simplest definition of forest health is the absence of significant diseases and pests or poor tree vigor.

Many authors express a belief that "a crisis in forest health" exists in the Inland West (Leonard 1993; Mutch 1993) and we agree. Several phenomena and symptomatic conditions are listed as proof, such as severe insect outbreaks, extensive root disease and increasing incidence and severity of catastrophic wildfire. Effective fire exclusion in past decades seems to have exacerbated these problems. It is typically pointed out that with the absence of fire, many inland forest types have developed away from open forest conditions previously maintained by frequent natural underburning. The presettlement forests usually were composed primarily of seral species. In contrast, closed canopy, multistoried stands, with a predominance of mid to late successional species are now common. The current stand conditions have a much higher potential for catastrophic stand replacement fires and increased incidence of insects and pathogens resulting in reduced tree vigor and increased mortality rates (Arno 1988; Hungerford et al. 1991).

Inland Western forests exhibited a far different range of conditions in the presettlement period than that observed today. Current forests are more extensive in area and exist on many previously nonforested sites, with more carbon stored as tree biomass (Harvey et al. 1992).

Role of Inorganic Plant Nutrients in Forest and Ecosystem Health

We believe, as do others (Kimmins 1977; Jurgensen et al. 1990), that significant evidence exists to support causality for stress induced by mineral nutrient deficiency and consequent implications for Inland Western forest ecosystem health. We agree that carbon and nitrogen are key nutrient elements (Harvey et al. 1994) but we also strongly suggest that other inorganic nutrient elements should be recognized as significant determinants of plant, and therefore forest ecosystem, vigor. We will describe results implicating potassium as an example of a critical "other element." We will also explore how human intervention, by design or not, can affect forest nutrient reserves and consequently inland forest health (Harvey et al. 1989). We presume that the nutritional status of inland forests is outside the historic range of variation since nutrient deficiencies are universal in the region (Mika et al. 1992) and humans have recently excluded fire from forest ecosystems and have often displaced or otherwise removed nutrients during harvesting or silvicultural activities. A more detailed understanding of all the relationships between mineral nutrient status and current forest health status is beyond the scope of this paper and would require more speculation than the authors are comfortable with. We therefore will concentrate on the research needed to improve our basis for understanding these relationships.

OVERVIEW OF NUTRITIONAL PROCESSES

Nutrient Requirements of Plants

An essential inorganic nutrient element for plants is defined as an element required to complete its life cycle. The common list of essential inorganic nutrient elements for green plants includes Carbon(C), Hydrogen(H), Oxygen(O), Nitrogen(N), Phosphorus(P), Potassium(K), Sulfur(S), Calcium(Ca), Magnesium(Mg), Iron(Fe), Manganese(Mn), Molybdenum(Mo), Copper(Cu), Boron(B), Zinc(Zn), Cobalt(Co) and Chlorine(Cl). Other elements such as Sodium (Na), Silicon(Si) or Selenium(Se) have also been demonstrated to be necessary for some plant and animal species.

Concepts of Nutrient Deficiency and Sufficiency

Plants fundamentally are "fixers" of atmospheric carbon. Plants create building block carbon, hydrogen and oxygen molecules and in the process "split" water, releasing oxygen. Forest process models typically use nitrogen reserves in forest ecosystems as a key variable to represent the potential for plant and ecosystem function, vigor and productivity. The relationship or ratio between inorganic nutrient elements such as nitrogen to the quantity of carbon (biomass) that can be fixed (Running and Gower 1991) is used along with other primary growth factors such as light, oxygen, water, and heat in these process models to predict the ability of a site to fix carbon. In addition to carbon/nitrogen, the relationships between all essential inorganic nutrient elements can be expressed as critical nutrient ratios (Ingestad 1979). Later in this paper we will give examples of the nitrogen/potassium relationships we have been exploring. Since quantities and ratios of 16 or more elements can influence carbon fixation, it is reasonable that inorganic elements can strongly influence plant, and therefore ecosystem, function in a profound manner. Typically though, only a few inorganic nutrient elements limit carbon assimilation and allocation on most individual sites.

Inorganic Plant Nutrients in Inland Forest Ecosystems

The nutrient capital of a forested site is contained in the soil, living and dead plants and animals (Tomlinson and Tomlinson 1990; Buol et al. 1989). The soil forming factors are climate, parent materials, organisms, relief and time, and maybe subsequently indentified human activity. Soils of the Inland West are predominantly young in the time scales of soil formation, primarily because of a relatively recent geologic history of glacial, interglacial, and volcanic activity. Because of the relatively cool and dry climate, mineral weathering and soil development processes occur slowly and the steep terrain limits soil profile development. Current soil properties of the Inland West, including the ability to supply nutrients, are therefore predominantly a function of the accumulated soil organic matter and the relatively unaltered soil parent material reflecting the rock, minerals and geomorphic processes from which the soil profiles were developed. Nitrogen is the only essential inorganic element which is fixed from the atmosphere by plants and microorganisms predominantly on site. Other removals and additions to the nutrient capital occur mostly through precipitation, wind transport, leaching and soil movement.

Brockley et al. (1992), state that ". . . inadequate nutrition is characteristic of many forests of the Interior Northwest. Nitrogen deficiencies are widespread and serious enough to dominate nutrient management concerns. On many sites, forest management activities and recurring wildfires may exacerbate existing nutritional problems. Because a large proportion of site nutrients may be contained in the forest floor and surface mineral soil, many interior forest soils–those where organic matter is produced, recycled, or replaced very slowly (e.g., dry, cold or fire-damaged sites)–are vulnerable to harvesting or site preparation practices that remove or displace surface organic layers and nutrient rich topsoil (Jurgensen et al. 1990)." Also, the removal of needles and branches in whole-tree harvesting operations will likely have a negative impact on soil nutrients and long-term soil productivity (Kimmins 1977).

Nutritional Implications of Ecosystem Disturbance and Recovery

In the discussion above, location and compartmentalization of inorganic nutrients in the living and dead biomass and soil was described. This is a dynamic process with nutrient transfers occurring in a process termed nutrient cycling (Cole et al. 1967). It is clear that among other effects, nutrient cycling often occurs in ways uncommon in presettlement conditions. These effects can be characterized, at least at this time, as abnormal lack of disturbance. Large volumes of standing live and dead biomass contain much of the inorganic nutrients previously held in the soil, or in the case of nitrogen, there may be higher quantities stored than in conditions receiving regular disturbance (McDonald 1993; Cochran and Hopkins 1991).

Fire–Fire is the primary carbon and inorganic nutrient recycling agent for the majority of forest types in the Inland West (Harvey et al. 1992; Arno 1988). Most nutrient elements are non-volatile; that is, they remain behind in ash (except perhaps in very hot burns of long duration). Nitrogen is volatilized during combustion and is returned to the atmosphere primarily as nitrogen oxides. The greatest risks to nutrient reserves as a consequence of fire are excessive nitrogen volatilization and the potential erosion of nutrient rich ash after very intense fire events.

Windthrow–Windthrow itself does not have a large impact on nutrient reserves since nutrients are returned through decomposition. The danger in the Inland West, though, is that a fire in the resulting downed and heavy fuels can be very intense, leading to excessive soil heating and to the loss of nitrogen and soil organic matter.

Harvesting & Slash Disposal–Effects of harvesting and slash disposal vary widely, depending on site conditions and the specific site treatments. The key point to remember is that nutrients are most concentrated in the fine branches and needles. Bole wood (logs) are relatively low in nutrient content. Whole tree harvesting is a process by which whole trees and attached crowns are transported to landings for processing and disposal. This process can result in significant export of nutrients, which ultimately may need to be replaced by fertilization. Follow-up slash disposal can have a great impact on nutrient capital as well, particularly if very intense slash fires volatilize nitrogen from the slash, duff and upper soil horizons. Dozer and windthrow piling, prior to burning, localizes nutrients, making access to nutrients for the next stand unevenly distributed. Brockley et al. (1992) and Kimmins (1977) provide comprehensive reviews of these topics.

Insects–Insect effects depend on the severity of the infestation and consequent level of tree mortality. In a mass mortality event, such as by pine bark beetles, effects on nutrient capital can be compared to windthrow mortality. Defoliation episodes result in accelerated nutrient recycling as nutrient rich foliage is returned to the forest floor as fecal material. Again, no severe loss of nutrient capital may occur, particularly if understory plants and recovering trees quickly reabsorb nutrients.

Root disease–Root disease effects on nutrient capital are relatively undramatic. Mortality occurs at a relatively low rate and is often quite localized such as in *Armillaria* root disease centers.

Current Conditions and Studies

Nitrogen concentration in foliage of inland conifers is apparently inadequate for optimum forest growth, based on results in almost all studies completed (Moore et al. 1991; Shafii et al. 1989; Brockley et al. 1992). In addition to the prevailing nitrogen deficiency, potassium (Mika and Moore 1991), sulfur, and micronutrients–particularly boron, iron and magnesium–have also been suggested as deficient on some sites. This list will likely be expanded as investigations continue.

Nutrient and Water Stress Relationships:
Insights from Fertilizer and Herbicide Treatments

Soil moisture deficits are clearly another significant limiting factor for growth of inland forests. Data from Montana trials (Mandzak, unpublished), including combinations of herbicide and fertilization treatments, indicate that nutrient status influences water relations and that competition with understory plants affects both water status and foliar nutrient content. Figure 1 shows predawn moisture stress at a typical installation for herbicide removal of competing vegetation (PRONONE), fertilization with a "complete" fertilizer mix, a combined "weed and feed" treatment and no treatment options. Predawn moisture stress in a ponderosa pine (*Pinus ponderosa* Dougl.) stand with a well developed grass and low shrub understory is lowest for the herbicide and greatest for the control treatment, thus indicating substantial interspecies competition for water. It is especially interesting that moisture stress is low after the fertilizer treatment, in spite of substantial increases in conifer foliage and understory biomass. The increased potential for evapotranspiration resulting from the increase in leaf area is presumed to be offset by improvements in plant water control, such as through turgor pressure adjustments and stomatal control. Water potential was intermediate with the combined fertilizer and herbicide treatment. Figure 2 is a plot of 15 herbicide release installations of

FIGURE 1. Pre-dawn water potential at the Johnsrud site.

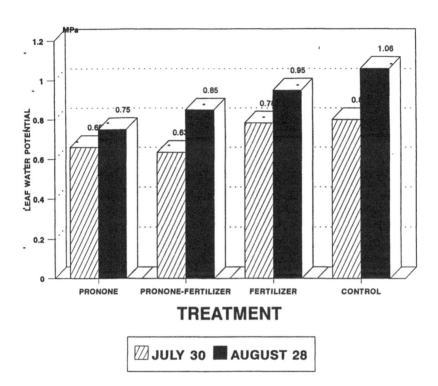

which four had the combined treatments described above. These installa-tions are displayed from left to right on the graph in ascending order of control foliar nitrogen. Note that herbicide control of the understory usual-ly resulted in increased foliar nitrogen concentration, suggesting a signifi-cant reduction in competition for nutrients. Substantial tree growth re-sponse was measured after two years for all treatments. Similar effects occur with other elements such as potassium. Smethurist et al. (1993) reported similar results for slash pine (*Pinus elliottii* Engelm. var. elliottii) in Florida. Understory plants may have a significant role in nutrient cycl-ing; however, other than for N-fixation, nutritional ecology of the under-story has been little studied.

Intermountain Forest Tree Nutrition Cooperative Studies

The Intermountain Forest Tree Nutrition Cooperative located at the University of Idaho was organized partially around the presumption that a

FIGURE 2. Ponderosa pine needle N concentrations for fifteen nutrition study sites in Montana, ranked in ascending order of control plot foliar nitrogen.

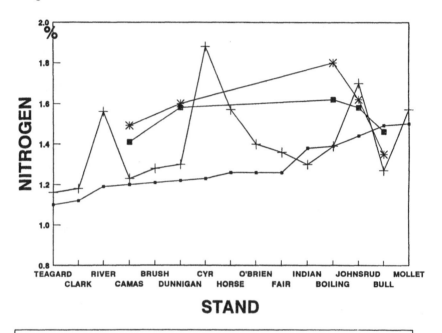

simple nitrogen deficiency of inland forests could be documented, thus producing the basis for growth enhancing silvicultural treatment. Figure 3 shows average dormant season foliar N concentrations for various combinations of geographic region and fertilizer treatments within the Inland West. The results show that the presumption of nitrogen deficiency was correct and that nitrogen fertilizer was taken up by the trees, resulting in increased foliar concentrations. Growth rates were also shown to increase significantly for all study regions. While a general dosage-dependent growth response was obtained, some installations produced little or no response, in spite of nitrogen uptake and deficiency.

In an effort to explain why some installations did not respond, pre- and post-treatment tree nutritional status was examined more closely. Evidence pointed to potassium as a nutrient potentially in low supply. Nitrogen application reduced K foliar concentrations and increased N con-

FIGURE 3. Average dormant season foliar N concentrations for the various combinations of geographic region and fertilizer treatment (1.4% N concentrations is inadequate for Douglas-fir).

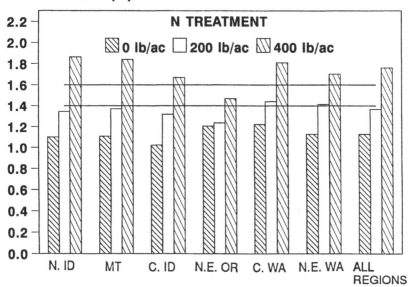

N CONC. (%)

REGION

centrations, thus the K/N ratios were reduced after treatment (Figure 4). Furthermore, post-treatment tree mortality was related to pre-treatment foliar potassium status. While gross growth response was about equal regardless of K nutrition status, net (after mortality) response was greatly reduced on sites with poor pre-treatment K status (Figure 5). This work is reported in detail in Mika and Moore (1991).

In 1987, six fertilizer trials were established in ponderosa pine stands in western Montana. By that time, interest had developed in the potassium-nitrogen relationship such that the IFTNC decided to substitute a nitrogen plus potassium treatment for the higher nitrogen rate treatment that had been standard for the regional ponderosa pine experiment. After four years, two very interesting responses resulted. First, mortality was higher in the plots treated with 200 pounds per acre of nitrogen as urea (200N) than in the 200N + 200K plots. In fact, three of the six installations had

FIGURE 4. The empirical cumulative frequency distribution of the ratio of foliar K and foliar N concentration for Douglas-fir in the Inland Northwest.

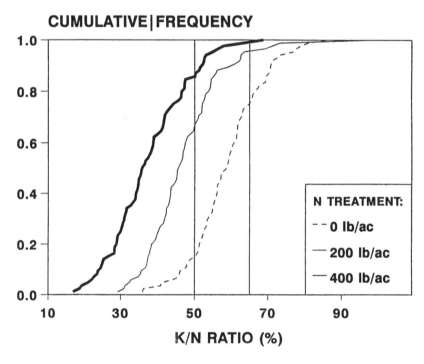

negative net growth response to nitrogen alone, but all six installations had positive net growth response to the N + K treatments (Figure 6). Even more interesting was that mountain pine beetles (*Dendroctanus ponderosa* Dougl.) were the primary cause of mortality, and their activities were concentrated in the nitrogen only plots. No trees were lost to beetles in the N + K plots (Figure 7). This work is reported in Mika et al. (1994). Follow-up investigations continue to elucidate what mechanisms "protected" the trees in the N + K plots. A pattern is emerging that indicates potassium deficiency causes inadequate chemical defense compounds to be produced when nitrogen alone is used. There is considerable evidence from agricultural studies documenting a link between plant nutrition, chemistry and incidence of insects or diseases (Huber and Arny 1985).

Other results indicate that it may be possible to alter tree root chemistry to the detriment of *Armillaria* root rot by manipulating tree nutritional status. Analysis of mortality patterns in nitrogen fertilized Inland Douglas-

FIGURE 5. Net basal area periodic annual response by 2 yr period for all combinations of fertilizer treatment and K status for Douglas-fir in the Inland Northwest.

NET BA PAI|RESPONSE (sq.ft/ac)

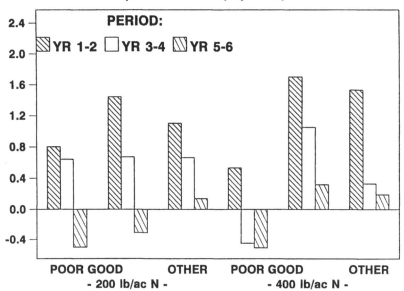

K STATUS and N TREATMENT

fir (*Psedotsaun menziesii var. glanca [Beissen] Franco*) installations showed that much of the mortality was associated with *Armillaria (Aemillaria ostoyae [Romasn.] Herink)* root rot (Figure 8). The mortality was greatest in installations with poor foliar K status and in plots with 400 N treatments when compared with those sites having good pre-treatment foliar K status.

Work by Entry et al. (1991) may explain the physiological basis for the root rot mortality and for favorable N + K fertilization results. They investigated some previously installed fertilizer trials in northern Idaho and inoculated trees located in the treatments with *Armillaria* root rot. They found that fertilizer-induced *Armillaria* infection was greatest on thinned and nitrogen fertilized plots and that it was related to root chemistry. Tree roots from thinned-only stands contained high concentrations of phenols and low sugar concentrations while nitrogen fertilized trees had high root sugar concentrations. The ratio of phenols to sugars was strongly corre-

FIGURE 6. The cumulative frequency distribution for net volume response to N or N + K fertilization for ponderosa pine in western Montana. The horizontal axis shows values of response obtained while the vertical axis indicates the proportion of all stands responding at that rate or less. The solid line shows the results for N fertilization while the dotted lines shows N + K fertilization results.

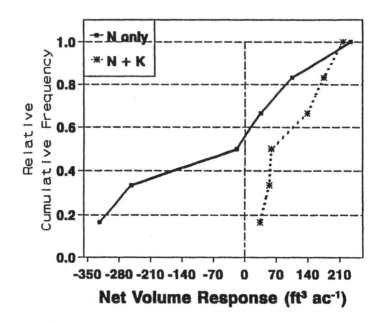

lated to incidence of *Armillaria* infection (Figure 9). Several IFTNC cooperators established nutrition experiments including combinations of nitrogen and potassium fertilizer treatments in Douglas-fir stands with active *Armillaria* infestation. Four years after the fertilizer treatments were applied, root samples were collected and analyzed using methods similar to Entry et al. (1991). The root phenol/sugar ratios by treatment are provided in Figure 10. These ratios are proportional to Entry et al.'s values shown in Figure 9. Potassium treatments significantly increased the root phenol/sugar ratio. Given that a high ratio is bad for the fungus and good for the trees, this experiment demonstrates that it may be possible to change tree root chemistry to the detriment of *Armillaria* by manipulating tree nutritional status (Moore et al. 1993).

Ylimarto (1991), in a study with Scots pine, found that scleroderris stem canker of seedlings was related to N and K fertilization regimes. Burdon (1991) described the selection pressure applied by fungal patho-

FIGURE 7. Percentage of trees dying in 4 years by fertilizer treatment and cause of mortality for ponderosa pine in western Montana.

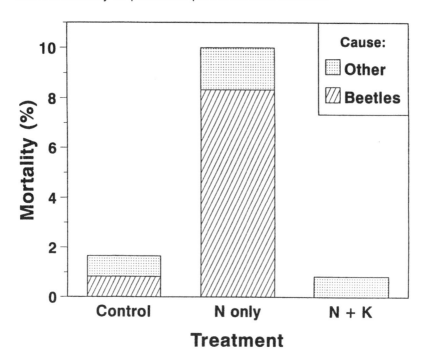

gens in plant populations. Waring et al. (1987) suggested that in mountain hemlock stands of the Oregon cascades, increases in nutrient availability and light, following death of the mature forest caused by the root rot fungus *Phellinus weirii* (Murr.) Gilbeatson, increased resistance of young replacement trees to infection by the pathogen. Dmitri (1977) stated that "As in human or veterinary medicine insufficient or excess nutrition of forest plants often leads to disease." Arbitrary fertilizer dressings, often applied without prior soil or needle analysis, may have more negative than positive results. But there is impressive evidence that stand improvement and fertilizer applications according to site, tree species and stand age can contribute to increasing "ecological resistance and health of the system." Garraway (1975) helped elucidate the effects of root chemistry on *Armillaria mellea* (Vahl:Fr.) Kummer thallus growth by demonstrating the superiority of glucose in growth media over other sugars and starch. Singh (1983) reported that nutrient deficient soil media of low pH produced

FIGURE 8. Percent basal area mortality six years after nitrogen fertilization by cause, treatment, and pre-treatment foliar potassium status for a region-wide Douglas-fir experiment.

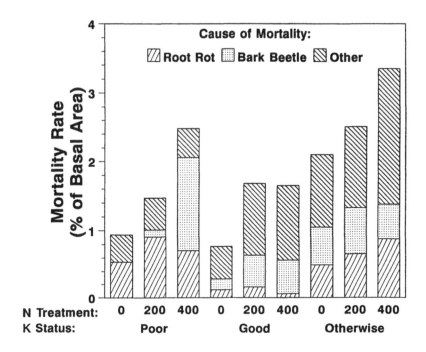

smaller, less vigorous conifer seedlings and the nutrient deficient seedlings incurred greater infection and mortality from *Armillaria*.

Other Investigations

Investigation of the relationships between defoliating insects and tree nutrient status and physiological defense mechanisms has been pursued to some extent. Mason et al. (1992) fertilized Douglas-fir and grand fir (*Abies grandis* [Dougl] Forbes) stands with nitrogen during a period of heavy infestation by spruce budworm (*Christoneura occidentalis* Freeman). They found that both populations and individual spruce budworm larvae increased in size. The trees, though, responded by producing more foliage and increased stem growth, thus indicating compensation by fertilization for defoliation effects. Waring et al. (1992) found that foliage enriched in amino acids due to nitrogen fertilizations increased

FIGURE 9. Incidence of *Armillaria ostoyae* infection of Douglas-fir versus the ratio of the energy available from root sugar concentrations. From Entry et al. 1991.

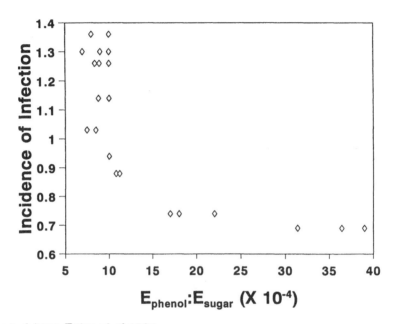

Adapted from Entry *et al* 1991

shoot growth more than enough to compensate for improved nutrition of budworm larvae and consequent increased feeding. Clancy et al. (1988) investigated the effect of seasonal and interspecific variation in foliar nutrients on spruce budworm performance. They suggest that foliar nutrition may be a key factor in budworm feeding behavior. Joseph et al. (1993), working with gypsy moth (*Lymantria dispar L.*) feeding on Douglas-fir in Oregon, found that increased foliar nitrogen improved larval growth and survival. However, high phenolics were shown to be lethal to gypsy moth larvae. Chiu et al. (1992) investigated Douglas-fir needles from treated and untreated IFTNC plots. Nitrogen and potassium treatments usually changed foliage color from green to a blue green (glaucous). It was found that the epicuticular wax changed to a different tubular, ornate morphology (waxes are phenolic compounds). Such structures may have beneficial effects on resistance to sunscald, to water loss and to pests.

FIGURE 10. Douglas-fir root phenolic/sugar concentration ratios four years after fertilization treatments.

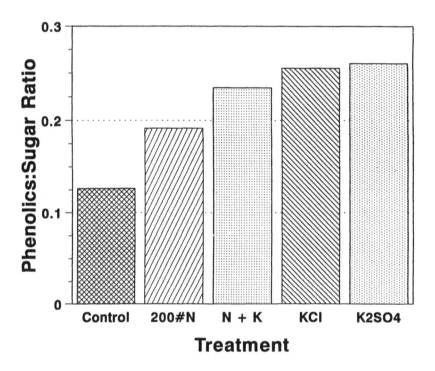

RESEARCH NEEDS

The following is a list of research focus areas that we feel will improve our understanding of forest nutrition's role in the health of Inland Western forests.

1. *Followups of Existing Nutrition Trials.* Much has been learned from IFTNC and other fertilizer trials in this and other regions. The original focus of these efforts was timber production, but they are well designed forest nutrition experiments and we can make many inferences for forest health ramifications of nutrition management.
2. *Designs for Future Nutrition /Forest Health Studies.* IFTNC has now developed installation designs that will help us understand the potential of forest nutrition management to deal with *Armillaria* root disease, bark beetles and other mortality causing agents. Optimum

nutrition trials (response surfaces) are a planned first step. This effort should be expanded.

3. *Inland Western Nutrient Cycling Studies.* Basic nutrient cycling studies are needed across a variety of sites, especially those experiencing health problems.

4. *Genetics in Forest Nutrition.* One issue in forest health is the loss of seral species from our forest communities. We need more information regarding how these species shifts affect nutrient cycling and the implications of those changes for forest health. Also, perhaps most importantly, it is recognized that there is substantial genetic control of nutrient uptake at both the species and intraspecific levels. More work is needed in this area of genetic/nutrition/forest health ecology.

5. *Collaborative Studies Between Forest Pathologists, Entomologists, Geneticists, and Physiologists with Forest Nutrition and Soil Scientists.* We feel that we have given some good examples of successful cross discipline efforts and strongly suggest that more be undertaken.

CONCLUSIONS

We must decide how, in managing forest ecosystems and their health in this Region, we can utilize what we have learned about forest nutrition. We have several suggestions for immediate application to forest management operations including forest harvesting, thinning, fertilization and prescribed fire. Nutrients should be conserved, particularly on nutrient poor sites, by learning tree foliage and branches on the site; fire should be reintroduced where possible and appropriate; and sometimes site nutrient capital will need to be restored through forest fertilization. We feel that a fundamental understanding of forest nutrition will be integral to developing effective solutions to the Inland Western forest health problems.

REFERENCES

Adams, D.L and L. Morelan Co-chairs. 1993. Forest Health in the Inland West. A Symposium. June 1-3, Boise, ID.

Arno, S.F. 1988. Fire ecology and its management implications in ponderosa pine forests. In: pp. 133-139. Baumgartner, D.M. and J.E. Lotan (Eds.). Ponderosa Pine: The Species and Its Management. Washington State University, Cooperative Extension. Pullman.

Brockley, R. P., R. L. Trowbridge, T. M. Ballard and A. M. Macadam. 1992. Nutrient management in Interior forest types. In: Chappell, H. N., G. F. Weetman and R. E. Miller (Eds.). Forest fertilization: sustaining and improving nutrition and growth of western forests. Institute of Forest Resources Contrib. 73. College of Forest Resources, University of Washington, Seattle.

Buol, S. W., F. D. Whole and R. J. McCracken. 1989. Soil genesis and classification. Iowa State University Press. Ames, IA. 446p.

Burdon, J.J. 1991. Fungal pathogens as selective forces in plant populations and communities. *Aust. J. Ecol.* 16:432.

Chiu, S., L.H. Anton, F.W. Ewers, R. Hammerschmidt and K.S. Pregitzer. 1992. Effects of fertilization on epicuticular wax morphology of needle leaves of Douglas-fir, pseudotsuga menzieziesii (pinaceae). *Amer. J. Bot.* 79(2): 149-154.

Clancy, K.M., M.R. Wagner and R.W. Tinus. 1988. Variation in host foliage nutrient concentrations in relation to western spruce budworm herbivory. *Can. J. For. Res.* 18:530-539.

Cochran, P.H. and W.E. Hopkins. 1991. Does fire exclusion increase productivity of ponderosa pine. In: p. 11. Proceedings-Management and productivity of western-montane forest soils. USDA Int. Res. Sta. Gen. Tech Rep. INT-280.

Cole, D. W., S. P. Gessel and S. F. Dice. 1967. Distribution and cycling of nitrogen, phosphorus, potassium, and calcium in a second growth Douglas-fir ecosystem. In: pp. 198-213. Symposium on primary productivity and a mineral cycling in natural ecosystems. University of Maine Press. Orono.

Dmitri. L. 1977. Influence of nutrition and application of fertilizers on the resistance of forest plants to fungal diseases. *Eur. J. For. Path.* 7:177-186.

Entry, J.A., K. Cromack, Jr., R.G. Kelsey and N.E. Martin. 1991. Response of Douglas-fir to infection by *Armillaria ostoyae* after thinning or thinning plus fertilization. *Phytopathology* 81:682-689.

Garraway, M.O. 1975. Stimulation of *Armillea mellea* growth by plant hormones in relation to the concentration and type of carbohydrate. *Eur. J. For. Path.* 5:35-43.

Harvey, A. E., M. F. Jurgensen and R. T. Graham, 1989. Fire-soil interactions governing site productivity in the Northern Rocky Mountains, In: pp. 9-18. Baumgartner, D. M. et al. (Comps.). *Prescribed fire in the Intermountain Region, Symposium Proceedings.* Washington State University, Cooperative Extension. Pullman. 171 pp.

Harvey, A.E., G.I. McDonald and M.F. Jurgensen. 1992. Relationship between fire, pathogens, and long term productivity in northwest forests. In: pp. 16-22. Kaufman, J.B. et al. (Coords.). Fire in Pacific Northwest Ecosystems: exploring emerging issues. Portland, OR, Jan. 21-23, 1992. Oregon State University. Corvallis.

Harvey, A. E., G. I. McDonald, M. F. Jurgensen and M. J. Larsen. 1994. Microbes: drivers of long-term ecological processes in fire-influenced cedar-hemlock-white pine forest in the Inland Northwest. In: Baumgartner, D. A. (Comp.). Symposium proceedings, Interior cedar-hemlock-white pine forests: ecology and management. Mar. 2-4,1993, Spokane, WA. Washington State University, Coop. Ext. Pullman. In press.

Huber, D. M. and Arney, D. C. 1985. Interactions of potassium with plant disease.

In: pp. 467-488. Munson, R.D. (Ed.). Potassium in agriculture. American Society of Agronomy. Madison, WI.

Hungerford, R. D., M. G. Harrington, W. H. Frandsen, K. C. Ryan and G. J. Niehoff. 1991. Influence of fire on factors that affect site productivity. In: pp. 32-50. Harvey, A. E. and L. F. Neuenschwander (Comps.). Management and productivity of western-montane forest soils. Symposium Proceedings. USDA For. Serv. Gen. Tech. Rep. INT-280.

Ingestad, T. 1979. Mineral nutrient requirements of Pinus silvestris and Picea abies seedlings. *Phsiol. Plant* 45:373-380.

Joseph G., R.G. Kelsey, A.F. Moldeke, J.C. Miller, R.E. Berry and J.G. Wernz. 1993. Effects of nitrogen and Douglas-fir allelochemicals on development of the gypsy moth, Lymantria dispar. *Journal of Chemical Ecology* 19(6): 1245-1263.

Jurgensen, M. F., A. E. Harvey, R. T. Graham, M. J. Larsen, J. R. Tonn and D. S. Page-Dumroese. 1990. Soil organic matter, timber harvesting and forest productivity in the Inland Northwest. In: pp. 392-415. Gessel, S. P., D. S. Lacate, G. F. Weetman and R. F. Powers (Eds.). Sustained productivity of forest soils: Proceedings of the seventh North American forest soils conference, June 25-28, 1988. University of British Columbia. Vancouver.

Kimmins, J. P. 1977. Evaluation of the consequences for the future tree productivity of the loss of nutrients in whole tree harvesting. *For. Ecol. Manage.* 1:169-183.

Leonard, G. M. 1993. Forest health and its relationship to ecosystem management. In: Forest Health in the Inland West. A Symposium. June 1-3, Boise, ID.

Mandzak, J. M. 1993 (unpublished). Montana chemical release and fertilization trials. Champion International Corporation. Milltown, MT.

Mason, R. R., B. E. Wickman, R. C. Beckwith and H. G. Paul. 1992. Thinning and nitrogen fertilization in a grand fir stand infested with western spruce budworm. Part I: Insect response. *For. Sci.* 38:235-251.

McDonald, G.E. 1993. Are current conditions temporary departures from long-term natural conditions? In: Forest Health in the Inland West. A Symposium. June 1-3, Boise, ID.

Mika, P.G., J.A. Moore. 1991. Foliar potassium status explains Douglas-fir response to nitrogen fertilization in the inland northwest, USA. *Water, Air, and Soil Pollution* 54:477-491.

Mika, P. G., J. A. Moore, R. P. Brockely and R. F. Powers. 1992. Fertilization response by interior forests: when where, and how much? In: Chappell, H. N., G. F. Weetman and R. E. Miller (Eds.). Forest fertilization: sustaining and improving nutrition and growth of western forests. Institute of Forest Resources Contrib. 73. College of Forest Resources, University of Washington. Seattle.

Mika, P.G., J.A. Moore and John M. Mandzak. 1994. Beetle caused mortality in ponderosa pine: induced by nitrogen fertilization but prevented by potassium amendment. *Plant and Soil.* In press

Moore, J.A., P.G. Mika, J.W. Schwandt and T.M. Shaw. 1993. Nutrition and

Forest Health. In: Baugmartner, D. A. (Comp.). Symposium proceedings: Interior cedar-hemlock-white pine forests: ecology and management. Mar. 2-4,1993. Spokane, WA. Washington State University Coop. Ext. Pullman. In press.

Moore, J.A., P.G. Mika and J.L. VanderPloeg. 1991. Nitrogen fertilizer response of Rocky Mountain Douglas-fir by geographic area across the inland northwest. *West. J. Appl. For.* 6(4):94-98.

Mutch, R.W. 1993. Unnatural stand density and species composition: Affecting the resilience of ponderosa pine in a fire environment. In: Forest Health in the Inland West. A Symposium. June 1-3, Boise, ID.

Running, S. W. and S. T. Gower. 1991. FOREST-BGC, A general model of forest ecosystem processes for regional applications. II. Dynamic carbon allocation and nitrogen budgets. *Tree Physiology* 9:147-160.

Shafii, B., J.A. Moore and J.R. Olson. 1989. Effects of nitrogen fertilization on growth of grand fir and Douglas-fir stands in northern Idaho. *West. J. Appl. For.* 4:54-57.

Singh, P. 1983. *Armillaria* root rot: Influence of soil nutrients and pH on the susceptibility of conifer species to the disease. Eur. J. For. Path. 1392:101.

Smethurst P.J., N.B. Comerford and D.J. Neary. 1993. Weed effects on early K and P nutrition and growth of slash pine on a spodosol. *Forest Ecology and Management* 60:15-26.

Tomlinson, G.H. and F.L. Tomlinson. 1990. Effects of acid deposition the forests of Europe and North America. CRC Press. pp. 51-82.

Waring R.H., K. Cromack, JR, P.A. Matson, R.D. Boone and S.G. Stafford. 1987. *Forestry* 60(2).

Waring R. H., T. Savage, K. Cromack, Jr. and C. Rose. 1992. Thinning and nitrogen fertilization in a grand fir stand infested with western spruce budworm. Part IV: An ecosystem/pest management perspective. *For. Sci.* 38(2):275-286.

Ylimartimo, A. 1991. The effect of nitrogen and potassium availability on scleroderris canker of scots pine seedlings. *Water, Air and Soil Pollution* 54: 307-313. Kluwer Academic Publishers.

Integrated Roles
for Insects, Diseases and Decomposers
in Fire Dominated Forests
of the Inland Western United States:
Past, Present and Future Forest Health

Alan E. Harvey

ABSTRACT. Forest ecosystems characterizing much of the Inland Western United States occupy precarious, changing environments that can be moisture, temperature, and/or nutrient limited. Rapid vegetative adaptations to inherent change are critical to both plant community stability and to the survival of individual species. Biological decomposition processes are often constrained and natural wildfires represent an important recycling agent. Recycling of resources is critical. It is proposed that native insect, disease and other decomposer activities, plus natural wildfire, historically provided coordinated biological and physical processes that were integral to carbon, nitrogen, and other nutrient cycling, and to rapid evolution and adjustment of native conifers (and of their ecosystems) in this dynamic environment. Current conditions, as imposed by traditional harvesting and fire control over the last 100 years, plus the introduction of white pine blister rust in the early 1930s, have changed many native vegetative and microbial systems. Endemic insects and diseases have responded to

---

Alan E. Harvey is Project Leader, USDA, Forest Service, Intermountain Research Station.

[Haworth co-indexing entry note]: "Integrated Roles for Insects, Diseases and Decomposers in Fire Dominated Forests of the Inland Western United States: Past, Present and Future Forest Health." Harvey, Alan E. Co-published simultaneously in the *Journal of Sustainable Forestry* (The Haworth Press, Inc.) Vol. 2, No. 1/2, 1994, pp. 211-220; and: *Assessing Forest Ecosystem Health in the Inland West* (eds: R. Neil Sampson and David L. Adams) The Haworth Press, Inc., 1994, pp. 211-220. Multiple copies of this article/chapter may be purchased from The Haworth Document Delivery Center [1-800-3-HAWORTH; 9:00 a.m. - 5:00 p.m. (EST)].

these changes by increasing activities. Their effects counter many of the destabilizing actions of site deterioration, fuel accumulation, changes in species and genetic compositions, increased stand densities and impairment of recycling processes. At least in the short term, many ecosystems are now highly vulnerable to potential damage from high fuel wildfire and perhaps to the momentum of alternative biological decomposition processes. Genetic resources and other diversity components may be at especially high risk. Many current trends place future values in increasing danger until the course is changed. Adjusting cycling processes, stand density and species composition will often be more important than controlling individual pests when managing forest health for this region in the future.

PERSPECTIVES FOR THE FUTURE

A Preemptive Approach

An appropriate perspective for the concept of future "forest health," from the above point of view, requires consideration of far more than pest management or integrated pest management or even ecological pest control (Norris 1988). Pest management has historically been limited because it was invoked only when there was a perceived problem. Pest management has largely been defensive in nature. Management of overall forest health should emphasize preemptive actions. For example, fiber (fuel) accumulation, species composition and stand density management should be utilized. Appropriate management of these factors has a high probability of regulating forest health without directly managing individual insect and disease agents. There is a higher probability of retaining at least moderately high levels of productivity in Inland Western ecosystems if they are allowed to develop within the natural envelope from which they originally evolved. This is a much sounder approach than waiting for a problem of a complex developmental nature to occur and then attempting to regulate a "causal organism."

We must also remember that these types of ecosystems are very young. In their current form many are less than 2,000 years old and cannot yet (if ever) be considered stable (Whitlock 1992). Thus, they are highly susceptible to change, particularly to novel change, and can be expected to react by becoming even less stable. These ecosystems should be considered uniquely and highly susceptible to damage and to resultant losses in productivity with associated increases in stress and endemic pest activities.

Introduced Pests

Introduced pests are a different matter than endemic insects and diseases. More introductions, like white pine blister rust or the gypsy moth, must be anticipated in the future. They are exceedingly destabilizing in the short-term and most ecosystems will require long recovery periods. Although movement of a pest is not unusual, with or without the involvement of humans, reducing the numbers and success of such introductions should remain one of our highest priorities in the ecosystem management forest health field.

Increasing Climate Variation

Another complication with the next forest rotation may be even more climatic variation and more rapid global and/or regional climate change (Peters 1990). Natural adjustment (and management) of future forests may demand hitting an even more rapidly moving environmental target.

Future Management Direction

Future actions should simulate natural disturbances and stimulate natural adjustment processes, or the effects they imposed, as they occurred during historical forest development. Current "out of phase" regional forests should be returned (gradually in most cases) to more ecologically balanced conditions. Future forest management should highlight ecosystem processes as keys for returning to a more balanced condition. Fiber (fuel/carbon) accumulation and cycling is the highest priority ecosystem process that must be returned to near historical levels. This function is the most out of phase with historical norms and directly affects density and species composition. Endemic insects and diseases should be recognized as an important and integral part of regulating all three. They also can be valuable indicators of imbalances and should be useful to, as well as a problem for, the natural resource managers of the future.

In any case, forests will continue to change and insects and diseases will continue to function within them. Depending on the nature of future changes (variation, amount, direction, duration) and the organism, insect and disease actions can either help stabilize or destabilize forest ecosystems. Human interaction with forests, and insect/disease agents will impact this process.

Choices

We do have choices. In some cases we may opt to leave things as they are, doing things much the same, because economics demands we do so . . . despite associated environmental costs. In others, we may be forced, by law, to intercede for the protection of non-forest (and forest) values, irrespective of cost. Hopefully, we can also choose to regulate many of our forests in more natural ways, and continue producing badly needed amenities and fiber while protecting forest components and processes.

An Opportunity

There is an obvious window of opportunity to harvest wood products as part of regulating fiber (fuel) accumulations. We should be able to regulate and use that fiber (sequester the carbon) while protecting basic ecosytem properties (maintain sufficient organic reserves) and processes (including fire), especially those that provide forest communities (and species) with the ability to change with their environment. Although compromises must be struck, and forestry practices and forest products processing will be more demanding and costly, we can accomplish appropriate fiber extractions without excessive loss of basic diversity, stability, productivity or value–a true challenge for the future.

A DYNAMIC ENVIRONMENT

An Environment Characterized by Change and Diversity

The Inland Western United States generally represents a region that is rugged, mountainous, highly dissected and that often contains young geological components. Typically, its climate is highly diverse and variable across both time and space. Most moisture arrives as winter snow. In recent geological time, major disturbances have been common. Some examples include ash depositions from Cascade Range volcanoes, like the heavy deposition of Mount Mazama ash 6,700 years ago (Fryxell 1965), and the climatic effects of continental glacier ice near Spokane, Washington only 14,000 years ago (Whitlock 1992). During the last 14,000 years, regional climates have progressed from cold/wet to cool/ dry to cool/moist, with accompanying changes in vegetation (Whitlock 1992). The ash depositions increased soil moisture storage in parts of the region, which brought about additional changes in effective climate (Geist and Cochran 1991). In some cases, this increased moisture storage allowed vegetation more characteristic of a moist, west coast climate or of a riparian site to occupy otherwise moisture limited areas (Meurisse et al. 1991). However, these forests are at especially high risk during extended dry periods.

Glacial scouring of valley bottoms created many thin, compacted soils with water and nutrient storage capacity that is extremely limited and, if physically disturbed or further compacted, that can easily lose much productivity potential (Meurisse et al. 1991). Many other shallow soils characteristic of the region are also fragile, especially where organic matter incorporated in upper soil horizons is limited (Harvey et al. 1993a).

In short, these are constantly changing environments where vegetation is often near its environmental limits for at least some species (especially climax conifers) in most forests. Many of our actions have the potential to rapidly alter them further.

Decomposition Processes

The typical temperature and moisture patterns of wet, cold winters and warm, dry summers can be severely limiting to biological decomposition (Olsen 1963; Harvey et al. 1979; Edmonds 1991). This allows accumulations of plant debris which, when combined with frequent lightning in summer rainstorms (that yield little water), repeatedly sets the stage for wildfire ignition (Arno 1980). The resulting fires are important. In the absence of fire, critical nutrients are tied up in plant debris, causing the site to become nutrient limited.

Thus, these forests are dependent on a combination of biological (primarily microbes and insects) and fire decomposition processes to regulate nutrient availability and cycling (Olsen 1981). Which of them dominates is dependent on site and climate (Harvey et al. 1993b). For example, fire dominates and occurs frequently in dry ponderosa pine forests and to a lesser extent with Douglas-fir, but biological decomposition dominates moist cedar/hemlock sites where fires occur much less frequently (Habeck and Mutch 1973; Arno 1980; Harvey et al. 1993b). The latter forest types usually occupy sites with deep, ash cap soils. Growth of vegetation on ash cap soils is less limited than on other soils because of the relatively high water and nutrient storage capacity. This likely further inhibits the ability of above-ground, more climate-limited biological decomposition processes to keep up with carbon accumulation.

ANALYZING THE PAST

Potential Insect and Disease Benefits

Characterizing this unique, dynamic environment permits analysis of past roles for insects and diseases. To reemphasize, historical environments were diverse and constantly changed so vegetation likely experienced frequent stress. Biological decomposition may have been constrained but recycling of nutrients was assured by fire (Olsen 1981). Fire-resistant species dominated sites that often burned. Most native "pests" were stress sensitive, i.e., they tended to attack unthrifty or stressed individuals (Waring 1987; Stoszek 1988). This probably generated the highest mortality in the poorest adjusted vegetation, an obvious benefit (Harvey et al. 1992). With fungal pathogens and at least some insects, it also probably accelerated decomposition, another obvious benefit (Mattson and Addy 1975). Localized centers of insect and disease activities created diversity in forest structure and species composition, also a benefit (van der Kamp 1991).

All of these helped to stabilize and diversify long-lived (100-400 years) tree communities that occupied potentially resource-limited sites with climates that changed over time periods which varied from days to several thousand years. Insects and diseases would have been integral to the development and function of these ecosystems (Martin 1988; Burdon 1991; Jarosz et al. 1991; Harvey et al. 1992).

THE PRESENT

Changing Conditions

In the present world, several things have altered the normal historical development and function of these forests. Perhaps most importantly, the numbers and types of disturbances are changing. Fires are often extinguished quickly, so forests burn less frequently (Baker 1992). However, wildfire effects can be severe in the presence of accumulated fuels and may have an associated loss of vegetation and organic matter storage (Brown 1983; Harvey et al. 1993a). Post-fire erosion of surface mineral soil horizons can also be a problem, especially with hot fires. As a result, representation of fire-adapted species in forest communities has been reduced. Non-fire-adapted species are invading areas where they were historically excluded by frequent fire (Monnig and Byler 1992; O'Laughlin 1994). Also, tree densities are vastly exceeding historical levels (Baker

1988; Monnig and Byler 1992; O'Laughlin 1994). Disturbances, in the absence of fire, such as harvesting, are increasing and may result in soil compaction, dislocation of surface horizons (especially reduction of ash cap depth) or inappropriate organic matter depletion (Harvey et al. 1993a). Tree removals associated with harvests have probably not improved adaptation of the community to the site, though they may not regularly reduce it (Wilusz and Giertych 1974).

In white and sugar pine country *(Pinus monticola* Dougl., *P. lambertiana* Dougl.) the introduction of white pine blister rust *(Cronartium ribicola* J.C. Fisch.) early in the century is reducing representation of these pines in many forest communities, to less than half of what they were even 40 years ago (Monnig and Byler 1992; O'Laughlin 1994). Fire-adapted species such as ponderosa pine and western larch (*P. ponderosa* Dougl. ex Laws., *Larix occidentalis* Nutt.), white and sugar pines, all tend to be broadly adapted species (Rehfeldt 1990) and appear relatively tolerant of many native insects and diseases (Monnig and Byler 1992). They are strongly reduced in the absence of fire. Other native conifer species are narrowly adapted and not very tolerant of changes in their environment or to many native insects and diseases. They are currently increasing rapidly (Monnig and Byler 1992; O'Laughlin 1994).

Ecosystem Response

The ecosystems have responded appropriately. They are now faced with reduced nutrient cycling, increased tree density, more shade tolerant late seral and climax species, less broadly adapted pioneer and early seral, "pest" tolerant species, with accompanying reduced vegetation adjustment to current site conditions. Stress levels have likely increased accordingly (McDonald 1990). Activities of "pests" have also increased (Baker 1988; Monnig and Byler 1992; Hessburg et al. 1993). These forests are adjusting (adapting) to prevailing conditions with the tools still at hand . . . insects, diseases, microbial decomposition, and sporadic, often severe wildfire.

An Accommodation

Where forest changes remain within reasonable natural variation, the ecosystems will eventually counter pest activities as they adjust. To facilitate that process, we should also adjust current human influences and restore many previously affected sites to better simulate situations in the past. No action, so long as fire regimes are not restored, will only serve to accelerate negative change. Current levels of insect and disease activity have common, sometimes complex environmental causes. Most have less to do with the individual organism than with present environmental conditions and the history of their development. Insects and diseases are a problem primarily because they are now extensive and highly active, creating an appearance and condition not in harmony with how society thinks it should be. However, real damage, in the form of severe wildfire effects, are increased by these conditions. Without modification of current actions and conditions, time lines for recovery of affected forests will continue to be extended.

REFERENCES

Arno, S.F. 1980. Forest fire history in the Northern Rockies. *J. of For.* 78:460-465.

Baker, F.A. 1988. The influence of forest management on pathogens. *Northw. Environ. J.* 4:229-246.

Baker, W.L. 1992. Effects of settlement and fire suppression on landscape structure. *Ecol.* 73:1879-1887.

Brown, J.K. 1983. The "unnatural fuel buildup" issue. In: pp. 127-128. Lotan, J.E. et al. (Eds.). Proceedings–symposium and workshop on wilderness fire. Gen. Tech. Rep. INT-182. USDA, Forest Service, Intermountain Research Station. Ogden, UT.

Burdon, J.J. 1991. Fungal pathogens as selective forces in plant populations and communities. *Aust. J. Ecol.* 16:423-432.

Edmonds, R.L. 1991. Organic matter decomposition in Western United States forests. In: pp. 118-128. Harvey, A.E. and L.F. Neuenschwander (Comps.). Proceedings–Management and productivity of western montane forest soils. USDA, Forest Service, Intermountain Research Station. Ogden, UT.

Fryxell, R. 1965. Mazama and Glacier Peak volcanic ash layers: relative ages. *Science* 147:1288-1290.

Geist, J.M. and P.H. Cochran. 1991. Influences of volcanic ash and pumice deposition on productivity of western interior forest soils. In: pp. 486-535. Harvey, A.E. and L.F. Neuenschwander (Comps.). Proceedings–Management and productivity of western montane forest soils. USDA, Forest Service, Intermountain Research Station. Ogden, UT.

Habeck, J.R. and R. W. Mutch. 1973. Fire-dependent forests in the northern Rocky Mountains. *Quat. Res.* 3:408-424.

Hessburg, P.F., R.G. Mitchell and G.M. Filip. 1993. Historical and current roles of insects and pathogens in eastern Oregon and Washington forested landscapes. In: pp. 486-535. Hessburg, P.F. (Comp.) Eastside forest ecosystem health assessment-Volume III: USDA, Forest Service, Pacific Northwest Forest and Range Experiment Station. Portland, OR.

Harvey, A.E., M.J. Larsen and M.F. Jurgensen. 1979. Fire-decay: interactive roles regulating wood accumulation and soil development in the Northern Rocky

Mountains. Res. Note INT-263. USDA, Forest Service, Intermountain Forest and Range Experiment Station. Ogden, UT. 4 p.

Harvey, A.E., G.I. McDonald and M.F. Jurgensen. 1992. Relationships between fire, pathogens, and long-term productivity in northwestern forests. In: pp. 16-22. Kaufman, J.B. et al. (Coord.). Fire in Pacific Northwest ecosystems: exploring emerging issues. Portland, OR. Jan. 21-23, 1992. Oregon State University. Corvallis, OR.

Harvey, A.E., J.M. Geist, G.I. McDonald et al. 1993a. Biotic and abiotic processes in eastside ecosystems: the effects of management on soil properties, processes and productivity. In: pp. 101-173. Hessburg, P.F. (Comp.). Eastside forest ecosystem health assessment–Volume III: USDA, Forest Service, Pacific Northwest Forest and Range Experiment Station. Portland, OR.

Harvey, A.E., G.I. McDonald, M. F. Jurgensen and M.J. Larsen. 1993b. Microbes: drivers of long-term ecological processes in fire-influenced cedar-hemlock-white pine forests in the Inland Northwest. In: Baumgartner, D.A. (Comp.). Symposium proceedings, Interior cedar-hemlock-white pine forests: ecology and management. March 2-4, 1993, Spokane, WA. Washington State University, Cooperative Extension. Pullman, WA. In press.

Jarosz, A.M., J.J. Burdon and W.J. Muller. 1991. Long-term effects of disease epidemics. *J. Appl. Ecol.* 26:725-733.

Martin, R. 1988. Interactions among fire, arthropods, and diseases in a healthy forest. In: pp. 87-91. Healthy forests, healthy world. Proceedings of the 1988 Society of American Foresters national convention. Rochester, NY. October 16-18, 1988. Society of American Foresters. Washington, DC.

Mattson, W.J. and N.D. Addy. 1975. Phytophagous insects as regulators of forest primary production. *Science* 190:515-522.

McDonald, G.I. 1990. Connecting forest productivity to behavior of soil-borne diseases. In: pp. 129-144. Harvey A.E., and L.F. Neuenschwander (Comps.). Proceedings–Management and productivity of western montane forest soils, USDA, Forest Service, Intermountain Research Station. Ogden, UT.

Meurisse, R.T., W.A. Robbie, J. Niehoff and G. Ford. 1991. Dominant soil formation processes and properties in western-montane forest types and landscapes–some implications for productivity and management. In: pp. 7-19. Harvey, A.E., and L.F. Neuenschwander (Comps.). Proceedings–Management and productivity of western-montane forest soils. USDA, Forest Service, Intermountain Research Station. Ogden, UT.

Monnig, G. and J. Byler. 1992. Forest health and ecological integrity in the Northern Rockies. USDA, Forest Service, Northern Region, FPM Rep. 92-7. 7p.

Norris, L.A. 1988. Forest pest management: an old problem looking for a new perspective. *Northwest Environ. J.* 4:185-197.

Olsen, J.S. 1963. Energy storage and balance of producers and decomposers in ecological systems. *Ecol.* 44:322-331.

Olsen, J. S. 1981. Carbon balance in relation to fire regimes. In: pp. 337-378. Mooney, H. A. et al. (Coord.). Fire regimes and ecosystem properties. Gen. Tech. Rep. WO-26. USDA, Forest Service. Washington, DC.

Peters, R.L. 1990. Effects of global warming on forests. *For. Ecol. and Manage.* 35:13-33.

Rehfeldt, G.E. 1990. Gene resource management: Using models of genetic variation in silviculture. In: pp. 31-44. Proceedings, genetics/silviculture workshop. Wenatchee, WA. Aug. 26-31, 1990. USDA, Forest Service, Timber Management Staff. Washington, DC.

Stoszek, K.J. 1988. Forests under stress and insect outbreaks. *Northw. Env. J.* 4:247-261.

van der Kamp, B. 1991. Pathogens as agents of diversity in forested landscapes. *Forest. Chron.* 67:353-354.

Waring, R.H. 1987. Characteristics of trees predisposed to die. *Biosci.* 37: 569-564.

Whitlock, C. 1992. Vegetational and climatic history of the Pacific Northwest during the last 20,000 years: Implications for understanding present-day biodiversity. *Northw. Environ. J.* 8.

Wilusz, W. and M. Giertych. 1974. Effects of classical silviculture on the genetic quality of the progeny. *Silv. Genet.* 23:127-130.

Assessing Forest Health Conditions in Idaho with Forest Inventory Data

Jay O'Laughlin

ABSTRACT. Periodic inventory measurements of forest characteristics published by the U.S. Forest Service are used to assess trends
in forest conditions in Idaho. Forest species composition, measured
by growing stock volume, has changed since 1952. Western white
pine and ponderosa pine have declined by 60% and 40%, respectively. True firs (mostly grand fir) increased by 60%, lodgepole pine by
almost 40%, and Douglas-fir, the predominant species throughout
the state, increased by 15%. Measurements of tree mortality across
the state and region from 1952 to 1987 establish a baseline regional
range for judging current conditions. Recent mortality data from
some Idaho national forests are much higher than the upper limit of
the baseline regional range. On the Boise and Payette National Forests in southwestern Idaho, annual mortality exceeds annual growth.
Recent inventories of national forests in northern Idaho show mature
stands have mortality well above the baseline regional range, which
projects into a negative net growth situation. Recent inventories of
private and other public forests in northern Idaho do not show similarly elevated mortality in mature stands.

INTRODUCTION

Forest health is a condition of forest ecosystems that sustains their
complexity while providing for human needs (O'Laughlin et al. 1994).

Jay O'Laughlin is Director, Policy Analysis Group, College of Forestry, Wildlife and Range Sciences, University of Idaho, Moscow, ID 83844-1134.

[Haworth co-indexing entry note]: "Assessing Forest Health Conditions in Idaho with Forest
Inventory Data." O'Laughlin, Jay. Co-published simultaneously in the *Journal of Sustainable Forestry*
(The Haworth Press, Inc.) Vol. 2, No. 3/4, 1994, pp. 221-247; and: *Assessing Forest Ecosystem Health
in the Inland West* (eds: R. Neil Sampson and David L. Adams) The Haworth Press, Inc., 1994, pp.
221-247. Multiple copies of this article/chapter may be purchased from The Haworth Document Delivery Center [1-800-3-HAWORTH; 9:00 a.m. - 5:00 p.m. (EST)].

221

Assessments of forest health require selecting appropriate ecosystem elements or processes as indicators to measure condition. These data can be used to establish standards against which measurements can be compared, and to provide a baseline for monitoring to identify changes in conditions. Efforts to do this nationwide are just beginning but have not been implemented in Idaho.

Forest inventory data on timber resources have been collected periodically on permanent plots established by the U.S. Department of Agriculture Forest Service on all forest ownerships since the 1950s. These data are used herein to determine changes in tree species composition in Idaho. Trends in growth and mortality data across large forest areas are used to establish a baseline for the Inland Northwest region and to compare more recent data on mortality rates. This analysis of forest conditions in Idaho is an important beginning for forest health assessment, but limited because it only deals with timber resources.

According to the U.S. Environmental Protection Agency, there are three high priority forest ecosystem response indicators, none of them ready for implementation in EMAP, the agency's Environmental Monitoring and Assessment Program (Riitters et al. 1990). Nonetheless, a wealth of data has been collected that relates to two of these indicators–visual symptoms of foliar damage and tree growth efficiency. There is a lack of adequate data for the third indicator–relative abundance of wildlife. The data on foliar damage are highly variable, making the determination and interpretation of a baseline range for Idaho data difficult (O'Laughlin et al. 1993). Inventory data on annual growth and mortality provide a rough measure of tree growth efficiency, and have been linked to forest health by Smith (1990), Marsden et al. (1991), and Norris et al. (1993).

TIMBER AND TIMBERLAND TRENDS

To put this analysis of conditions in Idaho's forests in context, it is instructive to review forest conditions nationwide and then compare Idaho to other states in the Inland Northwest region, as well as the nation as a whole.

Nationwide Trends

Prior to European settlement, forests covered roughly 950 million acres, or 42% of the United States (Figure 1). Approximately 850 million acres were timberlands, defined as forests physically and administratively

FIGURE 1. U.S. forest land area and standing sawtimber volume trends.

U.S. Forest Area and Standing Sawtimber

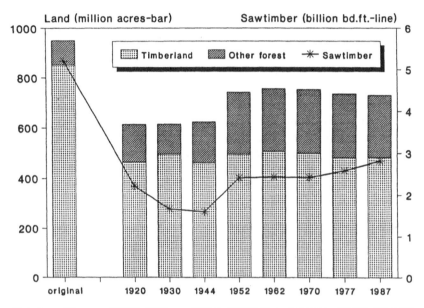

Data from Clawson (1979); Williams (1984); USDA Forest Service (1982); Waddell et al. (1989)

capable of producing at least 20 cubic feet of wood per acre per year. The original American forest contained an estimated 5 trillion board feet of sawtimber. Today roughly one-third of the nation is covered by forests, with most of that in the timberland category. Sawtimber volume is now approximately 60% of what it originally was.

Sedjo (1991) stated that by many measures America's forests are now in the best condition since the beginning of the twentieth century. The "favorable" condition of the nation's forests today is remarkable, considering the tremendous pressure people have placed on them. Many forests went from relatively undisturbed conditions in pre-colonial days through rapid conversion to agricultural lands during the mid-1800s and early 1900s. As Figure 1 illustrates, forest stocks have recovered since then, demonstrating both their serviceability and resiliency (Sedjo 1991).

Forests are vegetative systems capable of re-establishing themselves after disturbance, including fire, pests, logging, or grazing. In fact, destruction and restoration are components of the continuing natural cycle in these complex ecosystems. So, despite the stresses inflicted on forests by humans, American forests have shown an "amazing resilience." They have survived natural catastrophes in the past, and demonstrated the capacity to recover from the impacts of logging and agriculture (Frederick and Sedjo 1991). With management effort, the natural resilience of forests can be enhanced.

Tree mortality is a part of every living forest. The distribution of mortality across large areas is consistent and usually very predictable, except for periodic catastrophes. Loss of forest growing stock to mortality in 1986 was approximately 0.6% of the total growing stock volume in the United States. Mortality averaged 9 cubic feet per acre that year. On the Pacific Coast, mortality averaged about 15 cubic feet per acre in 1986; in the eastern regions, it ranged from 7 to 11 cubic feet per acre (USDA Forest Service 1990).

Inland Northwest Trends

Timberland area, timber growing stock volume by species, and growth and mortality data for the region are available from 1952 to 1987 (USDA Forest Service 1958, 1965, 1973, 1982; Waddell et al. 1989; Waddell 1992). Growing stock includes all live trees more than 5 inches in diameter at breast height. Comparable data prior to 1952 are not reliable and consistent.

Timberland and timber growing stock volume data for all ownership categories in Idaho, Montana, and the eastern portions of Oregon and Washington are displayed in Figure 2. For comparison purposes, trends for the entire United States during the same period are presented as tabular data at the bottom of Figure 2. The timberland area trends in Figure 2a show that Idaho timberland area, 14.5 million acres in 1987, declined from 15.5 million acres in 1952. Montana had approximately the same amount of timberland as Idaho in 1987. Eastern Oregon and Washington–that is, the portions of these states east of the Cascade Mountains taken together, or what the U.S. Forest Service calls the Ponderosa Pine region (Waddell et al. 1989) or the "eastside" region (Everett et al. 1993)–had slightly more timberland area in 1987 than either Montana or Idaho. These states are collectively known as the Inland Northwest region. The trend across the United States is similar, i.e., a slight decline in timberland from 1952 to 1987.

Also depicted in Figure 2b is the softwood timber growing stock volume in the same areas. Idaho has slightly more growing stock volume than Montana, and almost as much as the eastern Oregon and Washington region. Timber growing stock volume in Idaho increased by 12% from 1952 to 1987; volume in Montana remained relatively constant. Timber growing stock volume in eastern Oregon and Washington declined by 18% during that 35-year period. This indicates a situation in the "eastside" forests not currently evident in Idaho, Montana, or the rest of the United States.

CHANGES IN IDAHO FOREST TYPES

Idaho has 5 of 22 major forest types in the United States recommended by the Environmental Protection Agency for monitoring to assess forest health (Riitters et al. 1990). They are Douglas-fir (27.9% of the Idaho total), fir/spruce (27.6%), lodgepole pine (13.3%), ponderosa pine (9.6%), and western white pine (4.3%). These five forest types represent 82.7% of all Idaho growing stock volume (Benson et al. 1987).

Figure 3 illustrates changes in the composition of tree species in Idaho forests since 1952. The back row in Figure 3a shows that Douglas-fir has increased slightly and is the largest component of Idaho forests; Figure 3b shows Douglas-fir increased by roughly 1.2 billion cubic feet (top scale) or 15% (bottom scale). The second largest component in Figure 3a is the aggregation for Engelmann spruce, western larch, and other softwoods, primarily western redcedar and western hemlock. Taken together, spruce, larch, cedar, and hemlock increased by more than 30% from 1952 to 1987. The next component depicted in the illustration is true firs, consisting mainly of grand fir but including subalpine fir and a small amount of white fir. This component increased by 60%. Lodgepole pine increased almost 40% during the 35-year period.

Historically, the most important timber species in Idaho were ponderosa pine and western white pine. Both have declined since 1952, ponderosa pine by 40% and western white pine by 60% (Figure 3). Byler et al. (in press) estimated that the extent of western white pine may now be only 10% of what it was in 1900.

Based on these species changes, it is obvious that something significant has happened in Idaho's forests. Ponderosa pine has been reduced because it is a desirable timber species. Through the combined effects of fire exclusion and timber harvesting, Douglas-fir has invaded sites once occupied by ponderosa pine. Western white pine, also a desirable timber species, has been reduced primarily by the introduction of the exotic white pine blister rust fungus in the region in the early 1900s.

Figure 3b depicts the same data in a different way, and shows that growing stock volumes of both western white pine and ponderosa pine have declined by almost 2 billion cubic feet from 1952 to 1987. During

FIGURE 2a. Timberland area in Idaho, Montana, and the eastern portions of Oregon and Washington, 1952-1987, with U.S. data for comparison.

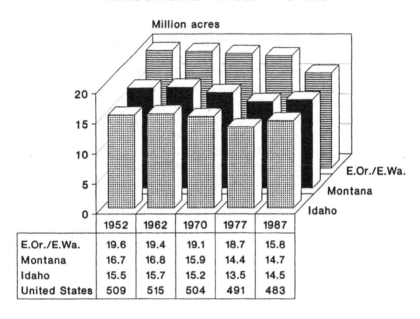

Timberland Area Trends

Million acres

	1952	1962	1970	1977	1987
E.Or./E.Wa.	19.6	19.4	19.1	18.7	15.8
Montana	16.7	16.8	15.9	14.4	14.7
Idaho	15.5	15.7	15.2	13.5	14.5
United States	509	515	504	491	483

that period, the true firs increased about 2.7 billion cubic feet. Spruce, larch, and other softwoods have increased approximately 1.7 billion cubic feet. Lodgepole pine and Douglas-fir have both increased more than 1 billion cubic feet.

Western white pine and ponderosa pine together have declined by almost 4 billion cubic feet while true firs and Douglas-fir have increased by a like amount. The increases in lodgepole pine and other softwoods thus represent about 3 billion cubic feet of net volume increase (or 12%) in Idaho's forests since 1952.

TREE GROWTH AND MORTALITY ANALYSIS

In order to insure scientific credibility, a quantitative approach toward ecological indicators is preferred when feasible. Indicators should also be clearly understandable by both the public and decision makers to be of value (Marshall et al. 1993). The relationship of forest growth and mortal-

FIGURE 2b. Softwood timber trends in Idaho, Montana, and the eastern portions of Oregon and Washington, 1952-1987, with U.S. data for comparison.

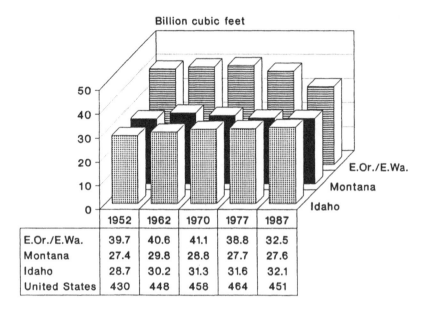

Softwood Timber Trends

Billion cubic feet

	1952	1962	1970	1977	1987
E.Or./E.Wa.	39.7	40.6	41.1	38.8	32.5
Montana	27.4	29.8	28.8	27.7	27.6
Idaho	28.7	30.2	31.3	31.6	32.1
United States	430	448	458	464	451

ity as a rough measure of tree growth efficiency is both quantifiable and understandable, and a national database on timber resource conditions dating back to 1952 exists.

Tree growth efficiency is one of the Environmental Protection Agency's three high priority response indicators for monitoring forest health. According to EMAP scientists (Riitters et al. 1990) "This is a measure of the overall ability of trees to maintain themselves in an ecosystem, which is an obvious but sometimes overlooked condition for the perpetuation of forests."

There are several ways to measure tree growth efficiency. Given adequate resources, tree growth efficiency can be measured either at the individual tree level or at the stand level (Marshall 1993). For large-scale health estimates, a more practical alternative may be to use the growth and mortality data collected by the U.S. Forest Service. The relationship of tree mortality and growth rates to forest health was made by Smith (1990):

FIGURE 3a. Trends in Idaho forest tree species composition, 1952-1987.

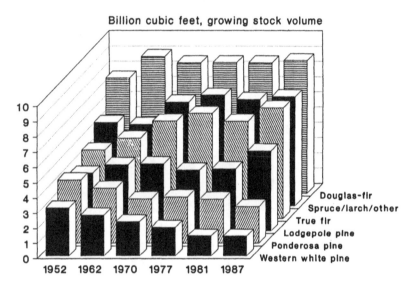

Idaho Forest Species Composition
Trends from 1952 to 1987

"One useful index of forest decline is reduced stand growth per unit area of ground."

Tree growth and mortality can be dramatically influenced by endemic and epidemic pest populations. Marsden et al. (1991) suggested using forest-wide data collected during periodic inventories that support the land management planning process on national forests. As they said, accurate quantitative measures of the presence, severity, and impact of pests are needed to make sound resource management decisions. The relationship of mortality to growth has also been suggested as an indicator of health by Norris et al. (1993), authors of the Society of American Foresters Task Force Report on Sustaining Long-term Forest Health and Productivity.

What rates of tree growth and mortality may be expected in the Inland Northwest? What are the "average" rates, and how do current rates compare to past averages? How do growth and mortality rates in Idaho compare to other regions? On the eastside of the Cascades–eastern Oregon and eastern Washington–and in Idaho and Montana, mortality ranged from 10 to 12 cubic feet per acre in 1986 (Waddell et al. 1989; Waddell 1992), or 11% to

FIGURE 3b. Trends in Idaho forest tree species composition with volume and percent changes by species.

Idaho Forest Species Composition
Change from 1952 to 1987

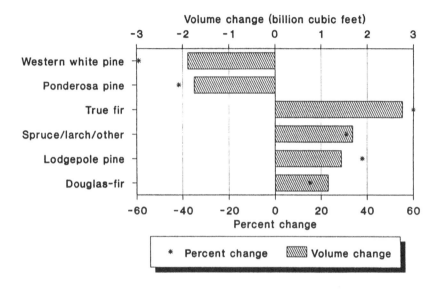

33% higher than the nationwide average. A baseline may be established by using a regional trend comparison approach similar to Figure 2 to analyze annual tree growth and mortality.

A Note on Statistical Reliability

The periodic timber resource inventories conducted by the U.S. Forest Service are sampled data and therefore have associated statistical standard errors. The scale of the estimate affects the standard error. As the area covered by the estimate becomes larger, more sample points are included and the standard error becomes smaller. Furthermore, estimates of land area made at the same scale as timber statistics–volume, annual growth, and annual mortality–have smaller standard errors. Timber volume estimates have slightly smaller standard errors than annual timber growth estimates. Annual mortality estimates have the largest standard error. In Idaho, mortality standard errors are two to three times that of annual growth (Benson et al. 1987; Wilson and Van Hooser 1993).

Table 1 displays the published information on standard errors from the state, regional, and national data presented in Figures 2 and 4. They are expressed as a percentage range of the estimate, computed with a 67% confidence interval. No standard errors for annual growth or mortality were given in the source document, nor were standard errors provided in more recent region-wide inventory reports.

Table 2 displays the published percent standard errors for timber statistics in Idaho, computed with a 67% confidence interval. No published stan-

TABLE 1. Percent standard error of timber statistics for the Inland Northwest, 1977. Data are from USDA Forest Service (1982).

	Idaho	Montana	Eastern Oregon	Eastern Washington	Rocky Mountain Region	United States
Timberland area	± 0.9%	± 0.7%	± 0.7%	± 0.9%	± 0.5%	± 0.1%
Growing stock volume	± 1.3%	± 1.9%	± 1.5%	± 1.4%	± 0.5%	± 0.3%

TABLE 2. Idaho softwood timber statistics and percent standard error on private and other public forest lands, 1987.*

	Private and other public forests		National forests		Total
	Timber volume	Percent standard error	Timber volume	Percent of total*	Timber volume
	(timber volumes in billion cubic feet)				
Net volume	8.648	+ 2.5%	23.440	73%	32.088
Net annual growth	0.244	+ 3.7%	0.463	66%	0.706
Annual mortality	0.038	+ 10.3%	0.120	76%	0.158

* Published standard errors for Idaho national forests are unavailable

Data from Benson et al. (1987); Waddell et al. (1989); Waddell (1992)

dard errors for individual national forests or aggregated national forest statistics for Idaho were located in the literature.

From the published standard error of ± 10.3% for mortality estimates on the private and other public lands of 4.5 million acres (33% of Idaho timberlands, the rest are national forests) it can be said that individual national forests will have a standard error of at least ± 10% for mortality estimates. Because mortality estimates have almost 3 times the standard error of growing stock volume estimates (Table 2), and volume estimates in Idaho had a standard error of ± 7.3% per billion cubic feet (USDA Forest Service 1982), standard errors for mortality on individual national forests may be at least ± 20%.

U.S. Forest Service plant pathologists Jim Byler and Sue Hagle (personal communication) believe that mortality figures in forest inventory publications and forest plans are systematically underestimated. It is difficult for professional foresters, let alone the temporary personnel who measure inventory plots, to tell whether or not dead trees have succumbed within the last five years. Based on their professional estimates of mortality using permanent plots that have been remeasured in the national forests in northern Idaho, Byler and Hagle feel that mortality figures in forest inventory publications and forest plans are underestimated by much more than 20%.

Annual Growth and Mortality Trends

As illustrated in Figure 4, mortality in Idaho has ranged from 115 to 202 million cubic feet per year during the six measurement periods from 1952 to 1987. No data have been compiled since 1987, the year the recent drought in the Inland Northwest began. During these same 35 years, net annual growth–a measure of growth with mortality deducted from it–increased more than 70%, from 412 to 706 million cubic feet per year. For comparison purposes, during the same time period annual removals–primarily harvests of useful timber products–increased 55%, from 233 to 360 million cubic feet per year. A similar pattern is evident in Montana. In eastern Oregon and Washington, the picture is somewhat different, as mortality actually declined slightly from 1952 to 1987. Net annual growth increased during the period, but removals increased much faster, almost doubling during the 35-year period. In 1977 and 1987 annual removals exceeded net annual growth of timber in the eastside forests of Oregon and Washington. In comparison, softwood trends across the United States reveal fairly constant mortality, a substantial increase of 66% in net annual growth, and an increase of 24% in timber removals (Figure 4). As a footnote, Idaho and Montana growth and mortality data for 1952, 1962, and 1970 in Figure 4

FIGURE 4a. Softwood timber annual net growth, mortality, and removal trends in Idaho 1952-1987.

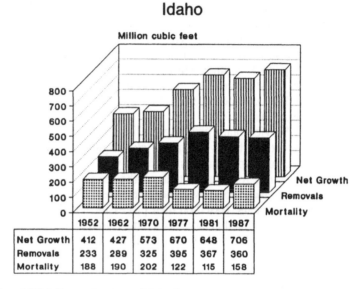

Idaho

Million cubic feet

	1952	1962	1970	1977	1981	1987
Net Growth	412	427	573	670	648	706
Removals	233	289	325	395	367	360
Mortality	188	190	202	122	115	158

Data from USDA Forest Service published reports

are estimates that were necessary because of revisions in published data (see explanation in USDA Forest Service 1982: 334).

Idaho National Forests–Idaho's national forests represent 40% of the land area in the state. These public lands, managed by the U.S. Forest Service, include two-thirds of the timberland acreage (67%) and almost three-fourths (73%) of the forest growing stock volume in Idaho (Waddell 1992). In Montana, 56% of the timberlands and 67% of the growing stock volume are in national forests. In eastern Oregon and Washington, national forests include 48% of the timberlands. It is not possible to tell from published statistics how much of the growing stock volume is in the eastside national forests. For the nation as a whole, national forests have almost 18% of the timberland acreage and 41% of the softwood growing stock volume (Waddell et al. 1989).

Each national forest conducts its own inventory on its own time schedule. Data on growing stock volume and annual growth and mortality are available in the forest plans for each of Idaho's national forests, and are used in this analysis. Inventory on other timberlands in Idaho is collected by the Research branch of the U.S. Forest Service.

FIGURE 4b. Softwood timber annual net growth, mortality, and removal trends in Montana, 1952-1987.

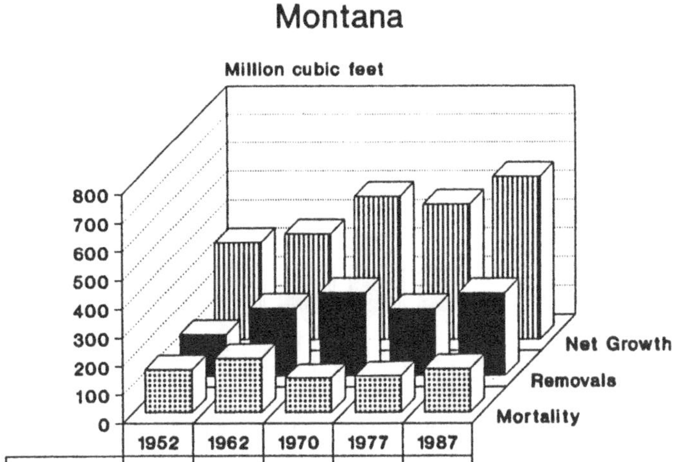

Montana

Million cubic feet

	1952	1962	1970	1977	1987
Net Growth	339	370	500	473	568
Removals	147	238	294	235	292
Mortality	150	190	120	128	153

Data from USDA Forest Service published reports

Public concern about forest health in Idaho is greatest in southwestern Idaho. The reason is revealed by analyzing the growth and mortality data from the forest plans (O'Laughlin et al. 1993). Averaged across all ten national forests, mortality was 18.3% of gross annual growth. Of the five national forests that have more than 2 billion cubic feet of growing stock volume, the Boise and Payette National Forests had, respectively, mortality at 31.3% and 24.9% of gross annual growth, significantly above the average. Of the five other national forests, the Targhee, with 1 billion cubic feet of growing stock volume, had mortality at 28.3% of gross annual growth.

The Boise and the Payette National Forests, along with Boise Cascade Corporation, are the largest timberland owners in southwestern Idaho. Growth and mortality trends for the Boise and Payette National Forests were obtained from past inventories and are depicted in Figure 5. For the Boise National Forest, four data points from 1954 to 1992 are depicted in the leftmost portion of Figure 5. The first three sets of data for 1954, 1979, and 1987 were provided from the inventory records on file in the Forest Supervisor's office. Because of the absence of current inventory data,

FIGURE 4c. Softwood timber annual net growth, mortality, and removal trends in the eastern portions of Oregon and Washington, 1952-1987.

Eastern Oregon and Eastern Washington

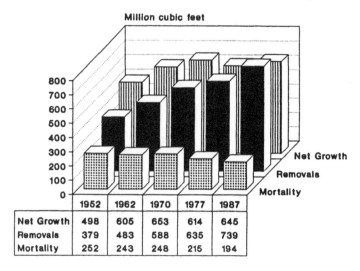

	1952	1962	1970	1977	1987
Net Growth	498	605	653	614	645
Removals	379	483	588	635	739
Mortality	252	243	248	215	194

Data from USDA Forest Service published reports

gross annual growth for 1992 was estimated statistically using least squares linear regression. The estimate for 1992 is slightly less than 1987 inventory data. Mortality trend data indicated a slight decrease from 1954 to 1987.

Mortality for 1992 was estimated by Boise National Forest staff. Consistent with the method for stand examinations of mortality when inventory is taken, estimates were made over a five-year period and then averaged for a 1992 estimate (Table 3). Insect-caused mortality was estimated by relating airborne visual estimates to land-based data in a geographic information system and verifying the estimates on the ground (Boise National Forest 1993). Fire-caused mortality was estimated by staff, based on a variety of records. Efforts were made by staff to be conservative with their estimates. Even so, these estimates are crude approximations. The next scheduled inventory on the Boise is 1995, when more accurate growth and mortality data will be collected.

Boise National Forest mortality data for 1992 (Figure 5) indicate a situation that best can be described as catastrophic, when mortality exceeds growth during a given period (McGuire 1958). Even without mortality estimates, significant growth reductions over time are evident in

FIGURE 4d. Softwood timber annual net growth, mortality, and removal trends in the U.S. 1952-1987.

United States

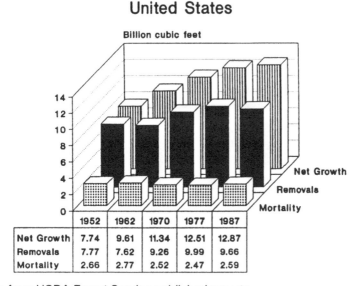

Billion cubic feet

	1952	1962	1970	1977	1987
Net Growth	7.74	9.61	11.34	12.51	12.87
Removals	7.77	7.62	9.26	9.99	9.66
Mortality	2.66	2.77	2.52	2.47	2.59

Data from USDA Forest Service published reports

Figure 5. The decline in gross annual growth alone indicates forest decline on the Boise National Forest. Annual growth declined by 39%, from 211.8 million board feet in 1954 to 129.8 million board feet 1987.

Declines in gross annual growth may result from overstocked young stands or from an aging forest. Whether either situation is good or bad is a value judgment. Some may say an aging forest is bad, because forest productivity, as measured by tree growth, declines as trees get older. Others may say it is good because the older a forest gets the closer it is to old-growth conditions. These data are for lands that have been identified in the forest planning process as suited for timber production. Growth reductions on those lands may represent a forest management problem, partly because 62% of the land base in the national forests in Idaho is not available for timber production according to forest plans.

For the Payette National Forest, unpublished data from inventories of the forest completed in 1979 and 1991 are depicted in Figure 5. Gross annual growth on suitable timberlands on the Payette declined 18% when sawtimber is the measure, and as shown in Figure 5, the decline was 36% for all timber growing stock volume. Along with growth decline, mortality

FIGURE 5. Timber annual growth and mortality trends in Idaho's Boise and Payette National Forests, 1954-1991.

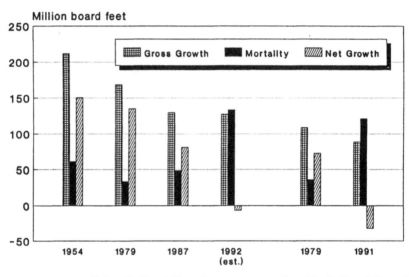

Growth and Mortality Trends
Southwestern Idaho National Forests

Data from unpublished forest inventories, except Boise National Forest staff 1992 estimates

tripled from 1979 to 1991. As on the Boise in 1992, mortality exceeded gross annual growth in 1991. This net loss of growing stock volume on the Payette (Figure 5) is even more significant when one considers that the inventoried land base identified as suitable for timber production has been reduced by 46%, from 795,980 acres in 1979 to 431,721 acres in 1991. Obviously, the situation on suitable timberlands in the Payette National Forest also meets the definition of catastrophic mortality.

The analysis that follows uses the annual growth and mortality data already presented, and is expressed in three ratios commonly used by the U.S. Forest Service to describe tree mortality: mortality per acre, mortality as a percent of growing stock volume, and mortality as a percent of gross annual growth.

Mortality Per Acre–One commonly used expression of mortality is on a

TABLE 3. Timber mortality estimates on Boise National Forest (by forest staff) 1988-1992 (million board feet).

	Suited lands*	Unsuited lands
Insect-caused		
Bark beetles	82,805	21,877
Tussock moth	222,543	36,634
Total	305,348	58,511
Fire-caused		
1989 fires	194,000	126,000
1992 fires	167,700	168,300
Total	361,700	294,300
Grand Total	667,048	352,811

* Suited lands are those identified in the forest plan as suitable for timber production. 55% of the total volume on suited lands volume was recovered in salvage sales (47% of insect-killed and 63% of fire-killed timber).

per-acre basis (Benson et al. 1987; USDA Forest Service 1990). Figure 6a depicts timber mortality on a per acre basis for the four regions used in the analysis. A range from approximately 7 cubic feet per acre per year to upwards of 13 cubic feet per acre per year is evident across the Inland Northwest region and the United States. This range of mortality can be used to represent the expected range of variability in timber mortality over large areas during the period 1952 to 1987. When mortality is outside this range, a forest health problem may exist, as tree mortality would fall outside the range of expectations. A word of caution: this range is based on limited data representing one-third or less of the life cycle of a forest stand.

The range of variability within the region, as depicted in Figure 6a, is used to determine if the mortality data for the Boise and Payette National Forests, depicted in Figure 5, lie outside this range. The lowest mortality rates in Figure 6a for 1952, 1962, and 1987 are for the United States. These data points were not used to define the regional range, which are the highest and lowest mortality rates for either Idaho, Montana, or eastern Oregon and Washington for each time period. Also presented in Figure 6b are the single data points for some of the other Idaho national forests. The Boise National Forest was at the upper limit of the regional range in 1954, but has fallen below it since then. Based on mortality per acre, the Boise did not appear to have an unusual situation until 1992, when estimates indicate

FIGURE 6a. Timber mortality per acre trends in Idaho, Montana, and the eastern portions of Oregon and Washington, 1952-1987, with U.S. trend for comparison.

Mortality per Acre
Inland Northwest Region and U.S. Trends

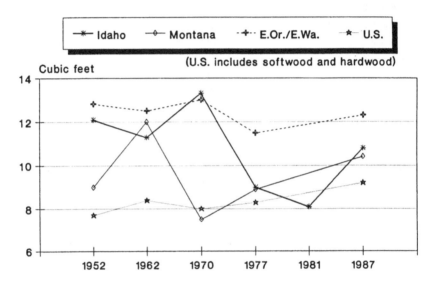

Data from USDA Forest Service published reports, and Land and Resource Management Plans for individual national forests.

twice the expected mortality. The Payette National Forest, with its high mortality rate in 1991, fell well outside the regional range of variability, having had four times the mortality that might be expected. Based on data presented in forest plans, the Clearwater National Forest fell just above the regional range in 1985. Other national forests were within the range or below it. Inventory data for 1990 on the 3.5 million acres of forests outside the national forests in northern Idaho (Wilson and Van Hooser 1993), which represent 23% of all Idaho timberlands, show 13.7 cubic feet per acre of mortality. This is slightly above the upper limit of the regional range for 1987 as shown in Figure 6.

Mortality as a Percentage of Growing Stock Volume–Using the same approach for another commonly used expression of mortality (USDA Forest Service 1990, Filip and Schmitt 1990), the range of variability lies

FIGURE 6b. Timber mortality per acre trends in Inland Northwest regional range and selected Idaho national forests.

Mortality per Acre
Regional Range & Idaho National Forests

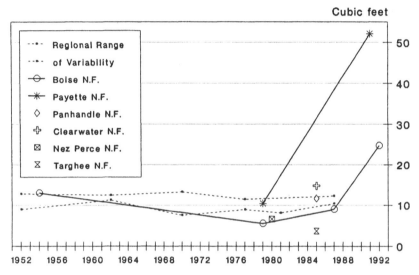

Data from USDA Forest Service published reports, and Land and Resource Management Plans for individual national forests

between .4 and .7 percent, as depicted in Figure 7a. Overlaying the data from the Boise, Payette, and other Idaho National Forests, the result in Figure 7b illustrates roughly the same relationship as in Figure 6b. The Boise National Forest was within the range until 1992, when mortality was twice what would be expected in one region. The Payette National Forest was within the range in 1979, but in 1991 mortality was almost four times above the expected range of variability. Other forests, based on the data reported in their forest plans, did not display anything unusual. However, based on analysis of permanent plots, national forests in northern Idaho are experiencing annual mortality of 3% to 4% in mature stands, with some plots as high as 10% (Hayes 1993, based on S. Hagle, personal communication). This level of mortality is even higher than the recent mortality data presented in Figure 7b for the Boise and Payette National Forests. Inventory data for 1990 on the 3.5 million acres of forests outside

FIGURE 7a. Timber mortality as a percentage of growing stock volume trends in Idaho, Montana, and the eastern portions of Oregon and Washington, 1952-1987, with U.S. trend for comparison.

Mortality as % of Growing Stock Volume
Inland Northwest Region and U.S. Trends

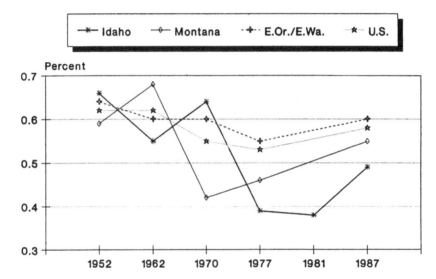

Data from USDA Forest Service published reports, and Land and Resource Management Plans for individual forests

the national forests in northern Idaho (Wilson and Van Hooser 1993) show that annual mortality is 0.63% of the growing stock volume, just at the upper limit of the regional range for 1987 in Figure 7.

Mortality as a Percentage of Gross Annual Growth–This is the measure suggested by the Society of American Foresters task force report on Sustaining Long-term Forest Health and Productivity (Norris et al. 1993). It has been used by the U.S. Forest Service to describe Idaho forest resources (Benson et al. 1987) and by Filip and Schmitt (1990) to describe root disease mortality on true fir species.

Figure 8a shows that the range of variability was from 15% to almost 35% of gross annual growth. The downward trend includes growth increases evident in Figure 4. Again, by overlaying this range of variability in the region with Idaho national forest data, the results in Figure 8b reveal

FIGURE 7b. Timber mortality as a percentage of growing stock volume trends in Inland Northwest regional range and selected Idaho national forests.

Mortality as % of Growing Stock Volume
Regional Range & Idaho National Forests

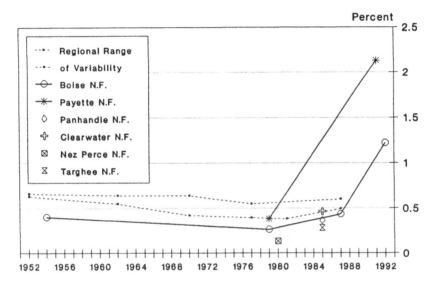

Data from USDA Forest Service published reports, and Land and Resource Management Plans for individual national forests

that the Boise National Forest was within the regional range in 1954 and 1979, and outside it in 1987 and 1992. In 1987 this was less a function of high mortality than it was low growth, but in 1992 the combination of high mortality and low growth resulted in mortality slightly in excess of growth. The Payette National Forest was in the range of variability in 1979, and in 1991 was well out of it, when mortality was 1.4 times gross growth on suitable timberlands. Mortality on the Targhee National Forest was 28% of gross growth in 1985, slightly higher than the regional range. Although not depicted, the Caribou National Forest had mortality at 25% of gross growth in 1985, at the upper limit of the regional range. Inventory data for 1990 on the 3.5 million acres of forests outside national forests in northern Idaho (Wilson and Van Hooser 1993), representing 23% of all Idaho

FIGURE 8a. Timber mortality as a percentage of gross annual growth trends in Idaho, Montana, and the eastern portions of Oregon and Washington, 1952-1987, with U.S. trend for comparison.

Mortality as % of Gross Annual Growth
Inland Northwest Region and U.S. Trends

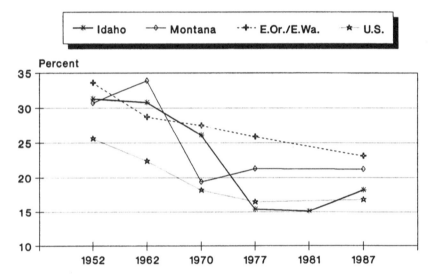

Data from USDA Forest Service published reports, and Land and Resource Management Plans for individual national forests

timberlands, show mortality as 17.1% of gross annual growth, well below the upper limit of the baseline regional range for 1987 in Figure 8.

Is There a Tree Growth and Mortality Problem in Idaho?

From the data that are available, tree mortality fell outside the expected regional range on the two national forests in southwestern Idaho (Figure 8b). Both the Boise and Payette National Forests have recently experienced levels of mortality that exceed gross annual growth. The forests also have declining gross annual growth, thus contributing to an unfavorable relationship between growth and mortality. To the extent that tree growth and mortality data reflect forest health, it may be said that both the Boise and Payette National Forests have a forest health problem on lands identified in forest plans as suitable for timber production. Similar analysis of

FIGURE 8b. Timber mortality as a percentage of gross annual growth trends in Inland Northwest regional range and selected Idaho national forests.

Mortality as % of Gross Annual Growth
Regional Range & Idaho National Forests

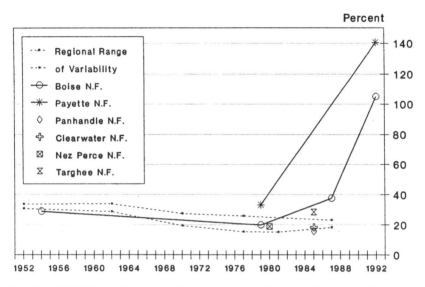

Data from USDA Forest Service published reports, and Land and Resource Management Plans for individual national forests

current data for other Idaho forests might also reveal symptoms of forest health problems.

DOES IDAHO HAVE A FOREST HEALTH PROBLEM?

According to Clark (1993), most of the scientists at a forest health symposium in Boise, Idaho, (June 1993) agreed that there is a forest health problem in Idaho. They commonly cited as a major underlying cause the conversion of once open forests, dominated by fire-resistant tree species, primarily ponderosa pine and larch, to dense forests dominated by fire-intolerant firs. Species conversion was caused by two factors acting independently or together: regeneration of firs on harvested sites, and invasion of firs as a result of excluding fire from performing its natural role. In some Idaho forests, this change in composition has led to increased insect and disease activity, tree mortality higher than tree growth, sometimes severe wildfires, and changes in wildlife populations (Clark 1993).

By analyzing existing tree growth and mortality data, a range of variability for mortality was defined. Using measurements of annual mortality in standard expressions related to acres, growing stock volume, or annual growth, it can be said that the Boise and Payette National Forests in southwestern Idaho have exhibited symptoms of a forest health problem. Both forests have higher than expected levels of mortality, in addition to growth decline, resulting in annual mortality in excess of annual growth.

What about the forests in northern Idaho? Although the recent drought situation was not as serious in northern Idaho as in the southern part of the state (O'Laughlin et al. 1993), because of the prevalence of Douglas-fir and root disease the long-term health of these forests is cause for concern.

On the Idaho Panhandle and Clearwater National Forests in northern Idaho, permanent inventory plots have been used to assess root disease mortality and its causes and effects since 1985. Several thousand trees on several hundred plots have been monitored. Mortality has been highest on grand fir and hemlock habitat types and lowest on Douglas-fir habitat types (Byler et al., in press; Hagle et al. 1992). According to Hayes (1993) and Hagle (personal communication), annual mortality averaged 3% to 4% of merchantable cubic foot volume in trees 80 years and older in the national forests in northern Idaho since 1985. On some plots annual mortality has varied from under 1% to more than 10%. Most mortality has been caused directly by root disease, especially in sawtimber size trees, or by stress attributed to root disease where the final cause of death was drought or insects. As Figure 7 shows, these data are much higher than those reported in the forest plans for these forests, but they do not represent the entire national forest. These plots were established after national forest plans were published in the mid-1980s. According to Byler et al. (in press), root disease has reduced timber volumes in mature stands by 40 percent in some areas of the Coeur d'Alene River basin in the Panhandle National Forest, and yield projections show negative net growth.

On private and other public forests in northern Idaho, recent timber resource inventory data do not show elevated levels of mortality in mature stands (Wilson and Van Hooser 1993). These two different sets of data were collected on permanent plots established for different purposes. As Marsden et al. (1991) put it, the problem of reconciling the quality of pest data in timber resource inventories remains.

In conclusion, this analysis demonstrates that tree growth and mortality data can be used as a measure of forest health condition and can be compared to existing time series data. The existing data are for 35 years, which may not be long enough to call an historic range. However, levels of mortality on national forests in southwestern Idaho are clearly at an elevated level–forests are dying faster than they are growing. Drought is partially responsible, but other factors–primarily species composition changes and dense stocking levels resulting from fire exclusion and past harvesting practices–have predisposed these forests to extreme effects of fire, insects, and disease. In northern Idaho, the prevalence of Douglas-fir and root diseases has been projected to have serious effects on long-term timber productivity. Existing inventories can be used to analyze forest conditions and document growth decline. Methods for acquiring new data on the causes of growth decline are worthy of attention by resource managers and researchers concerned about forest health. Immediate attention should be given to communicating forest health conditions to the public. The payoff will be a better understanding of forest health and collaborative efforts to design socially acceptable forest management strategies for protecting the long-term health and sustainability of our forests.

REFERENCES

Benson, R.E., A.W. Green and D.D. Van Hooser. 1987. Idaho's forest resources. USDA Forest Service Resource Bulletin INT-39. Intermountain Research Station. Ogden, UT. 114 p.

Boise National Forest. 1993. Insect mortality on the Boise National Forest, 1988-1992. Boise, ID. 5 p. + attached tables and appendices.

Byler, J.W., R.G. Krebill, S.K. Hagle and S.J. Kegley. Health of the cedar-hemlock-western white pine forests of Idaho. In: Interior cedar-hemlock-western white pine forests: ecology and management. Washington State University. Pullman. (In press).

Clark, L. 1993. Boise forest health symposium. *American Forests* 99(9/10):12.

Clawson, M. 1979. Forest in the long sweep of American history. *Science* 204: 1168-1174.

Everett, R., P. Hessburg, M. Jensen, B. Bormann, P.S. Bourgeron, R.W. Haynes, W.C. Krueger, J.F. Lehmkuhl, C.D. Oliver, R.C. Wissmar and A.P. Youngblood. 1993. Eastside forest ecosystem health assessment, volume I: Executive summary. USDA Forest Service unnumbered report. National Forest System and Forest Service Research. Portland, OR. 57 p.

Filip, G.M. and C.L. Schmitt. 1990. R_x for *Abies*: silvicultural options for diseased firs in Oregon and Washington. USDA Forest Service General Technical Report PNW-GTR-252. Pacific Northwest Research Station. Portland, OR. 34 p.

Frederick, K.D. and R.A. Sedjo. 1991. America's Renewable Resources: Historical Trends and Current Challenges. Resources for the Future. Washington, DC. 296 p.

Hagle, S., J. Byler, S. Jeheber-Matthews, R. Barth, J. Stock, B. Hansen and C.

Hubbard. 1992. Root disease in the Coeur d'Alene River basin: An assessment. USDA Forest Service, Idaho Panhandle National Forests, unnumbered publication. Coeur d'Alene. 16 p. + appendices.

Hayes, D. 1993. An estimate of timber mortality and potential for salvage, Nez Perce National Forest. USDA Forest Service, briefing paper, mimeo. Nez Perce National Forest. Grangeville, ID. 3 p.

Marsden, M.A., G.M. Filip and P.F. Hessburg. 1991. Using the forest timber inventory for sampling the occurrence of pests on interior Douglas-fir. In: pp. 109-114. Baumgartner, D.M. and J.V. Lotan (Eds.). Interior Douglas-fir and its management. Washington State Univ. Cooperative Extension. Pullman.

Marshall, J.D. 1993. The concept of health: trees, stands, and ecosystems. In: pp. 15-17. Proceedings, Forest health in the Inland West. USDA Forest Service, Boise National Forest, Dept. of Forest Resources. University of Idaho. Moscow.

Marshall, I.B., H. Hirvonen and E. Wiken. 1993. National and regional scale measures of Canada's ecosystem health. In: pp. 117-129. Woodley, S. et al. (Eds.). Ecological Integrity and the Management of Ecosystems. St. Lucie Press. St. Lucie, FL.

McGuire, J.R. 1958. Definitions. In: pp. 629-637. Timber resources for America's future. USDA Forest Service, Forest Resource Report No. 14. Washington, DC.

Norris, L.A., H. Cortner, M.R. Cutler, S.G. Haines, J.E. Hubbard, M.A. Kerrick, W.B. Kessler, J.C. Nelson, R. Stone and J.M. Sweeney. 1993. Sustaining long-term forest health and productivity. Task Force Report, Society of American Foresters. Bethesda, MD. 83 p.

O'Laughlin, J., R.L. Livingston, R. Thier, J. Thornton, D.E. Toweill and L. Morelan. 1994. Defining and measuring forest health. In: pp. 65-85. Sampson, R.N. and D.L. Adams (Eds.). Assessing Forest Ecosystem Health in the Inland West. Proceedings of the American Forests scientific workshop, November 14-20, 1993. Sun Valley, ID. The Haworth Press, Inc. New York.

O'Laughlin, J., J.M. MacCracken, D.L. Adams, S.C. Bunting, K.A. Blatner and C.E. Keegan, III. 1993. Forest health conditions in Idaho. Report No. 11, Idaho Forest, Wildlife and Range Policy Analysis Group. University of Idaho. Moscow.

Riitters, K.H., B. Law, R. Kucera, A. Gallant, R. DeVelice and C. Palmer. 1990. Indicator strategy for forests. In: Hunsaker, C.T. and D.E. Carpenter (Eds.). Ecological indicators for the Environmental Monitoring and Assessment Program. U.S. Environmental Protection Agency, EPA 600/3-90/060, Office of Research and Development. Research Triangle Park, NC. 14 p.

Sedjo, R.A. 1991. Forest resources: resilient and serviceable. In: pp. 81-120. Frederick, K.D. and R.A. Sedjo (Eds.). America's Renewable Resources: Historical Trends and Current Challenges. Resources for the Future. Washington, DC.

Smith, W.H. 1990. The health of North American forests: Stress and risk assessment. *Journal of Forestry* 88(1): 32-35.

USDA Forest Service. 1958. Timber resources for America's future. Forest Resource Report 14. Washington, DC. 713 p.

_____. 1965. Timber trends in the United States. Forest Resource Report 17. Washington, DC. 235 p.

_____. 1973. The outlook for timber in the United States. Forest Resource Report 20. Washington, DC. 367 p.

_____. 1982. An analysis of the timber situation in the United States: 1989-2030. Forest Resource Report 23. Washington, DC. 499 p.

_____. 1990. An analysis of the timber situation in the United States: 1989-2040. A technical document supporting the 1989 USDA Forest Service RPA Assessment. General Technical Report RM-199. Fort Collins, CO. 269 p.

Waddell, K. 1992. Forest statistics of the western states, 1987. USDA Forest Service unnumbered report, compiled for the Western States Legislative Forestry Task Force, Pacific Northwest Research Station. Portland, OR. 51 p.

Waddell, K.L., D.D. Oswald and D.S. Powell. 1989. Forest statistics of the United States, 1987. USDA Forest Service Resource Bulletin PNW-RB-168. Portland, OR. 106 p.

Williams, M. 1984. Predicting from inventories: A timely issue. *Journal of Forest History* 28(2): 92-98.

Wilson, M.J. and D.D. Van Hooser. 1993. Forest statistics for land outside national forests in northern Idaho, 1991. USDA Forest Service Resource Bulletin INT-80. Ogden, UT. 58 p.

Conceptual Origins
of Catastrophic Forest Mortality
in the Western United States

Allan N. D. Auclair
Julie A. Bedford

ABSTRACT. Estimates of changes this century (1890-1990) in the total mortality from natural causes (wildfire, pests) and harvesting in the Rocky Mountain and Pacific Coastal regions of the United States (excluding Alaska) were based on yearly national forest statistics. Total accrual per year doubled this century as a result of fire suppression and accelerated tree growth. Total depletion per year did not change significantly from 1920 to 1985; after 1985, extensive pest

Allan N. D. Auclair is Senior Scientist, and Julie A. Bedford is Environmental Scientist, Science and Policy Associates, Inc., 1333 H Street NW, Washington, DC 20005.

The authors thank J. L. Sarmiento (Princeton University, Princeton NJ) for providing the most recent estimates of the global net biospheric CO_2 flux; and the following for their most helpful commentary and suggestions on how to improve the text: C. C. Hardy (USFS, Missoula MT), A. E. Harvey (USFS, Moscow ID), R. W. Mutch (USFS, Missoula MT) and L. F. Neuenschwander (University of Idaho, Moscow ID).

Although the information in this document has been funded wholly or in part by the United States Environment Protection Agency, Office of Environmental Processes and Effects Research (Washington, DC) through EPA Contract No. 68-D1-0109 to SPA, Inc., it has not been subjected to Agency peer review process and therefore does not necessarily reflect the views of the Agency and no official endorsement should be inferred.

[Haworth co-indexing entry note]: "Conceptual Origins of Catastrophic Forest Mortality in the Western United States." Auclair, Allan N.D. and Julie A. Bedford. Co-published simultaneously in the *Journal of Sustainable Forestry* (The Haworth Press, Inc.) Vol. 2, No. 3/4, 1994, pp. 249-265; and: *Assessing Forest Ecosystem Health in the Inland West* (eds: R. Neil Sampson and David L. Adams) The Haworth Press, Inc., 1994, pp. 249-265. Multiple copies of this article/chapter may be purchased from The Haworth Document Delivery Center [1-800-3-HAWORTH; 9:00 a.m. - 5:00 p.m. (EST)].

and wildfire mortality sharply increased total losses. A model of accumulated fuel loads and shifts in forest density and species composition was used to account for this catastrophic jump in tree mortality concurrent with severe drought. The implications of a destabilized forest structure are discussed in terms of the needs to "defuel" many dominant forest types and to restore long-term sustainability.

INTRODUCTION

Forest conditions have seriously deteriorated in many regions of the western United States. Extraordinary levels of pests, wildfire, and drought over the past decade are killing millions of hectares of forest and woodland annually (Powell et al. 1993). There is a sense of catastrophic loss, and a corresponding management dilemma on how to resolve what appears to be a chronic and growing problem.

Forest scientists now recognize that many land use practices of the past may have inadvertently resulted in a serious alteration of forests and woodlands of the region. In particular, the short-interval, fire-adapted regimes of the early 1800s have been rigorously suppressed; in their place are less-frequent but intense, stand-replacement, crown fires with heavy fuels that can be very destructive of soils, site fertility, and watershed function.

There is an urgent need for a new management perspective, and an approach that is consistent with the natural ecological processes on which these ecosystems depend. To achieve this, there are important scientific questions that need to be more broadly tested and understood.

Some of the key questions include: What were the natural fire and pest conditions under which the dominant tree species evolved? To what extent are the forests at greater risk to wildfire and to pest outbreaks due to historical changes in tree species, stand densities, forest homogeneity, fuel connectivity, and accumulated fuel loading? Can the rise in pest-kill and fire extent since the mid 1970s best be explained by extraordinary climatic conditions? What practical silvicultural options exist to "defuel" the forests at high risk to fire?

The purpose of this paper is to develop an understanding of the historical antecedents leading to the current dilemma, namely the extraordinary levels of tree mortality and wildfire in the western U.S. forests. Specifically, our goal is to estimate the net volume of wood saved from burning by fire protection. Pyne (1988), consistent with earlier workers (Weaver 1943; Cooper 1961) hypothesized that continued fire protection was self-defeating since it resulted in the accumulation of tree litter (and other structural and species changes) that fueled catastrophic wildfires. We will examine the plausibility of this concept.

DATA AND METHODS

Approach

Accurate data on century-long changes in forest cover and productivity are not readily available. The most reliable estimates come from the national forest statistical database. In this study we compiled the statistics of the U.S.D.A. Forest Service for the Rocky Mountain and Pacific Coastal regions (excluding Alaska) into a standard format of annual total area (ha) and GMV or gross merchantable volume (m^3) of forest depletions. Depletion (wildfire, pests, harvest) records extended to early in this century, and all were consistent from 1950 to present. Extrapolations from the first date of record to 1800 were based on the average of the first decade of available depletion statistics (usually 1920-1929), except for harvesting which was estimated from literature summaries. Accruals (fire suppression, regrowth, accelerated tree growth) were estimated from the depletion summaries, and from Published chronologies of tree radial increment. All initial and final datasets, and any modifying assumptions to amend gaps or inconsistencies in the data, were fully documented (Auclair and Bedford 1993).

The annual balance between forest volume depletions and accruals, or "net forest volume," was calculated each year, 1890-1990, by subtracting the sum of forest volume depleted (wildfire, pests, harvest) from the sum of the forest volume accrued (regrowth, fire suppression, accelerated tree growth).

Net Forest Volume = <u>Total of Forest Depletions</u> – <u>Total of Forest Accruals</u>

Pest-Kill	Accelerated Tree Growth
Wildfire	Fire Suppression
Harvesting	Regrowth

Estimates of Forest Depletion

Loss to Forest Pests-Estimates of the total annual GMV loss to pests were based on the U.S.D.A. Forest Service statistics of the Pacific Coastal and Rocky Mountain regions. The periods included 1943-1952, 1953-1962, 1963-1970, 1971-1976, 1977-1986, and 1878-1992 (Powell et al. 1993,

Table 32). The average of 1952-1992 was used to represent the interval from 1942 to 1800. We assumed that 65% of the reported annual mortality of growing stock was pest-kill (Lewis 1993, personal communication).

Loss to Wildfires–Fire statistics in the United States reported annually the areal extent of wildfires from 1916 to present, and the fraction of wildfires in "non-forested watershed." The latter were areas containing scattered trees, shrubs or rangeland, and were arbitrarily assumed to contain a forest volume equivalent to 10% of a closed forest. There were no comprehensive continent-wide assessments of the extent of forest wildfires in the previous century. The suggestion has been made that the area burned annually was substantially more than at present (Wright and Bailey). In specific forest regions such as the mid-western U.S., charcoal data suggest wildfires were extensive and frequent compared to present fire regimes (Clark 1988). In the absence of better information, a conservative estimate of the levels of wildfire 1800-1915 was approximated using the 1920-1929 average areal extent of burns. Although fires in the previous century were more extensive and frequent, many were annual surface fires that consumed relatively little wood volume compared to the stand-replacement fires typical of this century (Pyne 1988).

The GMV loss annually to wildfire was estimated by multiplying the area burned by the GMV/ha of burnt forest. The latter was the average stocking of live trees on all forest lands in the Rocky Mountain and Pacific Coastal regions, multiplied by 0.77 which was the modifier for the average volume per hectare of burned stands (Auclair and Bedford 1993).

Forest Harvesting–Annual U.S.-wide estimates of removals in forest harvesting over the 1900 to 1990 interval were obtained from forest product summaries (USFS 1964, 1990). These national levels were then adjusted to the total GMV harvested in the Pacific Coastal and Rocky Mountain regions using the 1952-1992 forest statistics (Waddell et al. 1989, Table 29; Powell et al. 1993, Table 34). The estimates of harvesting from 1800 to 1899 were based on Clawson (1979). These national estimates were scaled downward to represent the western U.S. by assuming that harvesting in the Pacific Coastal and Rocky Mountain regions began in 1850 and had increased to 40% of the national total by 1900 (i.e., the same fraction as over the 1952-1992 period).

Estimates of Forest Accrual

Regrowth Following Depletion–Forest volume in regrowth was an estimate of the cumulative area depleted over time that had not reached age of maturity. To achieve this we backed our computer algorithm to year 1740 and used the average rates of depletion from 1890 to 1900. Time to maturity was estimated to be 150 years. By initiating the areal calculations in 1740, any lag in depletions 150 years hence (i.e., by 1890) were accounted for. Failure to use this algorithm resulted in an artifact, namely a sharp decrease in area of regrowth by the 1930-1940 decades. The volume of regrowth in any year was the product of the cumulative area of depletion (within each depletion process, biome, and year) multiplied by the rates of regrowth. Rates of regrowth used were the forest yields at maturity determined for each process, divided by the time to maturity.

Accelerated Tree Growth Rate–There was particularly convincing evidence of a significant increase in the rate of tree growth in most major forest biomes in North America (Auclair and Bedford, submitted).

Evidence of accelerated tree growth was based on a literature synthesis of dendrochronologies standardized for tree age and stand density. For purposes of comparison, we obtained an average growth rate increase in each major biome across North America by comparing the growth increment in 1890 with that in 1990, or with the most recent year of the chronology. In the western U.S., five tree chronologies were used to represent the temperate conifer forest biome (Brubaker and Graumlich 1987; Graybill and Idso 1993; Lamarche 1974; Lamarche et al. 1984; Parker 1987). To estimate the accelerated forest volume increment, the growth rate increase, 1890-1990, was multiplied by standard annual tree chronology increments of bristlecone pine (*Pinus longaeva* Bailey), 1890-1973 (Graybill and Idso 1993). This single species may not have accurately represented the regional trends. In the absence of standard increments on all the major species of the region, we improvised by using the 9-year running mean of the annual bristlecone pine values. We assumed no growth rate increase from 1974 to 1990 based on a synthesis of forest losses to crown dieback and growth rate decline (Auclair and Lindgren, in preparation).

Area Conserved by Fire Suppression–Two distinct phases of fire suppression were identified. The first occurred over the 1860-1890 period as a result of settlement and the introduction of large numbers of livestock. Grazing by cattle and sheep in woodlands had the effect of reducing ground cover (i.e., grasses) to levels insufficient to carry surface fires. Together with the active exclusion of fires (and inadvertent protection as forests were intersected by roads and croplands), the net effect was the ingrowth of trees and conversion of woodland (and low density stands) to forests of high tree density. The second phase occurred early in this century with the advent of systematic surveillance and mechanized fire fighting.

Actual fire statistics enabled us to estimate the efficacy of fire fighting over the past 7 decades. Forest volume saved by fire fighting was calculated using the decreasing volume burned from 1935 to 1955. The procedure involved computing a regression trend of the area burned over the 1920 to 1955 period. Horizontal lines were then projected from the regressed values of 1935 to 1990, and another from 1955 to 1990. The 1935-1990 line assumed that truly effective (i.e., mechanized) fire suppression capability in the U.S. began about 1935 (West 1992, and personal communication). The 1955-1990 line assumed that a high level of efficacy had been reached by the mid 1950s: The volume saved by fire fighting was the difference between the 1935-1990 line and the 1955-1990 line. The rationale for using a regression (and not the highly variable annual values) was that the capability to suppress wildfires was probably more or less consistent over time.

In the absence of systematic fire records in the 1800s, we made a number of provisional assumptions to approximate the volume of forest ingrowth and standing stock saved from burning as a result of agriculture and settlement. The area in question (74.3 million ha) was assumed to be the entire forest area in the western U.S. minus that occupied by moist stands of mixed conifers west of the Cascades in Washington and Oregon and the Sierra Nevadas in California. This area was multiplied by the standing stock (45 m^3/ha) of the 1860 woodlands. This was assumed to have been 40% of the present stocking levels (113 m^3/ha) (Birdsey 1992, Tables 1.1, 2.2). The ingrowth over the 1860-1990 period was the remaining fraction (60%), or 68 m^3/ha. Standing stock and ingrowth terms were calculated separately, since it could be argued that the standing stock would not have been at risk for fire-kill had the pre-1860 burning regimes been unaltered.

To assess the effect of fire protection on fuel loading (and hence risk to fire), we compared the cumulative forest volume saved by fire exclusion to the cumulative volume burnt annually. Both the 1860 and 1935 dates were used.

RESULTS AND DISCUSSION

Shifts in Forest Volume This Century

The forest volume lost to wildfire in the western U.S. decreased markedly from 1920 to notably low levels by 1955, followed by a surge in the mid 1980s. Volume losses to pests remained more or less constant until the mid 1980s, after which levels increased slightly.

Natural mortality (fire, pests) dominated depletions until about 1950, after which harvesting was increasingly dominant. Trends in wildfire and harvesting were reciprocal, so that total depletions, although highly variable year to year, did not show a distinctive trend until the surge of wildfires in the mid 1980 decade (Figure 1a).

Regrowth on formerly depleted sites was by far the strongest addition to forest volume. There was, nevertheless, significant gains from both fire suppression and accelerated tree growth, particularly from about 1950 on. Total accruals doubled over the century (Figure 1b).

The net forest volume showed unusually high levels of net accrual (i.e., net volume addition) over the 5 decades from 1930 to 1980 (Figure 2a). Much of this was due to rigorous fire suppression (Figure 3a). This pattern was concurrent with a similar trend in other temperate (and boreal) forests and may largely explain the drop in the global net CO_2 emissions from forest and other land ecosystems to a net uptake of atmospheric CO_2 (Figure 2b).

The sudden reversal in the global net biospheric CO_2 uptake in the late 1970s (Figure 2b) coincided with the surge of wildfires in the western U.S. and across the boreal zone (Auclair and Carter 1993). The persistent droughts in the western U.S. since the mid 1970s were symptomatic of similar climatic stresses across the temperate and boreal zone. All contributed to pest and wildfire surges to a degree significant enough to have influenced the global carbon balance. A part of the marked effect of wildfires on global carbon may be the great intensity of the burns (involving some combustion of coarse woody debris and soil organic matter) after a prolonged period of fire protection. It was noteworthy that the marked global warming from 1917 to 1926 (Hansen and Lebedeff 1987) and the severe droughts of the late 1920s and 1930s (i.e., the pre-suppression period) in the western U.S., and elsewhere in the hemisphere, produced only a minor fluctuation in net biospheric emissions.

Effects of Fire Suppression

The strong decreasing trend in areal extent of wildfires in the western U.S. had started by the 1920s and reached a low point about 1955. Fire-related mortality increased in the mid 1960s to mid 1970s decade, followed by a surge in the mid 1980s. The estimated volume saved from burning annually was about 50 million m^3/yr by 1955 (Figure 3a).

The cumulative forest volume saved by fire fighting exceeded the cumulative volume burned by the late 1950s (Figure 3b), and had reached a maximum of about 450 million m^3 (Figure 3c). The sudden surge of forest fires in the mid 1980s quickly reduced this volume saved to nil, suggesting that the overall effect of fire suppression on wood conservation had been minimal by 1990.

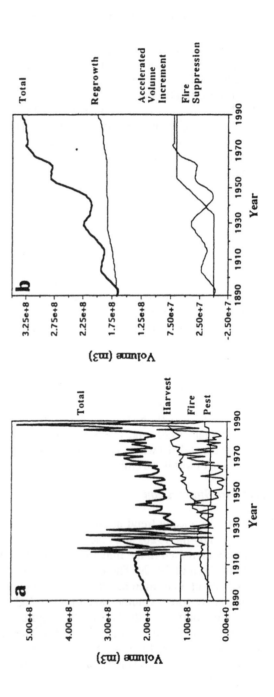

FIGURE 1. (a) Annual losses of forest volume in the western United States this century due to pests, wildfire, and harvesting. (b) Additions of forest volume over the same period due to fire suppression (since 1935), accelerated tree growth rate, and regrowth of depleted stands.

FIGURE 2. (a) The net forest volume trend in the western United States this century. (b) the global net flux of CO_2 from/to land (and marine) ecosystems this century (Sarmiento et al. 1992) showing the coincidence of the reversal in net uptake with a marked climatic discontinuity or "jump" in 1976 (Ebbesmeyer et al. 1992).

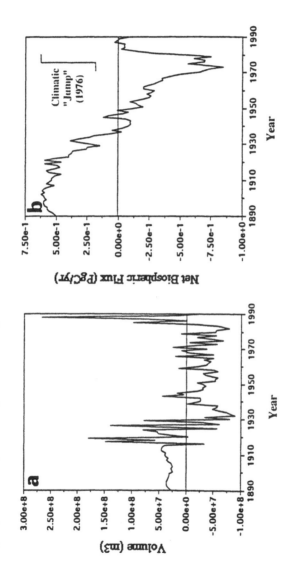

257

FIGURE 3. (a) The annual forest volume burned by wildfires in the western United States, and the volume saved by modern fire suppression annually. (b) The cumulative forest volume saved by fire suppression from 1935 to 1990 compared to the cumulative volume loss to forest wildfires in the same period. (c) The cumulative volume saved by fire suppression minus the cumulative volume burned.

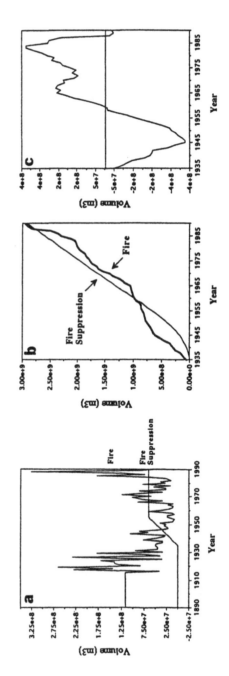

The exclusion of wildfires by settlement and grazing activities after 1860 had a potentially much greater effect on forest volume conserved than did modern fire fighting. The levels of increased forest volume due to ingrowth alone were two orders of magnitude greater than the forest volume saved by modern fire fighting this century (Figure 4a). The cumulative volume saved by fire fighting, 1935-1990, plus the ingrowth volume since 1860, greatly exceeded the cumulative volume lost to burning. Adding the cumulative volume harvested did not change the overall pattern (Figure 4b). The difference in the cumulative volumes provided a very rough measure of the accumulating fuel load and risk of catastrophic fires over the landscape in the absence of the natural (i.e., pre 1860) regimes. Fire was and remains the dominant factor in the forests of the U.S. West. Success in "damming" all wildland fires only served to impound tree litter into large reservoirs of fuels making possible future fires of catastrophic intensity. Viewed from this perspective, fire protection tends in some degree to be self-defeating (Pyne 1988: 31). This is a concept recurrent in the fire literature (Weaver 1943; Cooper 1961), but one inadequately practiced over the region (Mutch 1993).

The rapidity of the recent loss was striking (Figure 3c); any gains due to fire suppression over the 6-7 decades of fire fighting were eliminated in four consecutive fire seasons. A combination of factors best explain the suddenness of the decline.

Elevated Fuel Loads–The conversion of extensive areas of woodland and low density forest to high density forest stands in the latter half of the 1800s occurred under a more or less permanent regime of fire exclusion (Figure 4a,b). Mechanized fire fighting over the past 7 decades has been particularly effective (until recently) in suppressing wildfire. Fire can act as a "cleansing agent" where dry or cold conditions limit the natural rate of decomposition of tree and other woody litter. The accumulated litter ultimately fuels intense fires of sufficient severity to destroy organic components linked to soil fertility.

Increased Fuel Connectivity–This included high tree density and stand homogeneity (i.e., even age and complete crown cover) as trees seeded-in and established in the open spaces previously maintained by frequent fire. In ponderosa pine (*Pinus ponderosa* Dougl. ex Laws) types in the Southwest, tree density had increased 30-fold (from about 85 to 2500 trees/ha) with fire exclusion in the 1800s (Covington and Moore 1994). A shift to shade tolerant (i.e., moisture and fire sensitive) tree species also occurred in mixed conifer forest types. Species such as true firs (*Abies spp.*) increased 60% in Idaho over the past 4 decades alone, concurrent with a decrease in fire-resistant species such as western white pine (*Pinus mon-*

FIGURE 4. (a) The forest volume saved by modern fire suppression (1935-1990) compared to the ingrowth and 1860 standing stock volumes due to the exclusion of fires from western U.S. forest types over the 1860-1890 and subsequent period. (b) Estimates of the cumulative volume of forests in "ingrowth" plus the volume saved from burning by fire suppression, 1935-1990, and the cumulative extent of forest wildfires from 1860 to 1990. The difference is the volume of potential fuel accumulated in the landscape over and above that present under natural fire regimes prior to 1860.

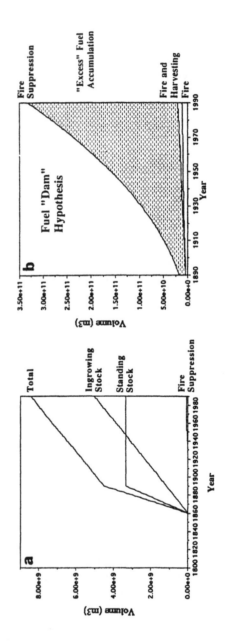

ticola Dougl. ex D. Don)(− 60%) and ponderosa pine (− 45%) (O'Laughlin 1994). With crowns often extending to the ground, true firs now provide a fuel ladder to the crown. The increases in the fraction of crown fires and average fire size over the past 40 years can partly be explained by the increase in fuel connectivity (Covington and Moore 1994).

Accelerated Tree Growth Rate–There is some evidence that tree growth rate in the western U.S. may have doubled over the past 150 years (Graybill and Idso 1993). This suggests that the trees today, with higher metabolisms than previously, may be more drought and temperature sensitive than in the past. Total annual precipitation may also have increased. If this can be substantiated, a case may be made that forests now are far more sensitive to droughts for physiological (e.g., large water demand) as well as structural reasons (e.g., high biomass, increased shoot/root ratio). True firs are particularly sensitive to cavitation injuries that result in crown dieback, desiccation, and insect attack (Tyree and Sperry 1989). In brief there is a need to consider seriously the possibility that an upward spiral of forces and conditions have culminated in forests that are more or less destabilized, and now respond in a hyper-sensitive fashion to stresses that would have elicited only mild damages in the 1930s, 1920s and in earlier periods. If true, this has management implications in that "defueling" forests by thinning and prescribed fires may not have the large salutatory effects expected.

Increased Landscape Synchronization–Because of extensive fire exclusion and cutting over a relatively narrow timeframe (1850-1900), many forests have simultaneously reached a point of maturity. This stage is typified by a transition phase to significantly lower forest volume and biomass as part of a natural stand dynamic that is largely independent of external factors (Bormann and Likens 1979). If the age to maturity of the dominant pine species is about 100-125 years, the timing of the current surge in pest-kill and wildfires may be "on schedule" consistent with the synchronization processes of the last century.

Climatic Shifts and Discontinuity–The climatic "jump" or discontinuity of 1976 (Ebbesmeyer et al. 1992) (Figure 2b) is especially noteworthy. This has been described as a "turned-on" El Niño (Kerr 1992), resulting in an almost continuous series of severe and persistent droughts over the western U.S. from the late 1970s to the present. The warming at the hemispheric and global levels since the mid 1970s has been extremely rapid (Hansen and Lebedeff 1987). The fact that wildfires surged over the 1976-1990 period in other geographic areas and forest types with contrasting management histories to that in the western U.S. forests lends credence to the view that climate has had a large role in the present dilemma.

GENERAL DISCUSSION

An historical and "big-picture" perspective is urgently needed to place the current dilemma in proper perspective, and to understand the importance of the abrupt and marked changes in forest function that occurred over a century ago. The historical antecedents to what are now extraordinary levels of risk to fire and pest outbreak are paramount in understanding the origins and effective solutions to the current problems of the Rocky Mountain and Pacific Coastal forest regions.

Our conclusions are that (1) the changes in the last century are far more important to the current problem than is generally recognized. Fire protection (including inadvertent fire exclusion with settlement and grazing) since the mid 1800s resulted in 40-60 times more fuel build-up than did modern (i.e., > 1935) fire fighting. (2) The concept of increased risk with fuel buildup is accurate in depicting the development of catastrophic fires under conditions of persistent fire protection. Our results for the post-1935 period indicate that recent wildfires have resulted in a sudden, rapid loss of virtually all the gains in wood volume saved by fire suppression over the last several decades. (3) Multiple, cumulative changes (5 factors above) have destabilized the western U.S. forests to the point they are now hypersensitive to climatic fluctuations.

Future studies to address the current problem of sustainability of forests in the Inland West must recognize that the region is particularly vulnerable to the kinds of global (and regional) climatic changes evident over the last decade (Mitchell et al. 1990). These changes are expected to accelerate in the near term (20-50 years), suggesting that any current sensitivities or imbalances could play an even greater role in the damage and demise of forests now under stress. Logic suggests that fuel loads far out of synchrony with normally frequent fires (as is presently the case in the majority of western long-needled pine forest types) will greatly compound the effects of severe drought.

Mutch (1993) has re-emphasized the need identified by fire scientists as much as half a century ago (Weaver 1943), but now in terms of a forest health emergency. The current problem stems not from balancing fire exclusion with an equally important program of prescribed fire on a landscape scale to sustain the health of fire-adapted ecosystems. He proposes nine practical steps to restore the balance.

A compelling case can be made for addressing the existing problems now rather than later. Silvicultural strategies to gradually "drain" the dammed-up fuel loads and lower the risk thresholds to catastrophic wildfires may prove very difficult to put into practice. These strategies will require the support of the public, especially where the national forests and parks are concerned. Moreover, other issues such as the design of forests (and the requisite silviculture) to ensure that forest vitality is maintained during the marked changes in climate expected in the near future, are likely to prove costly and complicated.

REFERENCES

Auclair, A.N.D. and J.A. Bedford. 1993. Forest depletion and accrual dataset: Area and volume estimates. Documentation File, Version 2. Office of Research and Development, U.S. Environmental Protection Agency. Washington, DC.

Auclair, A.N.D. and J.A. Bedford. Recent shifts in the wood volume balance of boreal forests: its implications for global CO_2. *Canadian Journal of Forest Research* (submitted November 3, 1993).

Auclair, A.N.D. and T.B. Carter. 1993. Forest wildfires as a recent source of CO_2 at northern latitudes. *Canadian Journal of Forest Research* 23: 1528-1536.

Auclair, A.N.D. and R. Lindgren. [In preparation]. Estimates of CO_2 flux from episodes of forest dieback in temperate and boreal forests of the Northern Hemisphere. Science and Policy Associates, Inc. Washington, DC.

Birdsey, R.A. 1992. Carbon storage and accumulation in United States forest ecosystems. General Technical Report WO-59. U.S.D.A. Forest Service. Radnor, PA.

Bormann, F.H. and G.E. Likens. 1979. Pattern and process in a forested ecosystem. Springer-Verlag, Inc. New York.

Brubaker, L.B. and L.J. Graumlich. 1987. 100-year records of forest productivity at high elevations in western Washington, USA. In: pp. 277-286. Kairiukstis, L., Z. Bednarz and E. Feliksik (Eds.). Methods of Dendrochronology I. Proceedings of the Task Force Meeting on Methodology of Dendrochronology: East/West Approaches. June 2-6, 1986. Krakow, Poland. International Institute of Applied Systems Analysis, and the Polish Academy of Sciences-Systems Research Institute. Warsaw, Poland. 319 p.

Clark, J.S. 1988. Effect of climate change on fire regimes in northwestern Minnesota. *Nature* 334: 233-235.

Clawson, M. 1979. Forests in the long sweep of American history. *Science* 204: 1168-1174.

Cooper, C.F. 1961. The ecology of fire. *Scientific American* 204(4): 150-160.

Covington, W. and M. Moore. 1994. Postsettlement changes in natural fire regimes and forest structure: Ecological restoration of old-growth ponderosa pine forests. In: pp. 153-181. Sampson, R.N. and D.L. Adams (Eds.). Assessing Forest Ecosystem Health in the Inland West. Proceedings of the American Forests scientific workshop, November 14-20, 1993, Sun Valley, ID. The Haworth Press, Inc. New York.

Ebbesmeyer, C.C., D.R. Cayan, D.R. McLain, F.H. Nichols, D.H. Peterson and K.T. Redmond. 1991. 1976 step in the Pacific climate: Forty environmental changes between 1968-1975 and 1977-1984. In: pp. 115-126. Betancourt, J.L.

and V.L. Tharp (Eds.). Proc. 7th Ann. Pacific Climate (PACLIM) Workshop, April, 1990. Interagency Ecological Studies Program Tech. Rpt. 26. California Dept. Water Resources. San Francisco.

Graybill, D.A. and S.B. Idso. 1993. Detecting the aerial fertilization effect of atmospheric CO_2 enrichment in tree-ring chronologies. *Global Biogeochemical cycles* 7(1): 81-95.

Hansen, J. and S. Lebedeff. 1987. Global trends of measured surface air temperature. *Journal of Geophysical Research* 92 (D11): 13345-13372.

Kerr, R.A. 1992. Unmasking a shifty climate system. *Science* 257:1508-1510. (Reference to studies of the 1976-77 climatic shift by Curtis Ebbesmeyer, Evans Hamilton Inc. Seattle, WA.)

Lamarche, V.C. 1974. Paleoclimatic inferences from long tree-ring records. *Science* 183: 1043-1048.

Lamarche, V.C., D.A. Graybill, H.C. Fritts and M.R. Rose. 1984. Increasing atmospheric carbon dioxide: Tree ring evidence for growth enhancement in natural vegetation. *Science* 225: 1019-1021.

Lewis, J.W. 1993. Average annual pest impacts on forests of the United States, 1972-1981. (Unpublished summary, and personal communication). Forest Pest Management, USDA Forest Service, FPM AB-2S. Washington DC. (Cited in Auclair and Bedford 1993).

Mitchell, J.F.B., S. Manabe, T. Tokioka, and V. Meleshko. 1990. Equilibrium climate change. In: pp. 131-172. Houghton, J.T., G.J. Jenkins and J.J. Ephraums (Eds.). Climate Change: The IPCC Scientific Assessment. Cambridge University Press. Cambridge, UK.

Mutch, R.W. 1993. Forest health and productivity: fire interactions. In: Proceedings of the Society of American Foresters National Convention on Forest Health and Productivity, November 8-11, 1993. Indianapolis, IN.

Parker, M.L. 1987. X-ray densitometry and image analysis as methods in dendrochronology in Canada and the United States. In: pp. 57-68. Kairiukstis, L., Z. Bednarz and E. Feliksik (Eds.). Methods of Dendrochronology I. Proceedings of the Task Force Meeting on Methodology of Dendrochronology: East/West Approaches. June 2-6, 1986. Krakow, Poland. International Institute of Applied Systems Analysis, and the Polish Academy of Sciences-Systems Research Institute. Warsaw, Poland. 319 p.

Powell, D.S., J.L. Faulkner, D.R. Darr, Z. Zhu and D.W. MacCleery. 1993. Forest statistics of the United States, 1992. Pre-publication Draft Tables. FIERR Staff Report, Forest Service, U.S. Department of Agriculture, Washington, DC.

Pyne, S.J. 1988. Fire in America: A cultural history of wildland and rural fire. Princeton University Press. Princeton, NJ.

O'Laughlin, J. 1994. Assessing forest health conditions in Idaho with forest inventory data. In: pp. 221-247. Sampson, R.N. and D.L. Adams (Eds.). Assessing Forest Ecosystem Health in the Inland West. Proceedings of the American Forests scientific workshop, November 14-20, 1993, Sun Valley, ID. The Haworth Press, Inc. New York.

Sarmiento, J.L., J.C. Orr, and U. Siegenthaler. 1992. A perturbation simulation of

CO_2 uptake in an ocean general circulation model. *Journal of Geophysical Research* 97: 3621-3645.

Tyree, M.T. and J.S. Sperry 1989. Vulnerability of xylem to cavitation and embolism. *Annals of Plant Physiology and Molecular Biology* 40: 19-38.

USFS 1964. The demand and price situation for forest products, 1964. Miscellaneous Publication No. 983. U.S.D.A. Forest Service. Washington, DC.

USFS 1990. United States timber production, trade, consumption, and price statistics, 1960-88. Miscellaneous Publication No. 1486. USDA Forest Service. Washington, DC.

Waddell, K.L., D.D. Oswald and D.S. Powell 1989. Forest statistics of the United States, 1987. Resource Bulletin PNW-RB-168, Pacific Northwest Research Station. USDA Forest Service. Portland, OR. 106 p.

Weaver, H. 1943. Fire as an ecological and silvicultural factor in the ponderosa-pine region of the Pacific Slope. *Journal of Forestry* 41: 7-14.

West, T.L. 1992. Centennial mini-histories of the Forest Service. USDA Forest Service, Rpt. FS-518. Washington, DC.

Wright, H.A. and A.W. Bailey. 1982. Fire ecology of the United States and Canada. John Wiley and Sons, Inc. New York.

Landscape Characterization:
A Framework for Ecological Assessment
at Regional and Local Scales

Patrick S. Bourgeron
Hope C. Humphries
Mark E. Jensen

ABSTRACT. A landscape characterization procedure is described as a first step in ecosystem management. Five attributes were used to characterize patterns at different scales of ecological organization from the plot to the region: climatic regions, ecoregions, biophysical environments, floristics, and vegetation. Examples of the characterization of selected attributes are presented for three western U.S.

Patrick S. Bourgeron is Regional Ecologist, and Hope C. Humphries is Ecological Modeler, The Nature Conservancy, 2060 Broadway, Suite 230, Boulder, CO 80302.

Mark E. Jensen is Regional Soil Scientist, USDA Forest Service, Northern Region, P.O. Box 7669, Missoula, MT 59807.

[Haworth co-indexing entry note]: "Landscape Characterization: A Framework for Ecological Assessment at Regional and Local Scales." Bourgeron, Patrick S., Hope C. Humphries, and Mark E. Jensen. Co-published simultaneously in the *Journal of Sustainable Forestry* (The Haworth Press, Inc.) Vol. 2, No. 3/4, 1994, pp. 267-281; and: *Assessing Forest Ecosystem Health in the Inland West* (eds: R. Neil Sampson and David L. Adams) The Haworth Press, Inc., 1994, pp. 267-281. Multiple copies of this article/chapter may be purchased from The Haworth Document Delivery Center [1-800-3-HA-WORTH; 9:00 a.m. - 5:00 p.m. (EST)].

study areas. Other aspects of the characterization process are illustrated with examples from the Northern Region of the Forest Service and from forested locations in the southwestern U.S.

Location of a study area within a climatic region provided an understanding of broad-scale climatic constraints operating on the biota. Assessment at the ecoregion level was used to examine finer scale environmental constraints due to landform effects. The ecoregion also provided a context for further analysis of biotic and environmental variability in a study area. Biophysical environments (combinations of environmental factors) were characterized within landscapes. The range of environmental variability in a landscape was compared to that of a larger region to determine the representation of regional environments in a landscape.

Biotic variability was characterized at a relatively coarse spatial scale by examining plant species distributions among floristic types. Such an analysis provided information about the long-term ecological and evolutionary pressures exerted on species in an area. Variability in vegetation was characterized at a variety of scales by examining the distribution of types within the levels of a hierarchical vegetation classification for the western U.S. In addition, biotic distributions along environmental gradients were described using canonical correspondence analysis to determine within-type variability. Temporal variability in vegetation was assessed within a hierarchical ecosystem-based framework.

Characterization of biotic-abiotic relationships is important in determining biotic responses to historical and current landscape conditions. We constructed statistical models to predict biotic responses to environmental factors using a powerful class of regression models, generalized linear models. Models developed for a plant community in the Southwest, the ponderosa pine series, revealed complex biotic responses to environmental factors. This result suggests that biotic-abiotic relationships warrant careful characterization, and that some commonly used simple models may not accurately predict biotic responses to environmental change.

INTRODUCTION

An important basis for ecosystem management is the identification, location, and description of the biotic and abiotic features of a landscape, as well as the characterization of biotic-abiotic relationships. Landscape features exhibit heterogeneity at a variety of scales of ecological organization (Turner et al. 1993). The characterization of this heterogeneity is accomplished by identifying relevant patterns and the processes which produce patterns in a landscape (Bourgeron and Jensen 1993). Therefore, the characterization process is a multiscaled approach conducted within a hierarchical framework (Bourgeron and Jensen 1993; Hann et al. 1993).

Organisms experience and are adapted to environmental variability induced by spatial and temporal heterogeneity, among other factors (Levin 1992). Human activities alter the range of natural variability in the environment. (Hereafter, we refer to the range of spatial heterogeneity as the range of variability in the environment, and the corresponding patterns in species and community distributions as the range of variability in the biota.) Management practices can change landscape configurations in ways that change species and community distributions. The impact of management practices needs to be understood in the context of the multiscaled organization of ecosystems.

Landscape characterization is the first step in implementing ecosystem management. An important part of the characterization process is the explicit definition of the relationships between the biota and the environment (Bourgeron et al. 1993b). The characterization of biotic-abiotic relationships allows for the prediction of species and community responses to landscape changes.

In this paper, we present an overview of a landscape characterization procedure consisting of five steps: the characterization of climatic regions, ecoregions, biophysical environments, floristics, and vegetation. In addition, we discuss characterization of biotic-abiotic relationships using predictive statistical models.

THE LANDSCAPE CHARACTERIZATION PROCESS

In characterizing large landscapes, a regional context must be provided. Such a regional context is the basis for (1) understanding broad-scale environmental constraints on the biota and (2) determining how well the biota and biophysical environments of an area represent the range of variability occurring in a larger region (Austin and Margules 1986). We have conducted landscape characterization using five attributes of the environment and vegetation (Bourgeron et al. 1993a; Engelking et al. 1993): Environmental variability is characterized at increasingly finer scales by defining climatic regions, ecoregions, and biophysical environments. Biotic variability is described using floristics and vegetation. Examples from three study areas are presented to illustrate different aspects of the landscape characterization process: the Gray Ranch, a Nature Conservancy preserve located in southwestern New Mexico; the Yampa River basin, in northwestern Colorado and south-central Wyoming; and the upper Blackfoot River region in western Montana. Studies differed in objectives and resources available; therefore, not all attributes were assessed in each study. In addition, the characterization of temporal dynamics in vegetation is illustrated using an example from the Northern Region of the Forest Service.

Climate

At a broad spatial scale, climatic patterns and constraints are understood by defining climatic regions, such as those delineated by Mitchell (1976) based on air mass boundaries for the western U.S. The Gray Ranch occurs in a climatic region characterized by a summer rainy season and infrequent intrusions of Pacific air masses in winter. The Yampa River basin is located in a region which is influenced by dry interior air masses in summer and infrequent intrusions of Pacific air in winter. The upper Blackfoot River region is in a climatic region that experiences frequent intrusions of Pacific air masses in winter. Species and ecosystems present in each of the three areas must possess features that can respond to the prevailing climatic regime, including the range of characteristic temperatures and the frequency, timing, and amount of precipitation.

Ecoregions

Within broad-scale climatic regions, climate is modified by the effect of landforms. Contiguous areas with similar climate, landforms, and vegetation have been delineated as ecoregions (Bailey 1980; Bailey et al. 1993). Environmental constraints on a landscape can be examined in the context of the ecoregion. In addition, ecoregions are important in landscape characterization as the context for the assessment of biotic and environmental variability.

The Gray Ranch is in Bailey's (1980) Mexican Highlands Shrub-Steppe ecoregion, characterized by high plains and north-south oriented mountain ranges. The Yampa River basin is located in portions of both the Wyoming Basin, with plains and low mountains, and the Rocky Mountain Forest ecoregion, containing high, glaciated mountains. The upper Blackfoot River region is in the Rocky Mountain Forest ecoregion.

Biophysical Environments

Environmental variability within landscapes is characterized by describing and mapping biophysical environments. For each area of interest, environmental factors are identified based on their ecological importance in determining ecosystem composition, structure, and function, as well as on spatial distribution patterns and availability of mapped information. Combinations of these environmental factors make up the biophysical environments. The representation of biophysical environments in a landscape was compared to the range of variability in environments in a larger region in order to determine how much of the regional variability was contained within a study area. Analyses were conducted for the Gray Ranch and the upper Blackfoot River region.

At the Gray Ranch, the factors selected to characterize environmental variability were geological substrate, soil suborder, and elevation. A biophysical environment represented a combination of a particular geological substrate, soil suborder, and elevation class. Geological substrate was used as an indicator of nutrient regime, soil suborder of soil water availability, and elevation of temperature and precipitation regimes.

Thirty of the 161 biophysical environments found in the Mexican Highlands ecoregion were present at the Gray Ranch. Twelve of the 30 biophysical environments occurred at both the Gray Ranch and elsewhere in the ecoregion. The shared environments were found to have a similar distribution among classes (G-test, Sokal and Rohlf 1981) and thus the Gray Ranch could be considered a representative "snapshot" of the ecoregion for these biophysical environments. Eighteen biophysical environments were found only at the Gray Ranch. Many of the environments unique to the Gray Ranch were associated with a playa lake. In general, the Gray Ranch contained mid-elevation, mesic biophysical environments but lacked some of the drier and colder environments found elsewhere in the Mexican Highlands ecoregion. The Gray Ranch has a high diversity of biophysical environments (19% of environments found in the ecoregion) considering that it makes up only 3% of the ecoregion's land area.

In the upper Blackfoot River region, biophysical environments were characterized to determine the range of environments present in the region's conservation network (wilderness and research natural areas) (DeVelice et al. 1993). The analysis used a rule-based model and a geographical information system to map biophysical environments based on soil water, temperature, and soil fertility limitations. The conservation network was found to represent cold environments well, but included little of the area characterized as water-limited.

Floristics

Characterization of biotic variability at a relatively coarse level of resolution is accomplished by describing the floristic distributions of species in a study area. At the Gray Ranch, species were classified by their patterns of geographical distribution into areal types (Whittaker and Niering 1965). Such an analysis indicates the long-term ecological and evolutionary pressures to which species in an area must respond (Whittaker and Niering 1965; McLaughlin 1986). Knowledge of the context for the evolution of species in an area can allow managers to provide conditions for the continuation of evolutionary processes as part of ecosystem management.

Twelve areal types were identified using the floristic subdivisions of McLaughlin (1986, 1989) and Whittaker and Niering (1965). Each of the 305 plant species occurring in plot samples at the Gray Ranch was assigned to an areal type based on information in Kearney and Peebles (1960), Whittaker and Niering (1965), Correll and Johnston (1970), and Wagner (1977). Although extensive floristic analysis has been conducted in the Southwest, phytogeographical categories have not been standardized (McLaughlin 1986, 1989). In our analysis, the determination of previous authors was used when there was consensus about species assignments to areal types. Otherwise, expert knowledge was the basis for assigning species.

At the Gray Ranch, a majority of species (65%) were associated with two areal types, the Southwestern and Madrean elements (Table 1). This result implies that most species at the Gray Ranch either have wide tolerance to variability in precipitation and hence are found throughout the entire Southwest, or they are adapted to transitional climate conditions characteristic of the area associated with the Madrean element, which occurs at the boundary between the precipitation regimes of the Sonoran and Chihuahuan deserts. Dependence on a particular precipitation regime, such as that of the Sonoran or Chihuahuan deserts, appears to be a less successful strategy for plants at the Gray Ranch. In general, southwestern floras (Madrean, Chihuahuan, Sonoran, Southwestern, and Southern Rocky Mountain elements) are well represented at the Gray Ranch, making up 77% of species.

An application of this approach to ecosystem management is the continued monitoring of an area of interest for additions and losses of species from areal types. Trends in such changes in species distributions could indicate the direction of change in broad-scale environmental constraints.

Vegetation

The assessment of variability in vegetation in a landscape has several aspects. Patterns of biotic variability are characterized at a range of spatial and temporal scales. In addition, biotic distributions are described along environmental gradients as a means of understanding within-class variability.

TABLE 1. Classification by areal type of 305 plant species occurring in plot samples at the Gray Ranch (Whittaker and Niering 1965).

Areal Type	Number of Species	Percent of total
Madrean	75	25
Chihuahuan	20	7
Sonoran	4	1
Southwestern	121	40
Southern Rocky Mountain	11	4
Western	26	9
Plains	10	3
Temperate	22	7
Northern	1	<1
Holarctic	7	2
Latin American	2	1
Global	7	2

Patterns in vegetation can be recognized at a variety of scales, and therefore characterization of vegetation can be usefully conducted within the framework of a hierarchical vegetation classification (Bourgeron and Engelking 1993). Plant communities defined within a classification represent recurrent patterns of co-occurring species (Austin and Heyligers 1989; Bourgeron et al. 1993b). A hierarchical vegetation classification has been developed in the western U.S. for use as a data layer for conservation planning and ecosystem management (Bourgeron and Engelking 1993). At the highest levels of the classification, Driscoll et al.'s (1984) physiognomic classification of U.S. vegetation was used, which is patterned after the UNESCO (1973) classification. Two floristic classification subdivisions, series and plant associations, were added under the physiognomic framework to characterize biotic variability at landscape and site scales. The rationale for coupling physiognomic and floristic systems has been justified on evolutionary, ecological, and management grounds (e.g., Westhoff and Van der Maarel 1978; Werger and Sprangers 1982). At all levels, existing natural vegetation was classified, provided that the types produced were maintained in space and time in the landscape.

Variability in vegetation was assessed at both the Gray Ranch and the Yampa River basin at all levels of the vegetation classification. The Gray Ranch was found to include a high percentage of regional vegetation types considering its relative size (Table 2). For example, 21% of the plant associations described in the ecoregion occurred at the Gray Ranch, although the ranch occupies only 3% of the ecoregion's land area. A large number of these plant associations have been documented only at the ranch. As part of a vegetation characterization of the Yampa River basin, 38 riparian plant communities were described, including five communities not previously reported in the western U.S., and 13 not previously reported in Colorado.

Occurrences of a plant community are not invariant. Species composition may show gradual changes along an environmental or geographic gradient. The same community in different locations may therefore respond differently to the same management treatment. As a consequence, it is important to characterize variability in composition. One way to quanti-

TABLE 2. Number of types within levels of a hierarchical physiognomic-floristic vegetation classification for the Gray Ranch and the Mexican Highlands Shrub-Steppe ecoregion.

Classification level	Number in ecoregion	Number in Gray Ranch	Number occurring only in Gray Ranch
Class	5	4	0
Subclass	9	8	0
Group	12	9	1
Formation	16	12	3
Series	71	26	9
Plant association	215	45	35

fy the variability in vegetation that is associated with environmental gradients is to use a gradient analysis technique such as canonical correspondence analysis (CCA) (ter Braak 1986) which has been found to be a robust and efficient method for direct ordination of species along environmental gradients (Palmer 1993).

Gradients in species and vegetation distributions were related to environmental attributes at the Gray Ranch and the Yampa River using CCA. In the ordination space defined by the first two axes, the environmental attributes most highly correlated with the distribution of species and vegetation types at the Gray Ranch were elevation, slope, and several parent material, soil, and landform categories. Further analysis resulted in determination of within-type variability along environmental gradients. This information was used to draw management boundaries at the Gray Ranch.

In the Yampa River basin, species associated with low elevation and marshy habitats were separated along the first two axes of the ordination space from species in montane and subalpine forests. Further analysis of willow communities revealed that environmental gradients which most strongly affected vegetation distribution differed with the location of occurrences in the landscape. Valley aspect appeared to be important in determining vegetation pattern in narrow, deep valleys at high elevations, but the shape of the valley was more important at moderate elevations.

The analyses described above resulted in assessments of spatial variability in vegetation patterns. It is also necessary to characterize temporal variability in vegetation. This has been accomplished using a hierarchical ecosystem-based framework in the Northern Region of the Forest Service as part of the process of assessing ecological condition (Hann et al. 1993). Successional stages are determined for each level of the hierarchy, including project, watershed, landscape analysis area, and higher levels. The forest successional stage classification system uses Oliver and Larson's (1990) stages: initiation, stem exclusion, reinitiation, and old growth. In addition, information on dominant species, number and kinds of cohorts regenerating after disturbance, and disturbance regimes may be provided.

At a given temporal and spatial scale, vegetation patterns are matched with the natural processes and functions that created them. Successional sequences can be characterized along each segment of an environmental gradient (see Tables 3 and 4 in Hann et al. 1993). Historical information is used to define the historical range of variability in plant community patterns within the area of interest (Swanson et al. 1993; Morgan et al. 1994). A variety of techniques exist for determining this range (Morgan et al. 1994).

BIOTIC-ABIOTIC RELATIONSHIPS

The landscape characterization process includes the description of patch structure resulting from the operation of processes such as disturbance. Plant and animal species respond to the configuration of environmental conditions in a landscape mosaic. Characterization of biotic-abiotic relationships is important in determining the nature of biotic response to patch structure, including the degree of fragmentation in the landscape experienced by plants and animals. In addition, characterization of biotic-abiotic relationships can be used to assess whether current environmental conditions in a landscape have been significantly altered from the historical range of variability. The number of patches available for native species may be reduced by human activities. Conversely, in a human-altered environment many more patches suitable for exotic species may be found today than were present historically.

We have developed predictive models of relationships between biotic entities and environmental factors over large areas in order to assess the condition of particular landscapes for species and communities. The approach assumes that the data encompass the full range of environmental combinations over which species and communities are distributed in a given area. This assumption necessitates the use of a regional data set to quantify species and community responses to key environmental variables.

A powerful class of regression models known as generalized linear models (GLMs) provide appropriate tools for predicting biotic responses to the environment (Nicholls 1989). Traditional statistical methods such as linear regression and analysis of variance are special cases of GLMs which can incorporate non-linear functions relating biotic responses to explanatory variables, categorical responses such as presence/absence and abundance classes, and non-normal error functions (McCullagh and Nelder 1989). All these conditions are commonly encountered in ecological data.

In GLMs, the shape of the curves describing biotic responses to environmental factors may take a number of forms including linear, bell-shaped, and skewed (Austin et al. 1990). Models can also contain interaction terms. The use of methods such as GLMs allows the construction and testing of complex models to accurately predict biotic responses for management purposes. Simple models for predicting biotic occurrences from the environment have been shown to be inadequate in some cases (Austin et al. 1984; Austin 1985; Austin et al. 1990).

The environmental factors selected for modeling species and community distributions should be the best possible estimates of niche and habitat dimensions, such as radiation, temperature, moisture, and nutrient regimes

(Mackey et al. 1988). Such information is not often collected in plot samples, but may be estimated using surface-fitting procedures (Busby 1991) or simulation models.

We have utilized GLMs to examine the relationship between a plant community, the ponderosa pine series, and its environment in the southwestern U.S., using data collected on approximately 1400 forested plots throughout the Southwest (Muldavin et al. 1990). Site presence or absence of the ponderosa pine series was used as the biotic response variable. A climate simulation model, MTCLIM (Hungerford et al. 1989), was employed to estimate temperature, precipitation, and radiation attributes at a site using site characteristics such as elevation, slope, and aspect, and weather data from a nearby weather station. Separate models were developed for each of four broad parent material categories (basalt, granite, other igneous rock, and sedimentary rock), since the ponderosa pine series was found to have different responses to each of the categories. The explanatory variables or predictors were temperature, precipitation, and radiation attributes.

Models differed in which climate attributes were significant predictors of ponderosa pine series response, but in all four models, annual radiation was a significant term, and all contained a single precipitation attribute and one or two temperature attributes (Table 3). For one of the most common parent material categories, other igneous rock, the response curve for annual temperature was skewed (contained a cubic function of the attribute) and the model included an interaction between annual temperature and annual radiation. The model developed for sedimentary also contained a temperature-radiation interaction term. The existence of these kinds of complex responses is evidence of the need for careful selection of methods to predict relationships between biotic entities and the abiotic environment.

A further result of the analysis was that, on a particular parent material category, the ponderosa pine series was predicted to occur over a range of temperature, precipitation, and radiation conditions associated with different positions in the landscape. In areas of high topographic relief, nearby occurrences of the community may have different environmental conditions, ranging from warm, dry sites to cool, moist sites. The implication of this result is that management treatments, such as prescribed burning or timber harvest, may have different effects on occurrences of a community within a landscape that differ in environmental characteristics.

CONCLUSIONS

Landscape characterization should include the description of patterns and processes at all relevant scales. To this end, a five step procedure for

TABLE 3. Significant model terms for ponderosa pine series models on southwestern forested plots. Sign of model parameters indicated with + or −. Order of polynomials indicated by superscript.

Parent material category	Model terms*
Basalt	$HT + Q4 - Q4^2 + AR$
Granite	$- CT - CT^2 + AP - AP^2 + AR$
Other igneous rock	$AT + AT^2 - AT^3 - CT - AP + AR - (AT * AR)$
Sedimentary rock	$HT - CT - CT^2 - Q3 + AR - AR^2 - (HT * AR)$

* HT is maximum July temperature
CT is minimum January temperature.
AT is average annual temperature.
AP is annual precipitation.
Q1 is first quarter precipitation (January-March).
Q2 is second quarter precipitation (April-June).
Q3 is third quarter precipitation (July-September).
Q4 is fourth quarter precipitation (October-December).
AR is annual radiation.

assessing spatial variability in the biota and the environment was described and examples were presented. The procedure includes a range of spatial scales from the plot level to large regions. Temporal variability is also characterized within a hierarchical framework at a variety of scales. Such descriptions of major ecosystem patterns and processes provides a basis for the assessment of the ecological condition of landscapes for conservation and management. The characterization of spatial and temporal variability can be used as a template for understanding cumulative effects of processes and management prescriptions. Landscape characterization can result in the identification of ecological limits in an area, as well as identifying the appropriate scales for planning and assessment efforts.

REFERENCES

Austin, M.P. 1985. Continuum concept, ordination methods and niche theory. *Annual Review of Ecology and Systematics* 16:39-61.
Austin, M.P. and P.C. Heyligers. 1989. Vegetation survey design for conservation: gradsect sampling of forests in north-eastern New South Wales. *Biological Conservation* 50:13-32.

Austin, M.P. and C.R. Margules. 1986. Assessing representativeness. In: Usher, M.B. (Ed.). Wildlife conservation evaluation. Chapman and Hall Ltd. London.

Austin, M.P., R.B. Cunningham and P.M. Fleming. 1984. New approaches to direct gradient analysis using environmental scalars and statistical curve fitting procedures. *Vegetatio* 55:11-27.

Austin, M.P., A.O. Nicholls and C.R. Margules. 1990. Measurement of the realized qualitative niche: environmental niches of five *Eucalyptus* species. *Ecological Monographs* 60:161-177.

Bailey, R.G. 1980. Descriptions of the ecoregions of the United States. Misc. Publ. 1391. U.S. Department of Agriculture, Forest Service, Intermountain Region. Ogden, UT.

Bailey, R.G., M.E. Jensen, D.T. Cleland and P.S. Bourgeron. 1993. Design and use of ecological mapping units. In: Jensen, M.E. and P.S. Bourgeron (Eds.). Eastside forest ecosystem health assessment–volume II: ecosystem management: principles and applications. U.S. Department of Agriculture, Forest Service, Pacific Northwest Research Station. Portland, OR.

Bourgeron, P.S. and L.D. Engelking. 1993. A preliminary series level classification of the western U.S. (Unpublished report). On file with: The Nature Conservancy, Western Heritage Task Force. Boulder, CO.

Bourgeron, P.S. and M.E. Jensen. 1993. An overview of ecological principles for ecosystem management. In: Jensen, M.E. and P.S. Bourgeron (Eds.). Eastside forest ecosystem health assessment–volume II: ecosystem management: principles and applications. U.S. Department of Agriculture, Forest Service, Pacific Northwest Research Station. Portland, OR.

Bourgeron, P.S., L.D. Engelking, H.C. Humphries and E. Muldavin. 1993a. Assessing the conservation value of the Gray Ranch: rarity, diversity and representativeness. (Unpublished report). On file with: The Nature Conservancy, Western Heritage Task Force. Boulder, CO.

Bourgeron, P.S., H.C. Humphries, R.L. DeVelice and M.E. Jensen. 1993b. Ecological theory in relation to landscape evaluation and ecosystem characterization. In: Jensen, M.E. and P.S. Bourgeron (Eds.). Eastside forest ecosystem health assessment–volume II: ecosystem management: principles and applications. U.S. Department of Agriculture, Forest Service, Pacific Northwest Research Station. Portland, OR.

Busby, J.R. 1991. BIOCLIM–a bioclimate analysis and prediction system. In: Margules, C.R. and M.P. Austin (Eds.). Nature conservation: cost effective biological surveys and data analysis. CSIRO. Australia.

Correll, D.S. and M.C. Johnston. 1970. Manual of the vascular plants of Texas. Texas Research Foundation. Renner.

DeVelice, R.L., G.J. Daumiller, P.S. Bourgeron and J.O. Jarvie. 1993. Bioenvironmental representativeness of nature preserves: assessment using a combination of a GIS and a rule-based model. In: Proceedings of the first biennial scientific conference on the Greater Yellowstone ecosystem. September, 1992. National Park Service. Mammoth.

Driscoll, R.S., D.L. Merkel, D.L. Radloff, D.E. Snyder and J.S. Hagihara. 1984.

An ecological land classification framework for the United States. USDA Forest Service Misc. Pub. 1439. Washington, DC.

Engelking, L.D., H.C. Humphries, M.S. Reid, R.L. DeVelice, E.H. Muldavin and P.S. Bourgeron. 1993. Regional conservation strategies: assessing the value of conservation areas at regional scales. In: Jensen, M.E. and P.S. Bourgeron (Eds.). Eastside forest ecosystem health assessment–volume II: ecosystem management: principles and applications. U.S. Department of Agriculture, Forest Service, Pacific Northwest Research Station. Portland, OR.

Hann, W., M.E. Jensen, P.S. Bourgeron and M. Prather. 1993. Land management assessment using hierarchical principles of landscape ecology. In: Jensen, M.E. and P.S. Bourgeron (Eds.). Eastside forest ecosystem health assessment-volume II: ecosystem management: principles and applications. U.S. Department of Agriculture, Forest Service, Pacific Northwest Research Station. Portland, OR.

Hungerford, R.D., R.R. Nemani, S.W. Running and J.C. Coughlan. 1989. MTCLIM: a mountain microclimate simulation model. Res. Pap. INT-414. U.S. Department of Agriculture, Intermountain Research Station. Ogden, UT.

Kearney, T.H. and R.H. Peebles. 1960. Arizona flora. University of California Press. Berkeley.

Levin, S.A. 1992. The problem of pattern and scale in ecology. *Ecology* 73:1942-1968.

Mackey, B.G., N.A. Nix, M.F. Hutchinson and J.P. MacMahon. 1988. Assessing representativeness of places for conservation reservation and heritage listing. *Environmental Management* 12:501-514.

McCullagh, P. and J.A. Nelder. 1989. Generalized linear models. 2nd edition. Chapman and Hall. New York.

McLaughlin, S.P. 1986. Floristic analysis of the southwestern United States. *Great Basin Naturalist* 46:46-65.

McLaughlin, S.P. 1989. Natural floristic areas of the western United States. *Journal of Biogeography* 16:239-248.

Mitchell, V.L. 1976. The regionalization of climate in the western United States. *Journal of Applied Meteorology* 15:920-927.

Morgan, P., G.H. Aplet, J.B. Haufler, H.C. Humphries, M.M. Moore and W.D. Wilson. 1994. Historical range of variability: a useful tool for evaluating ecosystem change. *Journal of Sustainable Forestry*.

Muldavin, E., F. Ronco and E.F. Aldon. 1990. Consolidated stand tables and biodiversity data base for southwestern forest habitat types. Gen. Tech. Rep. RM-190. U.S. Department of Agriculture, Forest Service, Rocky Mountain Research Station. Fort Collins.

Nicholls, A.O. 1989. How to make biological surveys go further with generalized linear models. *Biological Conservation* 50:51-75.

Oliver, C.D. and B.C. Larson. 1990. Forest stand dynamics. McGraw-Hill, Inc. New York.

Palmer, M.W. 1993. Putting things in even better order: the advantages of canonical correspondence analysis. *Ecology* 74:2215-2230.

Sokal, R.R. and F.J. Rohlf. 1981. Biometry. W.H. Freeman and Company. San Francisco.

Swanson, F.J., J.A. Jones, D.O. Wallin and J.H. Cissel. 1993. Natural variability-implications for ecosystem management. In: Jensen, M.E. and P.S. Bourgeron (Eds.). Eastside forest ecosystem health assessment-volume II: ecosystem management: principles and applications. U.S. Department of Agriculture, Forest Service, Pacific Northwest Research Station. Portland, OR.

ter Braak, C.J.F. 1986. Canonical correspondence analysis: a new eigenvector method for multivariate direct gradient analysis. *Ecology* 67:1167-1179.

Turner, M.G., R.H. Gardner, R.V. O'Neill and S.M. Pearson. 1993. Multi-scale organization of landscape heterogeneity. In: Jensen, M.E. and P.S. Bourgeron (Eds.). Eastside forest ecosystem health assessment–volume II: ecosystem management: principles and applications. U.S. Department of Agriculture, Forest Service, Pacific Northwest Research Station. Portland, OR.

UNESCO. 1973. International classification and mapping of vegetation. Switzerland.

Wagner, W.L. 1977. Floristic affinities of Animas Mountains southwestern New Mexico. M.S. thesis. University of New Mexico. Albuquerque.

Werger, M.J.A. and J.T.C. Sprangers. 1982. Comparison of floristic and structural classification of vegetation. *Vegetatio* 50:175-183.

Westhoff, V. and E. Van der Maarel. 1978. The Braun-Blanquet approach. In: Whittaker, R.H. (Ed.). Classification of plant communities. Dr. W. Junk Publishers. The Hague.

Whittaker, R.H. and W.W. Niering. 1965. Vegetation of the Santa Catalina Mountains: a gradient analysis of the south slope. *Ecology* 46:429-452.

Emphasis Areas as an Alternative to Buffer Zones and Reserved Areas in the Conservation of Biodiversity and Ecosystem Processes

Richard L. Everett
Paul F. Hessburg
Terry R. Lillybridge

ABSTRACT. Historically, buffer zones or reserved areas have been used to restrict the array of management actions within unique habitats or sensitive-species areas. This action, although necessary as an immediate protection measure, can create administrative fragmentation of the forest and associated problems over time. Rarely is the dynamic nature of the reserved or buffered area considered and the disturbance events that created and maintained these sites are not conserved.

Sizes of buffer zones and reserved areas generally reflect concern for a single management action, such as local timber harvest activities, and do not reflect adequate concern for other disturbances such

Richard L. Everett, Paul F. Hessburg and Terry R. Lillybridge, Center for Sustainable Forest and Rangeland Ecosystems; U.S. Department of Agriculture, Forest Service, Pacific Northwest Research Station, and Wenatchee National Forest.

The authors thank the following for their review of and comments on this paper: Maia Enzer, Dean Longrie, Bruce Marcot, Charlie McKetta, Tom Robinson, Bob Steele, and Dale Wilson.

[Haworth co-indexing entry note]: "Emphasis Areas as an Alternative to Buffer Zones and Reserved Areas in the Conservation of Biodiversity and Ecosystem Processes." Everett, Richard L., Paul F. Hessburg, and Terry R. Lillybridge. Co-published simultaneously in the *Journal of Sustainable Forestry* (The Haworth Press, Inc.) Vol. 2, No. 3/4, 1994, pp. 283-292; and: *Assessing Forest Ecosystem Health in the Inland West* (eds: R. Neil Sampson and David L. Adams) The Haworth Press, Inc., 1994, pp. 283-292. Multiple copies of this article/chapter may be purchased from The Haworth Document Delivery Center [1-800-3-HAWORTH; 9:00 a.m. - 5:00 p.m. (EST)].

as fire, livestock grazing, insect hazard, or flooding that involve much larger landscape scales. Boundaries are usually hard in the sense that few management activities are allowed. However, management may be required to maintain the unique habitat and adjacent landscapes in dynamic forest ecosystems of eastern Oregon and Washington.

We suggest a variable-source emphasis area that has multiple but softer boundaries that vary according to the spatial scale of the potential disturbance. Disturbances are evaluated at relative scales for their impact on the featured species or habitat and on sustainability of the associated landscape. Unique habitats or sensitive species are emphasized in management goals but required disturbance events are encouraged for long-term sustainability of the sensitive species, unique habitats, and associated landscapes.

INTRODUCTION

Immediate protection of sensitive species or unique habitats to prevent loss of species viability and biodiversity is required or implied by the Endangered Species Act of 1973 and subsequent amendments. In recent years, scientific panels have recommended reserved areas, buffer zones, and levels of management activities in the forest matrix to protect the spotted owl (Thomas et al. 1990), old growth ecosystems (Johnson et al. 1991), old-growth species, sensitive fish stocks and riparian areas (Thomas et al. 1993, Thomas and Raphael 1993). These efforts caused a re-evaluation of forest management practices and led to increased emphasis of all resources and sustainable ecosystems. However, we should not allow reserve areas suggested or established by these initial efforts to become set in stone. Each of these efforts called for gathering additional information and later modification of proposed reserved systems as required. The request to "avoid the paralysis and myopia fostered by boundaries" (Thomas and Raphael 1993) has applications in the implementation of reserve or buffer boundaries. Boundary flexibility is desirable if we are to conserve our future biological and socioeconomic options.

Landscapes of the interior west vary greatly in the type, frequency, and severity of disturbances that effect their sustainability over time. For landscapes that are frequently disturbed by natural events, we identified three major concerns associated with any future, long-term establishment of reserve or buffer areas: (1) the reserve area could isolate the species or habitat from the larger system in which it evolved and interacts, (2) reserve management may prevent disturbances required for species viability and sustainability of the landscape, and (3) declining forest health in reserve areas may jeopardize adjacent lands that are more readily managed to reduce insect/disease and fire hazard. Recent recommendations to allow management actions in reserve areas to reduce risk to sensitive species and their habitats (Thomas and Raphael 1993) are commendable, but the concept needs to be expanded to facilitate the reintroduction of disturbance levels required for sustainable landscapes and oil associated species, components, and functions (Everett et al. 1994). Also, we should consider expanding boundaries to address each of the larger-scale disturbance effects that may negatively or positively impact the unique habitat or sensitive species. We propose a multiple boundary approach that considers the spatial scale of each type of disturbance. Although the sensitive species or unique habitat is emphasized in each of these bounds, so is the sustainability of the associated landscape through the conservation of disturbance effects.

DYNAMIC NATURE OF ECOSYSTEMS

Disturbance events are required to create and maintain sustainable ecosystems and associated wildlife habitats and ecosystem processes. Disturbance is a major ecosystem component that must be considered in describing ecosystems (Cooper 1913; Sprugel 1991) and in prescribing management activities for the conservation of biodiversity. The lack of disturbance can be as damaging to regional diversity as excessive disturbance (Noss 1983). We are forced to re-evaluate the wisdom of creating static preserves (other than short-term) in dynamic systems (White and Bratton 1980).

Each type of forest community has a finite life span and window for contributing to larger-scale processes and structure. There is a feedback loop between the successional stage of the community and disturbance vulnerability associated with increasing tree longevity (White 1987). Fixed reserve locations may not be realistic in dynamic systems, and management should focus on maintaining sufficient habitat at various locations within a landscape over time. In equilibrium landscapes, where a dynamic steady state can be achieved, a shifting mosaic of reserve habitats is a viable objective. In nonequilibrium landscapes, characterized by large, low-frequency disturbances, the appropriate landscape scale for management may need to be increased if similar habitats are to be maintained in space and time.

Disturbances that maintain the landscape patterns may operate at scales significantly larger than reserve or buffer areas. To ignore the existence of these disturbance effects and attempt to maintain a static reserve system not only removes disturbances needed to maintain the unique habitat, but may also jeopardize sustainability of associated landscapes. As an example, fire suppression activities allowed large acreages of previously park-like ponderosa pine and mixed conifer stands to develop significant tree understories (Hessburg et al. 1993). Although these stands now have multicanopy layers and dense crown cover favorable for spotted owl habitat, they also have elevated fire, insect and disease hazards (Everett et al. 1992). Attempts to maintain these anomalous landscapes as reserves for the conservation of a single species must be weighed against the increased hazards to the associated landscape. Disturbances required for improved landscape sustainability may need to be re-instated outside and within reserve areas.

ADMINISTRATIVE FRAGMENTATION

Ecosystem management has been adopted by the Forest Service and other land management agencies as the basis for their management philosophy (Overbay 1992; Thomas and Raphael 1993). A hierarchical landscape approach has been suggested as the management framework to achieve sustainable ecosystems for the interior Pacific Northwest (Jensen and Bourgeron 1993). This hierarchical landscape approach prescribes desired conditions at multiple spatial and temporal scales. The system uses the coarse-filter (Hunter 1991) approach to conserving biodiversity by maintaining historical landscape patterns and the processes that create them. Historical landscape patterns define the environment in which indigenous species existed prior to European settlement and may serve as the first approximation of sustainable ecosystems that conserve both known and unknown components of biodiversity (Walker 1992).

To conserve biodiversity during the re-establishment of sustainable ecosystems, special consideration is required for sensitive species and unique habitats. The strategy used to conserve species or unique habitats should be consistent with the conservation strategy for all forest types and species within the landscape. Individual conservation strategies for multiple species and their required habitat can be in conflict (edge-dependent species vs. interior forest species) and individual species or unique habitat strategies may conflict with sustainability of the landscape (Irwin and Wigley 1993; USDA Forest Service 1993).

The hierarchical landscape ecology approach provides a common conservation strategy for all landscape components. The proposed emphasis area concept described below is an initial attempt to integrate the hierarchical landscape approach with the individual conservation needs of sensitive species or unique habitats. Habitat requirements of species and functional attributes of ecosystems need to be coordinated if both diversity and sustainability are to be achieved (Sharitz et al. 1992).

EMPHASIS AREA CONCEPT

Conserving biological diversity and sustainability of dynamic ecosystems requires the maintenance of disturbance effects and associated successional and structural stages within and adjacent to reserve or buffer areas. A management hypothesis to be explicitly tested over time is whether implementation of the emphasis area concept will facilitate the integration of the species or unique habitat into the conservation strategy for the larger landscape. Also, by emphasizing the maintenance of the sensitive species or unique habitat during this process, will biodiversity be conserved and conflicts between sustainability of ecosystems and individual components minimized?

The emphasis area concept, a derivative of Agee and Johnson's (1988) multiple ecosystem boundaries for primary ecosystem components, is relatively simple. Managers identify all major disturbance effects that are likely to impact the sensitive species or unique habitat. Multiple disturbance effects are defined at their respective spatial and temporal scales. Separate "soft" disturbance boundaries are established at the location of the species habitat and others at distances that reflect the larger area of influence of broader disturbance effects. For each disturbance effect the historical or desired disturbance regime is computed and its impact on the sensitive species, unique habitat, and sustainability of the landscape is described. Boundaries are "soft" in that disturbances which are required to maintain the species, unique habitat, or sustainability of the associated landscape are encouraged. Also, boundaries are flexible over time to reflect the shifting mosaic of habitat across the landscape. This approach is much different than the traditional "hard" permanent boundaries of reserves or buffer zones that have historically limited actions to custodial management, and more consistent with the increased management flexibility seen in more recent reserve systems (Thomas and Raphael 1993).

A Case Example

To illustrate the emphasis area concept we will examine the multiple disturbance-effect boundaries that were encountered in conserving a rare plant population of Wenatchee larkspur (*Delphinium veridescens*) and its associated landscape. This species occupies moist to seasonally wet sites adjacent to seeps, ephemeral streams or areas with high water tables. The multiple, soft disturbance boundaries would start at the plant population and the forest stand levels, where localized timber harvest could alter canopy cover and resultant plant shading (Figure 1). Prior research studies (Everett, unpublished data) have shown that this species has optimum growth and reproduction with 30 to 60 percent canopy closure. Thus, within this boundary (source area), disturbance levels that maintained the desired canopy would be encouraged if they are consistent with historical disturbance effects.

A second soft disturbance boundary would be established at the watershed drainage level. Both excessive tree cover that reduces available soil moisture and excessive loss of tree cover that elevates the water table

FIGURE 1. Variable-source areas and associated disturbance boundaries used in an example of an emphasis area approach to integrating individual species and landscape management.

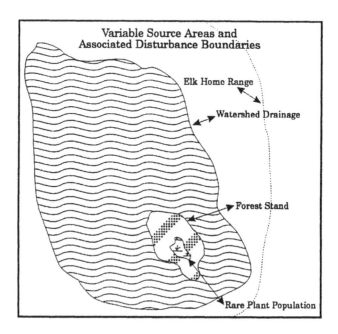

could adversely impact Wenatchee larkspur habitat. Historical disturbance frequency, extent, and intensity would provide initial estimates of the range of acceptable tree cover values for appropriate soil moisture and landscape characteristics. An analysis of the watershed hydrograph and the subsequent assessment of road construction and canopy removal on potential flooding or dewatering of the habitat would provide input into scheduling the extent or frequency of the desired level of disturbance. This information is needed to integrate Wenatchee larkspur viability with landscape sustainability.

A third and final soft boundary would occur at the home range of a growing elk herd that forages within the source area. Elk do not feed on the species except for nipping some flower stalks, but trampling can be a problem (Robson and Mehrhoff, unpublished data). Elk herd size and grazing/trampling effects currently exceed historical levels, but present no clear hazard to Wenatchee larkspur. However, increased canopy coverage and elk herd size could adversely impact the rare plant species with no-action under custodial management. In each of the soft boundaries the rare plant population is not isolated from the associated landscape and disturbance processes, but integrated into the overall management strategy to the level that information allows.

ADAPTIVE MANAGEMENT REQUIRED

In the above example, we rapidly encountered a point where scientific information was inadequate to prescribe needed disturbance effects. The emphasis area concept requires basic ecological information on systems and featured species. Where that information is lacking, a conservation approach suggests we proceed on an experimental basis and not widely prescribe treatments until more information is available. One option is no management–but systems will change with or without management actions. Adaptive management provides an approach to managing ecosystems when outcomes of management actions are uncertain (Walters and Hollin 1990). By stating our understanding of the system, the biology-ecology of the sensitive species, anticipated management effects, and by conducting appropriate landscape experiments, we will proceed faster in determining how to sustain all ecosystem components.

SOCIOECONOMIC ACCEPTABILITY OF EMPHASIS AREAS

Socioeconomic assessment of the emphasis area concept has not been defined. Individuals concerned with commodity output may object because the multiple boundaries impact larger areas, but this is offset by increased management options within all boundaries. Sustainable flows of resources may be enhanced because disturbance effects and opportunities for commodity extraction are encouraged if the actions benefit the desired habitat or species (Everett et al. 1993). Conversely, commodity extraction would decrease and restoration activities would need to increase if desired disturbance levels have been exceeded.

The long-term interdependence of economic and ecosystem health is becoming increasingly apparent (Ervin and Berrens 1993). The emphasis area approach may have wider social appeal than reserve areas because it allows for multiple rather than single use. The emphasis area concept integrates ecological and economic considerations by providing for the conservation of disturbance effects and associated resource flows in the maintenance of sensitive species, unique habitats, and sustainable landscapes.

SUMMARY

The dynamic functional components, as well as structural components, of reserve or buffer areas are conserved in the emphasis area approach. This concept differs from buffer zones and reserve areas in the conservation of disturbance effects within multiple "soft" boundaries that reflect disturbance effects at different scales. The multiple boundary approach integrates, rather than isolates, the unique habitat or sensitive species from the landscape in which it exists. This approach is consistent with the ecosystem management approach of managing landscapes as a whole rather than as separate components. The hazards associated with the emphasis area approach must be weighed against the hazards associated with the reserve system. Potential for loss of desired habitat resulting from unexpected events during a prescribed management entry in a managed emphasis area should be weighed against increased potential for loss associated with the accumulation of nonsustainable habitats over time in a managed reserve area. The emphasis area concept is as yet untested, much like ecosystem management, and both must proceed as experiments under an adaptive management process.

REFERENCES

Agee, J.K. and D. Johnson. 1988. Ecosystem management for parks and wilderness: a workshop synthesis. Institute of Forest Resources. Contribution Number 62. University of Washington. Seattle. 39 p.

Cooper, W.S. 1913. The climax forest of Isle Royale, Lake Superior and its development. *Botanical Gazette* 55:1-44.

Ervin, D.E. and R.P. Berrens. 1993. Critical economic issues in ecosystem management of a national forest. In: pp. 315-323. Eastside forest ecosystem health assessment–volume II: Ecosystem Management: Principles and applications. U.S. Department of Agriculture, Forest Service, Pacific Northwest Research Station. Portland, OR.

Everett. R.L. (Unpublished data). On file with the Forestry Sciences Laboratory, 1133 N. Western Avenue, Wenatchee, WA 98801.

Everett, R.L, S. Martin, M. Bickford, R. Schellhaas and E. Forsman. 1992. Variability and dynamics of spotted owl nesting habitat in eastern Washington. National Silviculture Workshop, Cedar City, UT, May 6-9, 1991.

Everett, R.L., P.F. Hessburg, M.E. Jensen and B.T. Bormann. 1993. Eastside forest ecosystem health assessment–Volume I: Executive Summary. U.S. Department of Agriculture, Forest Service, Pacific Northwest Research Station. Portland, OR. 57 p.

Everett, R.L., P.F. Hessburg, J.F. Lehmkuhl, M.E. Jensen and P.S. Bourgeron. 1994. Old forests in dynamic landscapes-dry site forests of eastern Oregon and Washington. *Journal of Forestry* 92(1):22-25.

Hessburg, P.F., R.G. Mitchell and G.M. Filip. 1993. Historical and current roles of insects and pathogens in eastern Oregon and Washington forested landscapes. In: pp. 485-536. Hessburg, P.F. (Comp.) Eastside forest ecosystem health assessment–Volume III: Assessment. U.S. Department of Agriculture, Forest Service, Pacific Northwest Research Station. Portland, OR.

Hunter, M.L., Jr. 1991. Coping with ignorance: the coarse-filter strategy for maintaining biodiversity. In: pp. 266-281. Kohn, K.A. (Ed.). Balancing on the brink of extinction–the Endangered Species Act and lessons for the future. Island Press. Washington, DC.

Irwin, L. and T.Wigley. 1993. Toward an experimental basis for protecting forest wildlife. *Ecological Applications* 3:213-217.

Jensen, M.E. and P.S. Bourgeron. 1993. Ecosystem management: Principles and applications. In: Eastside forest ecosystem health assessment-volume II. U.S. Department of Agriculture, Forest Service, Pacific Northwest Research Station. Portland, OR. 397 p.

Johnson, K.N., J.F. Franklin, J.W. Thomas and J. Gordon. 1991. Alternatives for management of late-successional forests of the Pacific Northwest: A report to the Agriculture Committee and Merchant Marines Committee of the U.S. House of Representatives. 59 p.

Noss, R.F. 1983. A regional landscape approach to maintain diversity. *BioScience* 33:700-706.

Overbay, J.C. 1992. Ecosystem management: National workshop on taking an ecological approach to management. April 27, 1992, Salt Lake City, UT. U.S. Department of Agriculture, Forest Service. Washington, DC.

Robson, K.A. and L. Mehrhoff. (Unpublished report). A report on the rare Wenatchee larkspur, *Delphinium veridescens,* in comparison with a common, sympartric relative, Western monkshood, *Aconitum columbianum.* USDA Forest Service, Pacific Northwest Research Station. On file with the Wenatchee

Forestry Sciences Laboratory, 1133 N. Western Ave., Wenatchee, WA 98801. 13 p.

Sharitz, R.R., L.R. Boring, D.H. Van Lear, and J.E. Pinder. 1992. Integrating ecological concepts with natural resource management of southern forests. *Ecological Applications* 2:266-2

Sprugel, D.G. 1991. Disturbance, equilibrium, and environmental variability: What is "natural" vegetation in a changing environment? *Biological Conservation* 58:1-8.

Thomas, J.W., E.D. Forsman, J.B. Lint, E.C. Meslow, B.R. Noon and J. Verner. 1990. A conservation strategy for the northern spotted owl: a report of the Interagency Scientific Committee to address the conservation of the northern spotted owl. U.S. Department of Agriculture, Forest Service, U.S. Department of Interior, Bureau of Land Management, Fish and Wildlife Service, and National Park Service. Portland, OR. 427 p.

Thomas, J.W. and M.G. Raphael. 1993. Forest ecosystem management: An ecological, economic and social assessment: A report of the Forest Ecosystem Management Assessment Team. U.S. Department of Agriculture; U.S. Department of Commerce; U.S. Department of Interior; and Environmental Protection Agency. Portland, OR.

Thomas, J.W., M.G. Raphael, R.G. Anthony et al. 1993. Viability assessments and management considerations for species associated with late-successional and old-growth forests of the Pacific Northwest. U.S. Department of Agriculture, Forest Service. Portland, OR. 523 p.

U.S. Department of Agriculture. 1993. White Sand–Landscape level integrated resource and ecosystem analysis. USDA, Forest Service, Powell Ranger District, Clearwater National Forest. Lolo, MT. 115 p.

U.S. Public Laws, Statutes, etc.; Public Law 93-205. [S.1983], Endangered Species Act of 1973. Act of December 28, 1973. In: United States statutes at large, 1973. 16 U.S.C. sec 668, et seq. (1976). U.S. Government Printing Office: 884. Vol 87. Washington, DC.

Walker, B.H. 1992. Biodiversity and ecological redundancy. *Conservation Biology* 6:18-223.

Walters, C.J. and C.S. Hollin. 1990. Large-scale management experiments and learning by doing. *Ecology* 71:2060-2068.

White, P.S. 1987. Natural disturbance, patch dynamics and landscape pattern in natural areas. *Natural Areas Journal* 7:14-22.

White, P.S. and S.P. Bratton. 1980. After preservation: the philosophical and practical problems of change. *Biological Conservation* 18:241-255.

A Process for Improving
Wildlife Habitat Models
for Assessing Forest Ecosystem Health

Larry L. Irwin

ABSTRACT. Vertebrate wildlife will probably continue to be a primary surrogate for assessing biological diversity in forested ecosystems. However, assessment tools such as wildlife-habitat models generally have proved to be poor predictors of wildlife population responses to landscape-scale changes in forest ecosystems. Forest ecosystem assessment therefore will require improved models. To improve modeling capabilities, scientists must clarify the primary determinants of wildlife habitat selection, which is a behavioral process that links wildlife populations with ecosystem processes. Wildlife populations respond to functional redundancies caused by multiple interactions among landforms, soils, and vegetation. Therefore, probing wildlife habitat selection responses to attributes of landforms, soils, and vegetation should result in improved wildlife-habitat models. In this paper, radiotelemetry data from a study on northern spotted owls *(Strix occidentalis caurina)* are used to illustrate how remote sensing and geographic information systems (GIS) analysis might clarify basic determinants of habitat selection.

Larry L. Irwin, National Council for Air and Stream Improvement, Corvallis, OR 97339.

The author thanks John Erickson, Hope Humphries, and Dale Toweill for providing comments that improved this paper.

[Haworth co-indexing entry note]: "A Process for Improving Wildlife Habitat Models for Assessing Forest Ecosystem Health." Irwin, Larry L. Co-published simultaneously in the *Journal of Sustainable Forestry* (The Haworth Press, Inc.) Vol. 2, No. 3/4, 1994, pp. 293-306; and: *Assessing Forest Ecosystem Health in the Inland West* (eds: R. Neil Sampson and David L. Adams) The Haworth Press, Inc., 1994, pp. 293-306. Multiple copies of this article/chapter may be purchased from The Haworth Document Delivery Center [1-800-3-HAWORTH; 9:00 a.m. - 5:00 p.m. (EST)].

INTRODUCTION

Current trends toward large-scale conservation planning result from an awareness that both internal structure and spatial arrangement of forest stands affect ecosystem processes. Large-scale planning also stems from federal laws that require maintaining well-distributed, viable populations of animals. Further, regulations for implementing these laws require assessments of indicators of ecosystem structure and function. Vertebrate wildlife most likely will continue to be a primary surrogate for judging whether conservation plans maintain healthy forest ecosystems. Such judgments will require evaluations of animal species richness in relation to landscape conditions. However, assessing forest ecosystem health will require more than identifying vertebrate community richness–viable populations must also be maintained. Demonstrating that populations are viable will require improved processes for linking wildlife population dynamics with habitat conditions at the landscape scale (Rosenzweig 1991; Pulliam et al. 1992; Noon 1993).

Habitat selection influences population dynamics because choice of habitat affects the balance between costs and benefits associated with various environmental stimuli (Partridge 1978). Here, I propose an analytical process that could help clarify basic determinants of wildlife habitat selection. The information provides a step toward predicting population responses to conservation plans that apply at landscape scales. For an example, I describe how the analytical process might result in predicting habitat selection by northern spotted owls, an indicator of late-successional and old-growth forest (LS/OG) conditions in the Pacific Northwest (Thomas et al. 1990). I assume that reliable predictions of habitat selection and, subsequently, population responses to conservation plans, require an accounting of the influences of the physical environment on the expression of animal-vegetation relationships. Here, the term wildlife refers to terrestrial vertebrates, although the analytical process might also be applicable to other animals.

WILDLIFE-HABITAT ASSOCIATIONS

Previously, methods of incorporating wildlife into land management plans relied on correlative studies of wildlife-habitat association. Such studies emphasized vegetation because vegetation can be changed by management, mapped readily, and predicted for future conditions (e.g., Thomas 1979; Bruce et al. 1985). The use of vegetation patterns to describe wildlife habitat is well established in the ecological literature (e.g., Holmes

1981; Schamberger and O'Neil 1986; Short 1988). Further, vegetation structure (most frequently vertical complexity) has been associated with animal species richness (MacArthur and MacArthur 1961; Short 1988). For example, Scott et al. (1992) used models of animal-vegetation associations to identify areas with potentially high levels of biological diversity that are not protected (i.e., gap analysis).

MacArthur et al. (1962) reasoned that ecosystems with complex habitats offer a greater number of potential niches and should support a greater species diversity than simplified systems. Thus, through the 1980s, wildlife biologists advocated successional mosaics with high amounts of edges (e.g., Thomas 1979) to promote habitat complexity. However, fragmenting forest ecosystems into mosaics of successional patches may reduce habitat quality for species that are associated with continuous forest stands (Harris 1984; Wilcove 1985; Robbins et al. 1989). Thus, horizontal complexity has been added to wildlife-habitat models to account for the effects of spatial distribution of forest stands on species richness (Flather et al. 1992).

Large-scale applications of wildlife-habitat association studies generally used successional stage or vegetation type as surrogate measures for vegetation structure. Map-based conservation strategies were then developed that specify the amounts and distribution of successional stages so as to minimize the effects of fragmentation. For example, Thomas et al. (1990) and Thomas et al. (1993) used LS/OG forest stages as a surrogate indicator for mapping large conservation reserves that would protect northern spotted owls and biological diversity, respectively, in the Pacific Northwest.

The essence of useful landscape management and assessment is reliable predictability. However, wildlife-habitat models based on vertical habitat complexity have not provided reliable explanations for the variation in wildlife population performance (Noon 1993). Furthermore, models that include indices of horizontal patchiness, which might account for bird species richness (Flather et al. 1992), may be unable to account for variation in population dynamics. For example, recent studies failed to demonstrate that indices to fragmentation of LS/OG forests can be used to predict reproductive success among northern spotted owls (e.g., Lehmkuhl and Raphael 1993; Meyer et al., in preparation).

Conservation strategies that advocate reserves of LS/OG forests assume that open-canopy species are "weedy" species that always do well in disturbed landscapes outside of reserves (Noss 1983). However, groups of vertebrate wildlife species respond differently to forest fragmentation due to numerous differences in life-history patterns (Hansen and Urban 1992; Hansen et al. 1993). For example, some species of birds that are open-canopy specialists in the Pacific Northwest are sensitive to afforestation (Hansen and Urban 1992), and some of these species are declining in abundance (Sharp 1990). Thus, although there probably is a link between the dynamics of some wildlife populations and forest fragmentation, reliable predictions about how habitat changes affect wildlife population performance apparently require more detailed information than vegetation structure and seral stage distribution.

PHYSICAL INFLUENCES
ON BIOLOGICAL RELATIONSHIPS

Habitat selection is a multi-scaled behavioral process that links wildlife populations with ecosystem processes (Rosenzweig 1991). Thus, developing an understanding of the primary determinants of habitat selection might help chart a reasonable course for improving wildlife-habitat models. In general, wildlife habitat results from interactions among physical and biological factors. These physical and biological factors interact with each other and with disturbances (e.g., fire, windstorms or logging) in numerous ways, creating functional redundancies within ecosystems that influence the distribution and abundance of wildlife (Partridge 1978). Multiple interactions among landforms, soils, and vegetation provide options for animals to use their repertoire of adaptations for finding specific resources (e.g., food, nest sites) and for avoiding enemies and inclement weather. Because attributes of topography and soils have strong effects on wildlife habitat selection, understanding their interaction with animal-vegetation relations might improve wildlife-habitat models.

Surprisingly few investigators have included measures of soil or topographic influences on wildlife-habitat relationships, despite intuitive recognition that physical factors influence wildlife distribution and abundance. Hill (1972) described soil fertility factors that influenced populations of cottontail rabbits *(Sylvilagus floridanus)*. Irwin and Cook (1985) found that interactions among vegetation, weather and topography influenced population dynamics of pronghorn antelope *(Antilocapra americana)*. Keith et al. (1993) described the effects of terrain on habitat selection in moose *(Alces alces)*, and Rice et al. (1993) incorporated precipitation and soil factors as well as attributes of vegetation to predict bobwhite *(Colinus virginianus)* population responses to management.

The papers described above suggest that more information than spatial patterns of successional stages is required to predict wildlife population responses to forest ecosystems managing. Thus, the next generation of wildlife-habitat models might be predicated upon a more complete understanding of the influences of the physical environment on biological relationships. The following discussion suggests a possible direction for clarifying those interactions.

SUGGESTED MODEL-BUILDING PROCESS

Ecologists must understand the interactions among wildlife (and plant) population dynamics and environmental variables at resolution scales that match those of the process of habitat selection and other life-history processes (Hansen and Urban 1992). Holling (1992) suggested using body-mass/territory size relations to define the scale for various vertebrate species groups (Figure 1). For example, ecosystem assessments must account for small-bodied or relatively immobile species as well as large, wide-ranging species. Scale-dependent assessments will be valuable because animals select habitats via hierarchial processes (Partridge 1978; Rosenzweig 1985).

Spotted Owl Example

Spotted owls are consistently associated with LS/OG forest stages (Thomas et al. 1990), but reproductive success is not predicted by either the amount or spatial arrangement of LS/OG forests (Lehmkuhl and Raphael 1993; Meyer et al., in preparation). Variation among physical factors, such as annual precipitation, and attributes of soils and topography, in addition to structure and patterning of successional stages, might account for much of the variation in reproductive success among northern spotted owls.

An animal's territory is that portion of its annual home range that is defended against conspecifics, and probably provides exclusive supplies of food. Territory size has not been reported for northern spotted owls. Core areas, or relatively small locations within home ranges that receive disproportionate use over an annual period (Samuel et al. 1985), might approximate spotted owl territories and/or that portion of a spotted owls' home range that most influences reproductive success. Millers et al.'s (1992) study provided information from 19 radio-tagged spotted owls in intermediate-aged forests in western Oregon that can be used as an example of core-area analysis of habitat selection.

Miller et al. (1992) mapped nocturnal foraging locations of spotted owls throughout annual periods. Home ranges were determined using the adaptive-kernel technique (Ackerman et al. 1989; Worton 1989). The adaptive kernel algorithm provides contour-line polygons for user-specified percentages of annual or total relocations.

FIGURE 1. Relation between territory size and body size in predatory birds. Solid triangles indicate hawks, owls, and eagles; open squares indicate insect-eating birds.

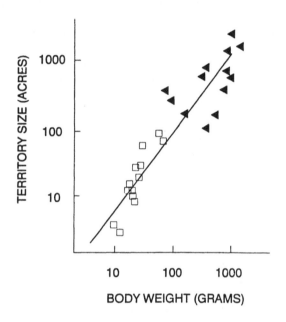

Spotted owls made intensive use of relatively small areas within their home ranges (Figure 2), also noted by Laymon and Reid (1986). Thirty percent of the locations were contained within 320 acres, or about 5% of annual home ranges. Sixty percent of the telemetry locations were contained within areas that encompassed slightly more than 700 acres, which averaged only 15% of the annual home ranges.

I propose that a core area of 700 acres be used as an analytical unit for predicting habitat selection in spotted owls. Several reasons support my proposal. First, the size of the core area is similar to the expected territory size based on body mass. Spotted owls, which range in body mass from about 550g to 660g (Forsman et al. 1984), would be expected to have breeding season territories of 600-725 acres, according to data in Holling (1992)

FIGURE 2. Concentration of use in home ranges by radio-tracked northern spotted owls, McKenzie River area, Oregon. Vertical bars indicate standard deviation.

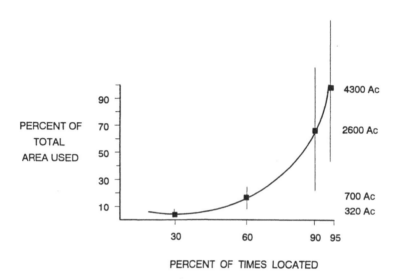

and Figure 1. Secondly, the 60% adaptive kernel polygons varied little among individual owls, although annual home ranges vary greatly. Third, such an analytical unit has ecological relevance, as demonstrated in empirical studies. For example, Meyer et al. (in preparation) found that site selection and occupancy by pairs of spotted owls in western Oregon were most strongly influenced by the amount of LS/OG forest in analytical circles that encompassed 500 acres. Fourth, an analytical unit for predicting habitat selection should be unconstrained by density-dependent effects (Rossenzweig 1991). Empirical data suggest that a 700-acre core area might be less constrained by density-dependent effects than annual home ranges because the core areas do not overlap.

The next steps toward building a biophysical model that relates to reproductive success would include the following: (a) predicting the probability of presence of an owl, or pair of owls, based on the set of physical and biological features of 70-acre core areas; and (b) comparing reproductive success among owl pairs that occur along gradients of important features that may be revealed in (a). The following paragraphs discuss some possibilities for constructing a tool that predicts habitat selection.

Remote Sensing

Recent studies have used remotely sensed habitat data as input to GIS to predict wildlife habitats (e.g., Hodgson et al. 1988; Clark et al. 1993). Homer et al. (1993) used GIS analysis of Landsat Thematic Mapper (TM) imagery to model structural and compositional features of sage grouse *(Centrocercus urophasianus)* habitats. Such studies traditionally created habitat layers by mapping *predetermined* wildlife habitat components, and were hampered by subjectivity, repeatability, and resolution problems (Manley et al. 1992). Even the best model could only take into account a simplified set of factors that determined a species' distribution.

I suggest *data-driven analyses* that combine a more complete, multidimensional set of physical and biotic factors. Such analyses incorporate GIS layers from an ecological land classification system that includes attributes of soils, landforms, and vegetation potential (Haufler and Irwin, in press). Then, Landsat TM or other satellite imagery could be used to create a map of current vegetation, based on successful applications by Cohen and Spies (1992) and Fiorella and Ripple (1993). The vegetation layer can be combined with a digital elevation model (DEM) for topographical modelling and a soils layer. Radio-tracking provides data for determining core areas. When classified Landsat imagery of vegetation is "draped" over a DEM, one can view a landscape in 3 dimensions. For example, Haufler and Irwin (in press) superimposed Miller et al.'s (1992) radio-telemetry locations on 3-dimensional images. Core areas were found to encompass concave topography that included the lower parts of east- and north-facing slopes. Use of lower slope positions suggests that deep soils may influence habitat selection, possibly through effects of shrubs on prey abundance.

Statistical Procedures

One potential statistical process for analyzing the apparent multi-factor influences on spotted owl habitat selection might first grid study areas into cells, whose size is based upon the accuracy limits of radio-tracking (which varies among studies). Logistic regression could then be performed, in which the sampling domain is the set of cells within the study area. Core-area cells with radio-locations receive a value of 1, and an equal number of randomly-located cells would be assigned a value of 0. Alternatively, random cells could be contained in a randomly located area equal in size to core areas.

Independent vegetation variables measured in each grid cell would include vegetation age-class or successional stage, stand size, understory vegetation, vegetation structure measures (e.g., snags and logs) and landscape metrics such as interspersion, contagion, and fractal dimension (Turner 1990). Then logistic regression or partial correlation analysis would allow investigators to statistically model the *joint spatial dependence* of the patterns of use of vegetation in core areas with physical variables such as attributes of landforms (slope, aspect, terrain shape) and soils (depth, nitrogen concentration, etc.). Lists of variables probably should first be reduced using principal components and cluster analysis, as suggested by Young et al. (1993), keeping in mind the cautions of Rextad et al. (1988) regarding multivariate analyses and those of White and Garrott (1986) regarding triangulation error in radio-tracking studies of habitat selection. Thus, the process should be considered exploratory rather than confirmatory.

Another statistical process would make use of the analytical capabilities of GIS. For example, using GIS overlay operations, stored attributes of soils, geology, and topography could be *intersected* with vegetation data and animal radio locations in core areas. Scores, or indices, reflecting the overall characteristics of each polygon that results from the overlays could be derived. By selectively weighting physical habitat properties and describing spatial variables such as forest patch size, shape, and arrangement with respect to other seral patches, both the *quantity* and *quality* of the habitat could be estimated. Such models are known as weighting or ordinal models (Rossi et al. 1992). These models can investigate the spatial dependence of core areas, or a similar ecologically significant portion of an animal's total use pattern, on interactions among physical and biotic features. This is potent, because it probes basic determinants of spatial use patterns (Rossi et al. 1992).

Mutual information analysis is another possible procedure for predicting habitat selection from core areas. Mutual information analysis provides a classification structure that is consistent with the well-accepted notion of scale-dependence and interactions among ecological factors (Davis and Dozier 1990). Of course, these relationships form the underpinnings of ecological patterns.

Mutual information analysis is a method for grouping samples that share a set of attributes, based on the association of those attributes with a categorical variable, such as a core-area. Mutual information analysis assumes that land surfaces are spatially ordered due to ecological interdependence among terrain variables (i.e., geology, topography, and vegetation are closely coupled). The method is predictive in the sense that land and vegetation classes are developed by searching samples for a pattern or combination that most closely resembles that of a mapped dependent variable (i.e., a core-area).

The resulting predictive capabilities of the geostatistical modeling would require subsequent testing to determine general reliability in predicting habitat selection. Further, comparative studies are required to determine how the conditions in core areas might be related to reproductive success. I suggest that this should be done using multiple study areas that are examined with similar techniques. Ultimately, scientists will want to explore experimental designs that probe how physical factors and vegetation conditions influence rates of survival and the process of dispersal.

DISCUSSION

Wildlife-habitat models can be constructed from data collected in extensive or correlative studies that suggest possible explanations of the observed phenomena; or they can be constructed from intensive or comparative studies that investigate key factors (i.e., test hypotheses) thought to influence reproduction and survival (Van Horne 1986). The information presented here relates to intensive, comparative efforts that might help build spatially-explicit wildlife-habitat models that help assess forest ecosystem health. The suggested analytical process, based on home range/body mass relations (Holling 1992), could be applied to different-sized species of similar trophic status (e.g., predatory mammals) that represent a gradient in home range size. This might provide a mechanism for drawing inferences across a hierarchy of scales. Such complex, scale-oriented models should improve the ability to predict vertebrate wildlife responses to forestry practices at the landscape scale.

Most importantly, large-scale planning that might employ such GIS-based models can incorporate the influences of the physical environment in addition to vegetation attributes. Such planning could also recognize that physical factors may alter an animal's successional affiliation. For example, bobwhite quail may respond positively to mid- to late-successional stages on unproductive sites but are considered early successional species on highly productive sites (Spears et al. 1993).

RESEARCH NEEDS

Scientific experimentation, our most valuable tool, should be brought to bear on the topic of forest ecosystem assessment. A validated understanding of biodiversity responses to ecosystem management will be manifest only on large geographic scales, long time frames (decades), or both. Because society is probably unwilling to wait decades, answers must come from experiments that are coordinated across large geographic areas (Irwin and Wigley 1993). Scientists and managers should join in developing archetypical experimental designs for focusing thinking about forest ecosystems within the context of ecological land classification, as proposed by USDA Forest Service (1993). Results from the suggested experiments would identify what is possible from managing various forest ecosystems and could help determine the scope of extrapolability based upon extant knowledge of underlying ecosystem processes.

Adaptive management experiments can probe wildlife habitat selection and population responses to the multiple interactions among the attributes of forest vegetation, landforms, local climate, and soils. Previous research considered topographical variables such as slope, slope shape or slope position as independent variables. Ecological theory, however, prompts us to recognize the multiple interactions, or dependencies, among environmental variables. Ecologically-based land classification systems could provide the supplementary information for more-reliable predictions of wildlife population responses to interdependent physical and biotic factors (Haufler and Irwin, in press).

REFERENCES

Ackerman, B.B., F.A. Leban, E.O. Garton and M.D. Samuel. 1989. User's manual for program home range. Contrib. No. 259, Univ. Idaho Forestry, Wildlife and Range Exp. Sta. Tech. Rep. 15. Moscow, ID.

Bruce, C., D. Edwards, K. Mellen, A. McMillan, T. Owens and H. Sturgis. 1985. Wildlife relationships to plant communities and stand conditions. Chapter 3 In: E.R. Brown (Technical Ed.). Management of wildlife and fish habitats in forests of western Oregon and Washington. USDA For. Serv. Pacific Northwest Region. Portland, OR.

Clark, J.D., J.E. Dunn and K.G. Smith. 1993. A multivariate model of female black bear habitat use for a geographic information system. *J. Wildl. Manage.* 57:519-526.

Cohen, W.B. and T.A. Spies. 1992. Estimating structural attributes of Douglas-fir/Western Hemlock forest stands from Landsat and SPOT imagery. *Rem. Sens. Environ.* 41:1-17.

Davis, F.W. and J. Dozier. 1990. Information analysis of a spatial database for ecological land classification. *Photogram. Engineer. Rem. Sens.* 56:605-613.

Fiorella, M. and W.J. Ripple. 1993. Determining successional stage of temperate coniferous forests with Landsat satellite data. *Photogram. Engineer. Rem. Sens.* 59:239-246.

Flather, C.H., S.I. Brady and D.B. Inkley. 1992. Regional habitat appraisals of wildlife communities: a landscape-level evaluation of a resource planning model using avian distribution data. *Landscape Ecology* 7:137-147.

Forsman, E.D., E.C. Meslow and H.M. Wight. 1984. Distribution and biology of the spotted owl in Oregon. *Wildl. Monographs* 87:1-64.

Hansen, A.J. and D.L. Urban. 1992. Avian response to landscape pattern: the role of species' life histories. *Landscape Ecology* 7:163-180.

Hansen, A.J., S.L. Garman and B. Marks. 1993. An approach for managing vertebrate diversity across multiple-use landscapes. *Ecological Applications* 3:481-496.

Harris, L.D. 1984. The fragmented forest: island biogeography theory and the preservation of biotic diversity. University of Chicago Press. Chicago.

Haufler, J.B. and L.L. Irwin. (In press). An ecological basis for forest planning for biodiversity and resource use. Proc. Internatl. Union Game Biol. 15-20 Aug, 1993, Halifax, Nova Scotia.

Hill, E.P., III. 1972. Litter size in Alabama cottontails as influenced by soil fertility. *J. Wildl. Manage.* 36:1199-1209.

Hodgson, M.E., J.R. Jensen, H.E. Mackey, Jr. and M.C. Coulter. 1988. Monitoring wood stork foraging habitat using remote sensing and geographic information systems. *Photogram. Engineer. Rem. Sens.* 54:61-68.

Holling, C.S. 1992. Cross-scale morphology, geometry, and dynamics of ecosystems. *Ecological Monographs* 62:447-502.

Holmes, R.T. 1981. Theoretical aspects of habitat use by birds. In: pp. 33-37. Capen, D.E. (Ed.). The use of multivariate statistics in studies of wildlife habitat. USDA For. Serv., Gen. Tech. Rep. RM-87.

Homer, C.G., T.C. Edwards, Jr., R.D. Ramsey and K.P. Price. 1993. Use of remote sensing methods in modelling sage grouse winter habitat. *J. Wildl. Manage.* 57:78-84.

Irwin, L.L. and J.G. Cook. 1985. Determining appropriate variables for a habitat suitability model for pronghorns. *Wildl. Soc. Bull.* 13:434-440.

Irwin, L.L. and T.B. Wigley. 1993. Toward an experimental basis for protecting forest wildlife. *Ecological Applications* 3:213-217.

Keith, R.J., D.A. Gauthier and D.G. Larsen. 1993. Home range and terrain suitability: spatial utilization of terrain by female moose in the southwest Yukon. Proc. Symp. on geographic informations systems in forestry, environment and natural resources management.

Laymon, S.A. and L.A. Reid. 1986. Effects of grid-cell size on tests of a spotted owl HSI model. In: pp. 93-97. Verner, J., M.L. Morrison and C.J. Ralph (Eds.). Wildlife 2000: modeling habitat relationships of terrestrial vertebrates. University of Wisconsin Press. Madison.

Lehmkuhl, J.F. and M.G. Raphael. 1993. Habitat pattern around northern spotted owl locations on the Olympic Peninsula, Washington. *J. Wildl. Manage.* 57:302-315.

MacArthur, R.H. and J.W. MacArthur. 1961. On bird species diversity. *Ecology* 42:594-598.

MacArthur, R.H., J.W. MacArthur and J. Preer. 1962. On bird species diversity II. Prediction of bird census from habitat measurements. *Amer. Natural.* 96:167-174.

Manley, T.L., K. Ake and R.D. Mace. 1992. Mapping grizzly bear habitat using Landsat TM satellite imagery. Fourth Forest Service Remote Sensing Conf. 4:231-240.

Meyer, J.S.., L.L. Irwin and M.S. Boyce. [In preparation]. Influence of habitat fragmentation on spotted owl site selection, site occupancy, and reproductive status in western Oregon. Humboldt State University. Arcata, CA.

Miller, G.P., D.F. Rock, L.L. Irwin and T.D. Wallace. 1992. Spotted owl occupancy and habitat use in intermediate-aged forests, western Oregon–1991 Progr. Rep. Unpubl. rep., USDI Bur. Land Manage. Eugene, OR.

Noon, B.R. 1993. Book review: wildlife-habitat relationships, concepts and applications. *J. Wildl. Manage.* 57:934-936.

Noss, R.F. 1983. A regional landscape approach to maintain diversity. *Bioscience* 33:700-706.

Partridge, L. 1978. Habitat selection. In: pp. 351-376. Krebs, J.R. and N.B. Davies, (Eds.). Behavioral ecology–an evolutionary approach. Sinauer Associates. Sunderland, MA.

Pulliam, H.R., J.B. Dunning, Jr. and J. Liu. 1992. Population dynamics in complex landscapes: a case study. Ecological Applications 2:165-177.

Rextad, E.A., D.D. Miller, C.H. Flather, E.M. Anderson, J.W. Hupp and D.R. Anderson. 1988. Questionable multivariate statistical inference in wildlife habitat and community studies. *J. Wildl. Manage.* 52:189-193.

Rice, S.M., F.S. Guthery, G.S. Spears, S.J. DeMaso and B.H. Koerth. 1993. A precipitation-habitat model for northern bobwhites on semiarid rangeland. *J. Wildl. Manage.* 57:92-102.

Robbins, K.V., D.K. Dawson and B.A. Dowell. 1989. Habitat area requirements of breeding forest birds of the middle Atlantic states. *Wildl. Monographs* 103:1-34.

Rosenzweig, M.L. 1985. Some theoretical aspects of habitat selection. In: pp. 517-540. Cody, M.L. (Ed.). Habitat selection in birds. Academic Press. Orlando, FL.

Rosenzweig, M.L. 1991. Habitat selection and population interactions: the search for mechanism. *Amer. Natural.* 137:55-528.

Rossi, R.E., D.J. Mulla, A.G. Journel and E.H. Franz. 1992. Geostatistical tools for modeling and interpreting spatial dependence. *Ecological Monographs* 62:277-314.

Samuel, M.D., D.J. Pierce and E.O. Garton. 1985. Identifying areas of concentrated use within the home range. *J. Anim. Ecology* 54:711-719.

Schamberger, M.L. and L.J. O'Neil. 1986. Concepts and constraints of habitat-model testing. In: pp. 5-10. Verner, J., M.L. Morrison, and C.J. Ralph (Eds.). Wildlife 2000: modeling habitat relationships of terrestrial vertebrates. University of Wisconsin Press. Madison.

Scott, J.M, F. Davis, B. Csuti, R. Noss, B. Butterfield, C. Groves, H. Anderson, S. Caicco, F. D'Erchia, T.C. Edwards, Jr., J. Ulliman and R.G. Wright. 1992. Gap analysis: a geographic approach to protection of biological diversity. *Wildl. Monogaphs* 123:1-41.

Sharp, B. 1990. Population trends of Oregon's neotropical migrants. *Oregon Birds* 16:27-42.

Short, H.L. 1988. A habitat structure model for natural resource management. *J. Environ. Manage.* 27:289-305.

Spears, G.S., F.S. Guthery, S.M. Rice, S.J. Demaso and B. Zaiglin. 1993. Optimal seral stage for northern bobwhites as influenced by site productivity. *J. Wildl. Manage.* 57:805-811.

Thomas, J.W. (Ed.). 1979. Wildlife habitats in managed forests: the Blue Mountains of Oregon and washington. U.S. Dep. Agric. For. Serv., Agric. Handb. No. 553. U.S. Govt. Print. Off. Washington, DC.

Thomas, J.W. (chairman), E.D. Forsman, J.B. Lint, E.C. Meslow, B.R. Noon and J. Verner. 1990. A conservation strategy for the northern spotted owl. Report of the Interagency Scientific Committee to address the conservation of the northern spotted owl. U.S. Gov. Print. Office. Washington, DC.

Thomas, J.W. (leader) and M.G. Raphael et al. 1993. Viability assessments and management considerations for species associated with late-successional and old-growth forests of the Pacific Northwest. The report of the Scientific Analysis Team. U.S. Gov't Printing Office: 1993-791-566 Reg. No. 10.

Turner, M.G. 1990. Spatial and temporal analysis of landscape patterns. *Landscape Ecology* 4:21-30.

USDA Forest Service. 1993. Summary: National hierarchial framework of ecological units for ecosystem classification (draft). USDA Forest Service. Washington, DC.

Van Horne, B. 1986. Summary: when habitats fail as predictors–the researcher's viewpoint. In: pp. 257-258. Verner, J., M. L. Morrison, and C.J. Ralph (Eds.). Wildlife Habitat 2000. Modeling habitat relationships of terrestrial vertebrates. University Wisconsin Press. Madison.

White, G.C. and R.A. Garrott. 1986. Effects of biotelemetry triangulation error on detecting habitat selection. *J. Wildl. Manage.* 50:509-513.

Wilcove, D.S. 1985. Nest predation in forest tracts and the decline of migratory songbirds. *Ecology* 66:1211-1214.

Worton, B.J. 1989. Kernel methods for estimating the utilization distribution in home-range studies. *Ecology* 70:164-168.

Young, J.A., W.A. Heise, K.B. Aubry and J.F. Lehmkuhl. 1993. Development and analysis of a landscape-level GIS database to assess wildlife use of managed forests. Symp. geographic information systems in forestry, environment and natural resources management.

An Ecological Framework for Planning for Forest Health

Jonathan B. Haufler

ABSTRACT. Concerns for maintaining biodiversity have led to the adoption of ecosystem management as the paradigm for federal land management. This approach will identify desired future conditions as the goal for management, based on ecological objectives for a given landscape. Some management efforts attempt to identify desired future conditions based on existing successional stages as defined by a classification of overstory vegetation types. Such an approach ignores most of the underlying ecological parameters of the landscape, and is inadequate for identifying past disturbance regimes and future successional pathways. An assessment of desired future conditions based on an ecological classification system is essential to overcome these inadequacies.

The strategy proposed in this paper uses an appropriate ecological land classification, based on either ecological land types or habitat types, included in a broader hierarchical classification system. It also uses a vegetation map of existing overstory vegetation. These two maps are overlaid to generate polygons of ecological units that can then be used to create an ecosystem diversity matrix. Each polygon (stand) can be evaluated as to its composition and structure relative to its possible placement within the ecosystem diversity matrix through comparisons with historical ranges of variability.

The overall ecosystem diversity matrix can then be examined in

Jonathan Haufler is Manager of Wildlife and Ecology, Boise Cascade Corporation, P.O. Box 50, Boise, ID 83728.

The author thanks L.L. Irwin for providing considerable input to this manuscript, and G. Aplet, R. Steele, and D. Ferguson for providing thoughtful reviews.

[Haworth co-indexing entry note]: "An Ecological Framework for Planning for Forest Health." Haufler, Jonathan B.. Co-published simultaneously in the *Journal of Sustainable Forestry* (The Haworth Press, Inc.) Vol. 2, No. 3/4, 1994, pp. 307-316; and: *Assessing Forest Ecosystem Health in the Inland West* (eds: R. Neil Sampson and David L. Adams) The Haworth Press, Inc., 1994, pp. 307-316. Multiple copies of this article/chapter may be purchased from The Haworth Document Delivery Center [1-800-3-HAWORTH; 9:00 a.m. - 5:00 p.m. (EST)].

307

terms of the distribution of successional stages within each habitat type or ecological land type. The goal should be to maintain at least adequate ecological representation of all successional stages within each habitat type that occurred historically, based on past disturbance regimes. Adequate ecological representation is defined as sufficient size and distribution of inherent ecosystems to maintain viable populations of all endemic species dependent on these ecosystems. This approach can maintain and enhance regional biodiversity, but also maintain flexibility in land management options.

INTRODUCTION

Ecosystem management has recently been accepted as the management approach to be implemented on Federal lands. One goal of ecosystem management is to maintain or enhance forest or ecosystem health. "An ecosystem is healthy if it maintains its complexity and resiliency" (U.S. Forest Service 1993). How do we know when ecosystems are maintaining their complexity and resiliency? Complexity may be measured by the maintenance of the overall biodiversity of a region. Resiliency can be measured by determining if ecosystems are following a predictable sequence of successional change following a perturbation, based on identifiable past sequences. Implementing management programs that incorporate measures of biodiversity and resiliency are therefore important to achieving the goal of maintaining or enhancing forest health.

Biodiversity can be defined as the full range of variety and variability within and among living organisms and the natural associations in which they occur. Biodiversity includes the consideration of ecosystem diversity, species diversity, and genetic diversity. As such, conservation of biodiversity potentially involves all landtypes and ownerships. A premise for the maintenance or enhancement of biodiversity is that our native flora and fauna, the focus of biodiversity, are adapted to the naturally occurring complex of ecosystems that have existed due to the physical environment and the disturbance history of a region. Given this premise, how do we manage for biodiversity, and thus forest health, while still considering the diverse human demands for forest resources? Ecosystem management planning proposes to do this by determining desired future conditions, and developing management strategies to obtain these conditions. The question is one of setting desired future conditions in a way that will achieve the maintenance of biodiversity.

Resiliency implies that successional pathways, within certain bounds or limits identified by historical ranges of variability (Morgan et al. 1994), can be identified for any given stand of vegetation. Existing vegetation in a stand can then be evaluated relative to this expected range of stand conditions. To understand and assess stands for resiliency and landscapes for complexity, an ecological perspective of the landscape and its ecosystems is needed.

ECOLOGICAL FRAMEWORK

Vegetation typing has often been the focus of land management programs. This offers advantages, as existing vegetation can be measured, predicted for future growth, and modified by management. However, basing management decisions on vegetation types suffers from the inability to accurately identify successional pathways or natural disturbance factors typically influencing a given stand. Further, vegetation typing excludes the influences of the physical environment (e.g., climate, soils, topography) on the inherent diversity of the landscape (Haufler and Irwin 1993). Vegetation typing tends to lump many different ecosystems into single classifications, producing overly simplistic maps of complex ecological landscapes. For example, aspen (*Populus* spp.) an intolerant early successional species, can occur on a number of habitat types, but would be typed the same based on vegetation typing.

The foundation of a sound biodiversity conservation strategy, one that will maintain forest health, is the conservation of ecosystem diversity. If we can maintain adequate ecological representations of the diverse ecosystems inherent in our landscapes, then we can go a long way toward maintaining species diversity and genetic diversity. Hunter (1990) suggested this as a coarse filter approach to the conservation of biodiversity. An essential basis for the management of ecosystem diversity is an ecological land classification system that identifies inherent variability in the physical environment that then influences the plant and animal associations for any given site. This classification should be in conjunction with vegetation typing of the landscape.

There are a number of ecological land classifications including habitat typing, developed by Daubenmire (1968) and others, as well as the hierarchical ecological land classification system being developed by the U.S. Forest Service (ECOMAP 1993). Any of these systems can be utilized if the units of land are described in a hierarchical fashion, with each succeeding level becoming more homogeneous to satisfy increasing specificity (i.e., Table 1). Ecological land classifications (ELC) allow for future successional pathways to be identified, provide better estimates of successional disturbance factors which have affected specific areas, reduce variance

TABLE 1. Hierarchical framework for an ecological land classification system (Ecomap, 1993).

Level	Scale	Descriptors
Domain	>10000s km^2	Large climatic region
Sections	10s to 1000s km^2	Geological formations within domains
Landtype association	100s to 1000s ha	Local climate and geological influences
Landtype	100s ha	Soil and plant association similarity
Landtype phase	<100s ha	Site fertility and plant association similarity

estimates for understory vegetation variables, and allow for ecological assessments of the landscape.

To utilize an ELC to manage for ecosystem diversity, the various habitat types or ecological land types need to be identified and mapped. Each habitat type has an associated array of successional stages of potential vegetation and productivity, and will support a corresponding array of animal species. Each array of species and successional stages will be different, to at least some degree, from other habitat types.

A separate map of the existing vegetation will supply information on the current seral stage present at any given site. When the existing vegetation type map is overlaid on the habitat type map, the resulting polygons represent ecological units (Figure 1) capable of describing existing vegetation for both overstory and understory characteristics, predicting ecological processes of the site such as successional pathways, and allowing estimates of the effects of specific forest planning activities to be made (Roloff and Haufler 1992).

DEFINING DESIRED FUTURE CONDITIONS

One way of describing a desired future condition for ecosystem diversity is to create an ecosystem diversity matrix listing the areal extent of habitat types by their various seral stages within the landscape of interest (Table 2). This matrix allows for the quantification of total amounts of each seral stage for each habitat type occurring across the landscape of

FIGURE 1. Overlay of vegetation type map with a habitat type map to generate a map of ecological units for use in planning for forest health.

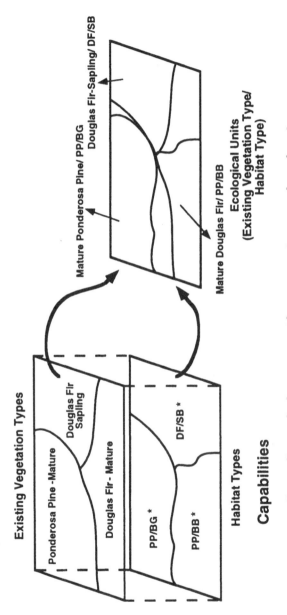

Existing Vegetation Types

Ponderosa Pine - Mature

Douglas Fir Sapling

Douglas Fir - Mature

DF/SB *

PP/BG *

PP/BB *

Habitat Types

Mature Ponderosa Pine/ PP/BG
Douglas Fir-Sapling/ DF/SB

Mature Douglas Fir/ PP/BB

Ecological Units
(Existing Vegetation Type/
Habitat Type)

Capabilities

1. **Describe existing vegetation, overstory and understory.**
2. **Predict successional change.**
3. **Identify natural disturbance factors.**

* PP/BG = Ponderosa Pine/Bunchgrass; PP/BB = Ponderosa Pine, Bitterbrush;
DF/SB = Douglas Fir, Snowberry

TABLE 2. Example of ecosystem diversity matrix for regional landscape analysis and planning. Stand ages are listed, but identifying specific seral stages for each habitat type would be better.

Stand Age (Seral stages)	Habitat Type							
	PP/BG[1]	PP/BB[1]	PP/WN[1]	DF/SB[1]	DF/MM[1]	GF/S[1]	GF/MM[1]	AF/BG[1]
01-10	2.38	*3.68	8.27	5.48	0.40	0.00	0.15	0.60
11-20	6.19	1.79	8.12	8.22	0.00	0.00	0.14	0.21
21-40	8.54	2.66	12.55	31.01	0.42	0.65	0.28	2.44
41-60	4.03	1.46	14.73	20.33	0.22	0.43	0.37	0.86
61-80	0.30	1.01	10.95	8.91	0.34	0.31	0.21	0.32
81-100	0.00	0.62	1.61	3.21	0.41	0.20	0.11	0.54
101-120	0.00	0.00	0.28	1.36	0.11	0.02	0.00	0.37
121-140	0.00	0.00	0.00	0.00	0.00	0.01	0.00	0.45
141-160	0.00	0.00	0.00	0.00	0.00	0.00	0.00	0.00
>160	0.00	0.00	0.00	0.00	0.00	0.00	0.00	0.05
TOTAL	21.44	11.22	56.51	78.52	1.90	1.62	1.26	5.84

*Numbers are thousands of acres.

1. PP/BG = Pondersoa Pine/Bunchgrass; PP/BB = Ponderosa Pine, Bitterbrush; PP/WN = Pondersa Pine, Western Needlegrass; DF/SB = Douglas-fir, Snowberry; DF/MM = Douglas-fir, Mountain Maple; GF/S = Grand Fir, Spirala; GF/MM = Grand Fir, Mountain Maple; AF/BG = Subalpine Fir, Beargrass. (Steele et al. 1981)

interest, and provides a basis for identifying the steps necessary to maintain or enhance ecosystem diversity. Ages of stands are represented in this table, although a better representation would be to list the various seral stages within each habitat type.

How will managers identify the desired future mixes of seral stages in order to maintain healthy and diverse ecosystems? An assessment of historical disturbance regimes (disturbance type, frequency, intensity, and extent) and the appropriate structural variation within these successional stages for each habitat type will provide the basis for defining desired future conditions. Information on the historical range of variability can be used to help identify successional stages that were in significant abundance. Habitat types that historically supported substantial acreages of old growth should be managed to provide adequate amounts and distributions of later seral stages to support obligate late successional species. Habitat types subject to frequent major disturbances, such as those supporting lodgepole pine *(Pinus contorta),* would need to be targeted to maintain sufficient acreage in early successional stages to provide for dependent species.

Watersheds containing high percentages of habitat types susceptible to frequent major disturbances should not be regulated to maintain substantial percentages in late successional stages, as this would not be justified using input from historical ranges. Watersheds need to display a range of successional compositions, consistent with historical ranges of variation, so that diversity of watershed conditions are also maintained. Not all watersheds need to be managed for historical ranges of variability, but not all watersheds should be targeted for uniform successional composition either.

How much of each seral stage for a particular habitat type in any landscape of interest needs to be maintained, and in what distribution, may still produce discussions and potential disagreements, but at least the discussion is focused on specific areas and features of the landscape with an appropriate ecological perspective. In this way, the ecosystem diversity matrix can be evaluated relative to a range of desired future conditions. The goal should not be to return to some previously occurring set of conditions or historical range of variability, but to provide at least adequate ecological representation of all successional stages in each habitat type to support viable populations of the species that depend on each type of ecological unit (Figure 1). If we plan correctly, we should provide for the habitat requirements of all of the native flora and fauna of the area, while still allowing flexibility in our management programs for multiple use benefits. The ecosystem diversity matrix can be used as a planning tool to produce desired conditions over time, with adequate consideration given to the dynamic nature of ecosystems and the range of possible management options. The matrix forces planners to recognize the dynamic processes at work in the landscape, and to incorporate a temporal component into the planning process.

Management for biodiversity through an ecosystem diversity approach should have the goal of providing adequate ecological representation of all of the seral stages that would naturally occur in significant acreages within each of the habitat types across the landscape. Once these representative ecosystems are adequately provided for, remaining acres within each habitat type could be managed for any of the alternative seral stages for that habitat type, thus maintaining management flexibility.

The second use of the ecosystem diversity matrix would be to define acceptable ranges of conditions in terms of compositions and structures for each ecological unit identified in the matrix. Stands within a habitat type that met appropriate criteria, as defined by an assessment of historical ranges of variability for that ecological unit, could be tallied in the matrix. Those stands that do not meet these criteria would be considered as outliers. These stands may have different management objectives in order to meet other societal needs, and therefore be in an acceptable condition, or they may be unhealthy, due to past influences on disturbance regimes. Those that are considered unhealthy could be manipulated to produce resilient conditions.

MANAGEMENT OPTIONS

Identification of disturbance regimes associated with each habitat type can help reveal appropriate management tools for satisfying ecosystem management and forest health objectives. These management tools need to include silvicultural options which, if judiciously applied, will produce ecological responses and patterns that simulate, if not mimic, those produced by natural disturbances within a habitat type. Assessment of the historical range of variability for conditions within an ecological unit can help provide the basis for determining appropriate methods of achieving ecological objectives. Some areas will need to be protected from disturbances and allowed to succeed to late-successional or old growth stages. In some of these areas, late-successional stages may be obtained more rapidly via silvicultural practices designed to produce a specified set of compositional and structural conditions.

Other areas will need to be disturbed to create or maintain early successional stages or healthy ecosystems. These disturbances may be by natural tools, such as fire, or may be substituted with other appropriate management tools. Silvicultural manipulations can be used to produce many ecological objectives in stands, while also providing for resource use. In fact, in some stands in the Inland West, silvicultural manipulations may be needed prior to the use of fire, due to past human alteration of fire regimes, if historical ranges of variability are to be restored (Agee 1993). Again, the historical ranges of structure and composition will allow for ecological guidelines for achieving desired stand characteristics. Such management actions need to be evaluated and adjusted through adaptive management experiments designed to reveal the best options for achieving the stated objectives.

Timber harvesting has been targeted as having negative effects on biodiversity (Norse 1990; Doppelt et al. 1993), and when used indiscriminately, it can. However, timber harvest can also be an effective tool for maintaining or restoring biodiversity and ecosystem health when used with both ecological and economic objectives in mind. Ecosystem management will require blending an understanding of natural disturbance regimes with appropriate management tools if we are to provide for both biodiversity and resource use. Historical ranges of variability provide essential information in the understanding of natural disturbance regimes, and for evaluating the status and health of existing stands of vegetation. With this understanding as a guide, but not a goal, and by using an ecologically-based approach to land management planning, we can plan for forest health while maintaining resource use for human needs.

REFERENCES

Agee, J.K. 1993. Fire ecology of Pacific Northwest forests. Island Press. Washington, DC. 493p.

ECOMAP. 1993. National hierarchical framework of ecological units. USDA Forest Service. Washington, DC. 18p.

Daubenmire, R. 1968. Plant communities: a textbook of plant synecology. Harper and Row. New York. 300p.

Doppelt, B., M. Scurlock, C. Frissell, and J. Karr. 1993. Entering the watershed. A new approach to save America's river ecosystems. Island Press. Washington, DC. 462p.

Haufler, J.B. and L.L. Irwin. (1993). An ecological basis for planning for biodiversity and resource use. Proc. Internatl. Union Game Biol. Halifax, Nova Scotia. In press.

Hunter, M.L., Jr. 1990. Wildlife, forests, and forestry. Principles of managing forests for biodiversity. Prentice Hall. Englewood Cliffs, NJ. 370p.

Morgan, P., G. H. Aplet, J. B. Haufler, H. C. Humphries, M.M. Moore and W. D. Wilson. 1994. Historical range of variability: A useful tool for evaluating ecosystem change. In: R. Neil Sampson and Dave L. Adams (Eds.). Assessing Forest Ecosystem Health in the Inland West. *Journal of Sustainable Forestry* 2(1/2):87-111.

Norse, P.H. 1990. Ancient forests of the Pacific Northwest. Island Press. Washington, DC. 327p.

Roloff, G.J. and J.B. Haufler. (1993). Forest planning in Michigan using a GIS approach. Proc. Internatl. Union Game Biol. Halifax, Nova Scotia. In press.

Steele, R., R.D. Pfister, R.A. Ryker and J.A. Kittams. 1981. Forest habitat types of Central Idaho. USDA Forest Service. Gen. Tech. Rept. INT-1 14. 138p.

U.S. Forest Service. 1993. Framework for ecosystem management. Draft. Washington, DC. 8p.

MANAGEMENT AND POLICY

Forest Health Management Policy:
A Case Study in Southwestern Idaho

Keith A. Blatner
Charles E. Keegan III
Jay O'Laughlin
David L. Adams

ABSTRACT. The two national forests in southwestern Idaho–the Boise and Payette National Forests–have both recently experienced five years of "catastrophic" timber mortality. That is, annual mortality exceeded annual growth on lands suitable for timber production.

Keith A. Blatner is Associate Professor, Department of Natural Resource Sciences, Washington State University, Pullman, WA 99164-6410.

Charles E. Keegan III is Director of Forest Industry Research and Associate Professor, Bureau of Business and Economics Research, University of Montana, Missoula, MT 59812.

Jay O'Laughlin is Director, Idaho Forest, Wildlife and Range Policy Analysis Group, and Adjunct Professor; David L. Adams is Professor; Department of Forest Resources, College of Forestry, Wildlife and Range Sciences, University of Idaho, Moscow, ID 83844-1133.

[Haworth co-indexing entry note]: "Forest Health Management Policy: A Case Study in Southwestern Idaho." Blatner, Keith A. et al. Co-published simultaneously in the *Journal of Sustainable Forestry* (The Haworth Press, Inc.) Vol. 2, No. 3/4, 1994, pp. 317-337; and: *Assessing Forest Ecosystem Health in the Inland West* (eds: R. Neil Sampson and David L. Adams) The Haworth Press, Inc., 1994, pp. 317-337. Multiple copies of this article/chapter may be purchased from The Haworth Document Delivery Center [1-800-3-HAWORTH; 9:00 a.m. - 5:00 p.m. (EST)].

Boise Cascade Corporation lands lie between and, in some cases, are interspersed with these two national forests, and did not experience such high levels of mortality. The major difference in management strategies is that the company practices stand density control on all its lands; the national forests have not historically placed the same emphasis on stocking control. The Payette and Boise National Forests developed different strategies for approaching their tree mortality problems. The Payette placed a priority on developing a comprehensive treatment plan for the affected acres in the roaded portions of the suitable base. The Boise opted for a three phase approach, which stressed rapid salvage efforts, followed by cultural treatments designed to improve the vigor of remaining stands, and a comprehensive public education program. In each case, the U.S. Forest Service was confronted with appeals or litigation due to the public mistrust of such efforts.

INTRODUCTION

Forest health conditions on the two national forests in southwestern Idaho–the Boise and the Payette–were such that annual mortality exceeded annual growth in the late 1980s and early 1990s (O'Laughlin 1994). Such a situation was defined by McGuire (1958) as catastrophic timber mortality.

Thousands of trees killed by defoliating insects, bark beetles and wildfire created this situation. Dead trees lose economic value rapidly. Because the best way to capture as much of the economic value as possible is to implement salvage logging operations as soon as possible, in recent years salvage sales largely replaced green timber sales on the Boise and Payette National Forests. Other cultural treatments, including the thinning of overstocked stands and slash removal, were combined with the salvage of dead material to enhance the health of the forest and to reduce the risk of costly wildfire.

The salvage of dead material and the application of cultural treatments were opposed by some environmental groups. These groups have interpreted the salvage sale programs as "business as usual" for the Forest Service, and just another scheme to "get the cut out." Salvage sales on the Payette National Forest were extensively appealed; the salvage efforts on the Boise National Forest are currently under litigation as this is being written.

In contrast, Boise Cascade lands lying between the two national forests offer a sharp contrast to adjoining national forest lands, with minimal losses due to insect attack. The sharp contrast between these lands appears to stem from the company's long-term emphasis on stocking control, the removal of trees at risk of insect attack, and the prompt salvage of mortality in these managed stands. These actions result in a younger age class structure than on the neighboring national forests, subjecting managers to criticism due the lack of an old-growth component and other factors.

This paper describes some of the recent forest conditions on the Payette and Boise National Forests and Boise Cascade Corporation lands in southwestern Idaho and the strategies used in addressing the salvage of dead and dying material and other efforts to confront forestry problems. Problems encountered in implementing these approaches are described and some suggestions are made for resource managers facing similar forest health problems. Information was obtained from U.S. Forest Service documents, other published sources, and through independent interviews with at least three representatives from each of the three organizations during the summer of 1992.

DESCRIPTION OF LAND AREAS

Figure 1 is a map of the land areas in southwestern Idaho managed by the three organizations featured in this case study. In some areas Boise Cascade lands are intermingled with national forest lands.

The Payette National Forest includes approximately 2.4 million acres. Almost 33% of these lands lie within wilderness areas. Of the remaining land, 135,900 acres are non-forested, and 530,900 acres are not physically suited for timber production, leaving 821,000 acres (or 34% of the land base) of tentatively suited forest land under the 1986 Forest Plan. The average annual allowable timber sale quantity for the Payette was initially projected at 80.9 million board feet, based on the intensive management of 431,100 acres of suited lands (Payette National Forest 1986). The inventory of the forest completed in 1991 was conducted only on these 431,000 acres (18% of the lands in the forest) because the remaining tentatively suited lands have effectively been removed from timber production through agreements with affected interest groups.

The Boise National Forest includes 2.65 million acres (USDA Forest Service 1992b). Two wilderness areas are located partly within the Boise's proclaimed boundaries, and account for 665,492 acres. The Boise also has 38 roadless areas totaling 1,209,000 acres. A projected annual allowable sale quantity of 85 million board feet is based on 656,114 acres of the Boise's 1,317,941 acres of lands identified as suitable for timber production (Boise National Forest 1990).

Boise Cascade Corporation owns 195,000 acres in southwestern Idaho

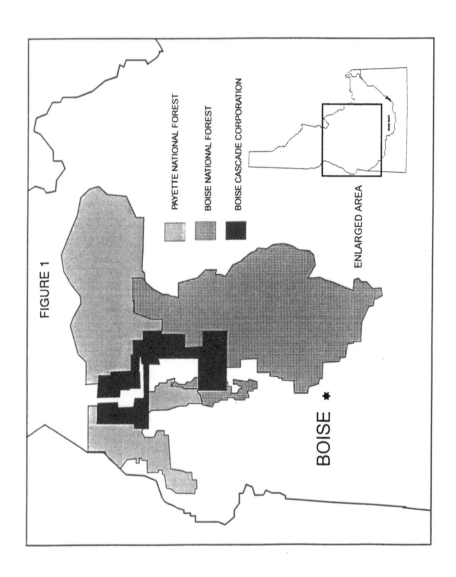

FIGURE 1

PAYETTE NATIONAL FOREST

BOISE NATIONAL FOREST

BOISE CASCADE CORPORATION

BOISE ★

ENLARGED AREA

in the vicinity of the two national forests (Figure 1). Forest types include 125,000 acres of grand fir (*Abies grandis*) or wetter habitat types and 70,000 acres of the drier Douglas-fir (*Pseudotsuga menziesii* var. *glauca*) habitat types. The company did not wish to release information on timber volume. The tree species by volume on Boise Cascade timberlands are Douglas-fir (30%), grand fir (30%), ponderosa pine (*Pinus ponderosa*) (30%), and smaller quantities of lodgepole pine (*Pinus contorta*), subalpine fir (*Abies lasiocarpa*), and western larch (*Larix occidentalis*) (10%).

It is not possible from national forest inventory data to similarly identify species composition. For example, the Payette National Forest uses a mixed conifer category that characterizes 66% of the suitable timberlands. By definition, 75% of the trees in a mixed conifer forest type are ponderosa pine, Douglas-fir, grand fir, or western larch. There are 3 subdivisions of mixed conifer based on productivity class; 18% of the suitable timberlands are high-site mixed conifer, 43% are medium-site, and 5% are low-site. Further subdivisions of productivity classes are the seven strata representing stand condition. The remaining 34% of the Payette's suitable timberlands are predominantly Engelmann spruce (6%), subalpine fir (6%), lodgepole pine (6%), and seedling/sapling/poletimber stands, clearcuts, and burned or undifferentiated areas (16%).

TREE MORTALITY

Outbreaks of western spruce budworm (*Choristoneura occidentalis*), mountain pine beetle (*Dendroctonus ponderosae*), western pine beetle (*Dendroctonus brevicomis*), Douglas-fir beetle (*Dendroctonus pseudotsugae*), spruce beetle (*Dedroctonus rufipennis* Kirby), Douglas-fir tussock moth (*Origyia pseudopsugata*), and fir engraver (*Scolytus ventralis* LeConte) have all occurred in southern Idaho over the past decade on both public and private forest land (O'Laughlin et al. 1993). Most of these insect species are normally present in the forests of southwestern Idaho, however, population levels remain at endemic levels until conditions are suitable for a major outbreak. Insect outbreaks on the two national forests in recent years have occurred primarily in concentrated locations corresponding to species mix, habitat type, and stand conditions.

Payette National Forest

The 1991 inventory of the Payette National Forest, confined to the 413,000 acres of suitable timberlands, showed recent mortality levels four times that of the late 1970s. The mortality from insect epidemics is expected to decline in the future, but the high mortality levels over the past five years have resulted in a loss in net growth, and a reduction in the volume of merchantable timber in the suitable timberland base. As might be expected from the analysis of growth and mortality rates on the Payette, a substantial proportion of the forest has mortality exceeding growth. In total, almost 46% of the suitable timberland on the Payette in 1991 was in this condition (O'Laughlin et al. 1993).

Forest officials expect spruce beetles to kill almost all of the commercial volume of Engelmann spruce. This is the predominant species on approximately 6% of the suitable timberlands. They expressed some concern for the health condition of the stands affected by high rates of mortality, but felt the mortality situation had peaked. They recognized the overstocked condition of stands had led to health problems, and suggested stand density index as an indicator of such potential situations.

Boise National Forest

Detailed inventory data for the Boise National Forest are not available because of the differences in the timing of individual forest inventories. A new inventory of the Boise is not scheduled until 1995. But because of the current insect and wildfire situation, Boise National Forest staff members have estimated that the average annual mortality volume from 1988 to 1992 exceeded annual growth. More than 400,000 trees died from bark beetle attacks on the forest from 1988 to 1991. In addition, the Douglas-fir tussock moth defoliated more than 240,000 acres during 1990 and 1991 (Boise National Forest 1992d). The increase in Douglas-fir tussock moth attacks tend to weaken the trees and make them more susceptible to the Douglas-fir beetle. The predominance of older stands of Douglas-fir on the Boise continues to make Douglas-fir beetles a major concern. Wildfires have also killed substantial number of trees (O'Laughlin 1994).

Boise Cascade Corporation

The company estimated their mortality losses, from all causes, as ranging from 75,000 to 150,000 trees per year during the period 1987-1992, with salvage efforts recovering 8 to 10 million board feet, or roughly 45,000 to 56,000 dead and dying trees per year during the same period.

It is difficult to compare mortality losses across all three ownerships because of differences in the inventory systems used and the timing of measurements. Boise Cascade could estimate the mortality volume, but was unable to provide a breakdown of mortality from insect attack as opposed to other causes. Among those interviewed, it was generally agreed that Boise Cascade lands had experienced the lowest levels of mortality from insect attack. However, there are important differences among the three organizations. The national forests are managed for multiple uses, determined under a forest plan as required by the National Forest Management Act (NFMA) of 1976. Some of these lands are designated as suited for timber production, where timber can be the dominant use, but all other uses and amenities are still important. Boise Cascade lands are managed primarily for timber production, but other uses and amenities are important, including visual considerations, raptor sites, public use, and water quality.

RESPONSE TO RECENT INSECT OUTBREAKS

Individual national forests throughout the West have tended to develop their own responses to forest insect outbreaks, with direction from their forest plan, the regional office, and the Forest Pest Management section of State and Private Forestry in the U.S. Forest Service. This approach is consistent with the agency's decentralized planning process and allows each forest to respond to specific situations such as high levels of timber mortality. Many factors govern the response of a national forest to management issues, including access to the affected areas (that is, the extent of roading), the views of various interest and user groups, ecological factors, environmental laws (including the Endangered Species Act), professional judgment, and the forest plan. Boise Cascade's response to insect outbreaks has been to immediately salvage dead and dying trees to minimize losses in value resulting from degradation. Again, the difference in approaches reflects the management objectives of the organizations and the underlying laws governing the management of public and private forest lands.

Payette National Forest

Beginning in 1988, the Payette made a major shift in its harvest program to salvage accessible Engelmann spruce mortality on the suitable timberland base and by the end of 1993 expected to have salvaged 200 million board feet. Salvage efforts on the Payette followed normal timber sale procedures, including the preparation of an environmental impact statement (EIS). The salvage program is now nearly complete in the roaded areas, but portions of the unroaded timber base have extremely heavy mortality and most of it will not be harvested.

The Payette placed a high priority on developing a comprehensive silvicultural and fuels treatment plan for the affected areas, and on the development of the forest's transportation system as part of their salvage program. This comprehensive prescription called for not only the salvage of dead and dying timber, but also the harvest of timber and other vegetative treatments under the selected regeneration system. Slash treatment and the construction of roads were also a part of the effort. Given the magnitude and speed of the epidemic and the constraints faced in preparing timber sales, resource managers felt that only a small part of the infested acreage could be effectively treated. The Payette, therefore, concentrated their salvage program on their roaded timber base.

The rationale for this approach stemmed, in part, from the Payette National Forest Plan. Based on public input during the development of the forest plan, the forest chose to maximize back country recreation opportunities and concentrate timber production on the best sites. An analysis of yield tables indicated the forest could sustain historic harvest levels on about 431,100 acres, or about 18% of the entire forest. The plan was to take a tree farm approach, managing timber as the dominant use on those lands, with most of the rest of the forest to be managed primarily for nontimber product outputs (Payette National Forest 1986). However, many of the acres designated primarily for timber production were unroaded, with about 30% of the first decade's timber sale program planned to come from roadless areas. Hence, intensive timber management and access were considered crucially important to fulfill forest plan goals and objectives.

Planners felt the preparation of the EIS's needed for entry into the roadless areas of the timber base and the likelihood of appeals or litigation would have led to considerable delay between the time the sale was proposed and the time the timber could be harvested. The reduction of such probable delays was viewed as crucial for the economic feasibility of salvage sales, especially because of the rapid value loss associated with Engelmann spruce relative to other species in the region. The loss in value and increased cost of entering roadless areas would in all likelihood have resulted in these sales being substantially below cost, given stumpage prices during most of the salvage period.

In attempting to implement this approach to combat its mortality situation, the Payette encountered two complicating factors. First, critics have focused much of their attention on the tree farm approach for managing the forest's suitable lands, as outlined in the forest plan and approved in 1986. Virtually every timber sale is now appealed. Second, because a relatively large proportion of the 431,100 suitable acres are to be managed primarily for timber production, it is difficult to prepare salvage sales or harvests of green timber at the levels projected in the Forest Plan, without violating the standards and guidelines for water quality set forth in the forest plan.

Boise National Forest

Given the large volume of dead and dying trees on the forest and the problems encountered by other national forests attempting to effectively address forest health problems, the Boise developed a forest health strategy in 1992 with three major elements (Boise National Forest 1992a):

1. . . . a short-term measure to salvage dead and dying trees from lands suitable for commercial timber harvest to recover their economic value.
2. . . . restore and improve forest health by reducing the number of trees competing for water. This longer-term measure will require thinning living trees over large acreages of commercial forest.
3. . . . share information about forest health and methods to restore and improve the resilience of the forest ecosystems.

Salvage Program–While implementing the first part of their forest health strategy, the Boise elected to shift the 1992 sale program from the harvest of green trees to dead and dying trees, and salvage 67.5 million board feet of insect-killed trees with an estimated value of $8.5 million. In 1992,87 million board feet of salvage timber was sold. To speed up the salvage of dead and dying material, the forest requested and received an exemption from appeals under 36 CFR 217.4 (a)(11). This regulation states:

> Decisions related to the rehabilitation of National Forest System lands and recovery of forest resources resulting from natural disasters or other natural phenomena, such as wildfires, severe wind, earthquakes and flooding when the regional forester or, in situations of national significance, the Chief of the Forest Service determines and gives notice in the federal record that good cause exists to exempt such decisions from review under this part.

Resource managers viewed this approach as crucial to avoid an estimated value loss of approximately 50% if the trees were left standing for one additional year on the site pending the preparation, review, and possible appeal of documents. The Boise National Forest (1992d) developed a position paper on the timber salvage and forest rehabilitation program, based in part on the support of federal legislation, specifically the Appropriations Act of 1992, which states:

> The managers urged the Forest Service to pursue a timber salvage program which will allow for the removal of maximum salvage volumes while protecting the full range of environmental values.

The Boise National Forest (1992d) concluded that the following policies and procedures reflecting the Forest Plan intent were considered applicable when implementing the insect damage recovery and rehabilitation:

* The implementation of recovery and rehabilitation activities may only occur following site specific project analysis as stipulated by the National Environmental Protection Act.
* The implementation of recovery and rehabilitation activities may occur on both suited and unsuited lands on the BNF, in occurrence with standards and management area direction contained in the Forest Plan (pages IV-57 and IV-61). The estimates are that more than 90% of the timber salvage and reforestation will occur on suited lands.
* The implementation of recovery and rehabilitation activities may occur within inventoried roadless areas, as well as roaded areas when current Forest Plan management activities provide for timber management activities.
* Salvage volume offered for sale under the Forest's policy to recover damaged timber resources will count toward the Forest's ASQ and will substitute for green volume harvested from suited lands. The Forest Plan ASQ (850 million board feet) will not be exceeded.

To gain public support for the salvage operation, the Boise National Forest made a substantial effort to involve the public in designing and reviewing the proposed action, including public meetings, letters to interested parties, and personal contacts. Although there was considerable support for salvaging the affected timber and rehabilitating the area, some interest groups and individuals expressed their preference for leaving most of the utilizable timber in the woods for long-term soil productivity and wildlife habitat purposes. Based on the scoping efforts undertaken, resource managers felt the salvage sales would be at risk of appeal, so the exemption procedure was used. The exemption process only eliminated the possibility of an appeal under the Forest Service appeals procedure; it did not eliminate the possibility of litigation based on the National Environmental Policy Act (NEPA) or other laws and regulations pertaining to the harvest of timber and its associated impacts.

Boise National Forest planners found it necessary to develop a streamlined approach for the preparation of NEPA documents and timber sale procedures if the salvage efforts were to be completed in a timely manner. Site-specific Environmental Assessments (EAs) were prepared for each of the affected areas, some of which spanned multiple drainages with scattered pockets of dead and dying timber.

The South Whitehawk Salvage Sale (Boise National Forest 1992b) provides an example of how planners dealt with small pockets of dead timber across large areas. Approximately 5 million board feet of dead and dying trees on 1,100 acres were scattered throughout the 9,500 acre Whitehawk and No Man Creek drainages. Although two inventoried roadless areas contained 202 of the 1,100 acres, no new roads were needed because the entire sale was logged by helicopter. However, 12 landings (6 new and 6 reconstructed) were needed to support the yarding operations. These landings were located outside roadless areas in order to preserve roadless characteristics. In addition to the salvage harvest, 350 acres were designated for replanting. Insect baiting and trapping would be used in and around harvest areas to reduce insect populations. The number of snags were specified: for every 100 acres, 180 snags at least 12 inches in diameter at breast height (dbh) and 11 snags larger than 20 inches dbh were to be left after logging to meet wildlife requirements and to contribute to the structural diversity of the forested ecosystem. No trees or snags were harvested from old-growth stands dedicated in the Forest Plan within the salvage area (Boise National Forest 1992b).

Dead and dying trees were not marked prior to the sale, in order to reduce the time required to prepare the salvage sales. Instead, sale contracts were designed to provide the needed information for designating trees to harvest within a specific sale area. For example, the Cayuse Point Salvage Sale (Boise National Forest 1992c) included an estimated 5 million board feet of timber located on 4,300 acres. Bidders were provided with a map of the sale area and details of the sale specifications. For example:

- In logging the subdivisions numbered 1-3 on the Sale Area Map, all dead trees (slough bark and no foliage) meeting Utilization Standards, standing and down, are designated for cutting.
- During the period June 1 through December 31:
  1. all trees retaining no green or red foliage are designed for cutting.
  2. all trees with red foliage and which exhibit bark beetle infestation as indicated by boring dust from ground level to at least 6 feet

high on the bole and/or brood galleries in the stump are designated for cutting.
3. all dead trees with sloughing bark and no foliage are designated
for cutting, except for Douglas-fir and subalpine fir with red foliage and no evidence of bark beetle infestation.

Costs and time savings during the timber sale preparation phase were at
least partly offset by increased sale administration costs and personnel
needs during the actual salvage of the timber (Boise National Forest
1992d).

Thinning Program–The second part of the Boise National Forest strategy is to restore forest health by reducing the number of trees. This will be
accomplished by thinning.

Will this program work? One benefit of a thinning program coupled
with prescribed burning became evident after the Foothills Fire in August
1992. The lesson was described by U.S. Forest Service fire scientist Bob
Mutch (1993) during an interview conducted by the Association of Forest
Service Employees for Environmental Ethics:

> The Foothills Fire burned from late August into the middle of Sep-
> tember. The fire wound up at something like 257,000 acres. It was a
> wildfire-started on BLM land and burned into the Boise National
> Forest Service. This fire, after it burned into Forest Service lands,
> burned primarily in the ponderosa pine type that also had this very
> significant encroachment of Douglas-fir over the past 50 to 60 years.
> The Foothills Fire burned at very high intensity and killed not only
> young-age stands, but killed old-age stands of ponderosa pine as
> well because it had that characteristic Douglas-fir ladder fuel present
> almost everywhere.
>
> But there was one very significant lesson that was learned on the
> Foothills Fire. There's a small 2,500-acre drainage in the Foothills
> Fire [called Tiger Creek] that had previously been thinned to im-
> prove the density of the stand to make it not quite so thick in young
> coniferous trees. And following the thinning operation, it had also
> been prescribed-burned. And so here was the opportunity to look at
> what thinning coupled with prescribed-burning did in terms of bring-
> ing the health of that stand through a very serious wildfire. That
> entire 2,500 acres came through the wildfire practically intact with
> very little mortality because the fuels had been removed through the
> thinning. Other fuels had subsequently been removed by prescribed-
> burning that left the stand in a very natural, more open condition that

we would have expected to see there in pre-settlement times. That's the kind of lesson we would like to apply on a larger scale.

From a forest health standpoint, the lesson of Tiger Creek appears straightforward. Fewer widely-spaced trees receive more sun, water, and nutrients, and appear to survive wildfire better than their more crowded counterparts in fuel-heavy forests (McLean 1993). In addition to being thinned and burned with low-intensity prescribed fire, dead and dying timber in the managed forest in the Tiger Creek drainage had been salvaged (Idaho Forest Products Commission 1993).

Sharing Information–In November 1991, American Forests and the BNF brought together a large group of citizens, agencies, user groups, industry representatives, and others to discuss a strategy for dealing with the growing concerns about the health of the BNF. The result was a letter of intent signed by a variety of partners to investigate the causes and to gather and share information.

While the partnership was taking shape, the BNF created a Forest Health Coordinator position to work with the partnership and to coordinate forest health management efforts. The first Forest Health Strategy brochure was developed in May of 1992 as a tool for informing the public about the Boise's strategy for confronting forest health issues. More than 30 presentations were made to local service groups (such as Rotary and Optimist Clubs), congressional staff persons, state and local agencies, environmental groups, industry associations, and professional societies. The staff on the BNF actively worked to get the message out to local media with newspaper and magazine articles, television segments, and radio spots. Field tours were planned and more than a dozen were conducted for interested groups.

When the partnership was formed, American Forests put together an advisory group of local citizens, interest groups, state agencies and other interested parties, including the University of Idaho, to help the partnership frame appropriate questions to guide the information gathering and sharing process. The partnership developed and sponsored a symposium in June 1992, attracting over 200 participants from federal and state agencies, interest groups and individual citizens. The partnership also developed and sponsored a workshop in November 1993 titled: An Assessment of Forest Health in the Inland West.

Other Considerations–The third part of the Boise National Forest Strategy is public education. Views of selected representatives of the environmental community are presented in the next section and indicate it may take more than education to convince some people that forest health presents problems are in need of solutions. As long as a forest health strategy, is presented as a timber management strategy it is likely to encounter strong opposition from some quarters.

In an effort to overcome the economic losses resulting from delaying salvage harvests, Rep. Larry LaRocco (D-ID) has proposed forest health legislation, in 1992 and again in 1993, that would give managers additional flexibility and resources to cope with these situations, but Congress has not yet enacted the bill into law.

In October 1993, a coalition of conservation groups filed suit to enjoin the Forest Service from exempting salvage sales from appeals (*Idaho Sporting Congress et al. v. U.S. Forest Service* CIV 93-0390-S-EJL). Although the U.S. Forest Service has been allowed to continue ongoing salvage sales for the present, the litigation has not been settled. However, it is now a moot point. The option of exempting salvage sales from appeal has been eliminated under the recently revised NFMA planning regulations.

Boise Cascade Corporation

Forest management on Boise Cascade lands is dramatically different than on Payette and Boise National Forest lands, for a number of reasons. First, unlike national forests, the company manages land with efficient timber production as the dominant use objective. Second, the company completed an extensive road network on their lands during the 1950s, thus allowing salvage logging of even relatively small volumes on a continuous basis. Third, the predominant harvesting system has been individual tree selection since the late 1940s (Boise Payette Lumber Co. 1947).

Harvest and leave tree selection criteria were adapted from Keen's (1936) classification system for ponderosa pine in the Black Hills region, which drew on the earlier work of Dunning (1928) in the Sierra Nevada. Keen specified four age groups and four vigor groups for a total of 16 classes. Keen's classification system was designed to reduce insect losses, allowing the forester to move through the forest quickly, marking high risk trees for removal.

Boise Cascade adopted the use of Keen's system beginning in the 1960s and developed similar guidelines for Douglas-fir and grand fir. Boise Cascade has also developed individual tree selection rules to aid in achieving desired stand structures to be used in combination with Keen's classification.

In the early 1970s the company developed stocking guidelines by habitat type to maximize growth and protect against bark beetle attack. These guidelines are achieved by selective harvesting, commercial thinning, and precommercial thinning. The company works to create a diversity of tree species wherever possible to reduce the risk of insect attack. Company forest managers believe this approach to tree selection and stocking control is a key factor in the relatively low levels of insect attack on their lands. The company's ability to rapidly respond to insect outbreaks over a well-developed road network is also an important factor to be considered in evaluating the relative effectiveness of this approach.

VIEWPOINTS ON FOREST HEALTH

Forest Resource Managers

Discussions with forest managers indicated a definite perceived difference in management direction and levels of forest health problems. When interviewed by Garber (1992), Dave Van De Graaff, resource manager for Boise Cascade Corporation in southwest Idaho, expressed more concern about the future condition of the forest than acquiring a short-term glut of dead timber. He expressed pessimism about the future, because he doubts the U.S. Forest Service will be able to protect enough of southwest Idaho's forest to produce timber on a sustained basis.

John Kwader, a unit forester with Boise Cascade, said he saw one crucial difference between what was done on company lands and national forests. The company trains its timber markers to select trees for harvesting, and sends them out with a paint gun only when they are experienced. It is the timber marker who will determine what the future forest will be like. The U.S. Forest Service sends part-time seasonal employees out to mark its trees with specific written instructions. Many of these seasonal employees have experience, but many do not.

Chief Regional forester for Boise Cascade, Herb Malany stated that "If you cut the old, weak, and wounded, you have a pretty good forest left." This philosophy, along with a greater emphasis on stand density management, describes the essence of the forest health management strategy for Boise Cascade. The company relies heavily on natural regeneration and, as Malany said, "with the exception of a couple of habitat types, natural regeneration is successful."

According to Malany, thinned areas near Cascade, Idaho, were "graphically different" from unthinned areas where mountain pine beetle was a problem. Indeed, Boise Cascade has a photograph of a boundary between their land and an adjacent national forest tract. Trees on the national forest side of the line are red—obviously defoliated or dead; those on the company side are green. In the Idaho City area, unthinned stands of Douglas-fir suffered much more from Douglas-fir beetle. Malany said, "We see more problems on Forest Service land" in the Idaho City area–probably due to less thinning and less logging in overstocked stands.

The company has experienced less mortality from the fir engraver beetle in thinned areas. However, they have seen greater bark beetle mortality in stands that were opened up too fast; the grand firs are damaged by sun scald, then fir engraver beetle damage increases. Company policy is to not remove more than 40% of the basal area in the first entry. This is done to reduce sun scald, and also helps to reduce windthrow.

The impression from talking with Herb Malany, Steve Fletcher (Forest Health Coordinator, Wallowa-Whitman National Forest), and several foresters on the Boise National Forest is that the U.S. Forest Service is constantly in a reactive mode, trying to deal with the aftermath of recent outbreak situations. Partially because of staffing problems, it takes all their time to prepare for actions in the problem areas, leaving no time to get ahead of the problems. The U.S. Forest Service also has much more mature and over-mature timber to deal with; Boise Cascade has cut more of their older age classes, and therefore has less material in susceptible conditions.

Selected Views from the Environmental Community

Environmental groups have reduced the forest health management issue to salvage logging. Francis Hunt, lobbyist with the National Wildlife Federation, described forest health "as nothing more than a thinly veiled attempt by industry to increase the timber cut and weaken environmental laws" (Swisher 1992).

The benefits of salvage logging would seem to be a win-win situation from a forest health perspective. Quantities of flammable fuel are removed, reducing the potential of costly wildfires, and economic value is recovered. But not everyone agrees.

Some environmental groups cite several objections to salvage logging. Some even say that salvage logging is just an excuse to help beleaguered national forest managers achieve their assigned allowable cuts. According to Gray (1992), environmentalists are worried that the U.S. Forest Service will use salvage logging as an excuse to enter roadless areas. Craig Gehrke (1993), Idaho State Director of the Intermountain Regional Office of the Wilderness Society, agreed and also expressed concern that the agency can avoid doing environmental impact statements and administrative appeals of timber sales by using salvage logging to "get the cut out." Lawson (1992) said, "I'm cynical enough to think that this recent editorial effort ["more and more stories concerning forest health"] is appearing because there is a need to sell timber salvage sales to the public as a necessary operation in order to maintain or enhance forest health."

Mike Medberry, an employee of the Idaho Conservation League, was reported to have said (Garber 1992) that the U.S. Forest Service doesn't have a clue what to do, so the agency does the only thing it knows how to do–cut down trees. In the same *Idaho Statesman* article, Ron Mitchell of the Idaho Sportsmen Coalition was said to believe that the U.S. Forest Service looks at dying trees and only sees two-by-fours going to waste. Both were said to believe the forest health issue is a ploy to open up roadless areas that the Idaho Conservation League has proposed for designation as wilderness.

Gregory Aplet (1992), Forest Ecologist with the Wilderness Society in Washington, DC, believed the forest health situation in the Inland West is being exaggerated in order to accelerate timber harvesting in the name of ecological restoration to compensate for harvest reductions on the westside of the Cascade Mountains. Aplet acknowledged a compelling case for restoration and recognized catastrophic fire risk potential, but did not see any role for salvage logging. Because logging and fire management created the problem, he said it is "a bit hasty to conclude that logging and fire management will provide the solution." Larry Tuttle, former Regional Director of the Wilderness Society based in Oregon, said "You can never remove enough [dead timber] to change the fire picture (Swisher 1993)." Similarly, Gehrke (1993) stated "Most conservationists remain unconvinced that salvage logging makes an overall healthier forest . . . a quick and dirty fix of salvage and thinning might very well be the last thing [stressed forest ecosystems] needed.

Roy Keene (1993), a consulting forester and director of the Public Forestry Foundation in Eugene, Oregon, said there really is a forest health problem, and the real debate should focus not on whether there is a problem, but on what to do about it. He asked "Can salvage logging cure the sick forest?" He didn't say yes or no, but offered six "well-proven historical standards" to guide salvage logging. He didn't say so, but these guidelines remind one of the physician's adage: do no harm.

- Forest health salvage logging should rescue, save, or heal the site, not impact it further.
- Salvage activities should focus on preserving living and growing timber and promote younger growth, particularly in shade-intolerant species.
- Salvage activities should maintain or increase productivity as well as the capital value of the stand.
- Salvage harvesting should be efficient, yet not too rushed.

- Salvage activities should help reduce fuel levels and fire hazards.
- Salvage activities should not draw further from the forest without first protecting it.

CONCLUSIONS

It is difficult to evaluate the merits of these three different approaches for addressing forest health conditions in southwestern Idaho. These management responses involve specific situations encountered during different periods and in different places. Individual forest plans, national forest policy, and the management objectives of a private corporation underlay the development of three different sets of management guidelines. When comparing the approaches employed by the three organizations, it is important to remember that each ownership faced a unique set of issues. Because all three shared the common goal of a healthy forest, something can be learned in each case.

The salvage program on the Payette National Forest preceded that on the Boise National Forest by five years, with several key differences. The stumpage values for Engelmann spruce during much of the salvage period on the Payette were very low, generally under $50 per thousand board feet, compared to 1992 stumpage values at least 4 times as much. The low stumpage prices did not allow the use of helicopter logging for entry into roadless areas without building roads. The Payette, unlike the Boise, did not seek exemptions from appeals on their salvage program because the regional office did not support the use of exemptions until such time as the Payette was well along in their salvage program. The Payette did receive added funding from the salvage share fund to cover increased costs of the salvage program.

All the resource managers interviewed expressed the need to get out ahead of the current tree mortality situation through an aggressive stand density management program. They felt current forest stocking levels were too high on many sites, given the drought conditions experienced over the past six years in southern Idaho and that density management was the key tool in controlling future insect epidemics, with or without drought conditions. All managers interviewed felt better stocking control could be accomplished through wider application of thinnings, prescribed burning, salvage logging, and other related cultural practices. Boise Cascade's use of a modified form of Keen's (1936) classification system to guide individual tree selection combined with an extensive road network may have significantly reduced the incidence of insect activity on their lands. Conversely, stocking control will not solve all of the region's forest health problems on every site. Prescriptions must be site specific and consistent with existing or potential forest health problems.

Those interviewed said it was important to start building public support for management practices designed to promote the health and vigor of forests. At the present time, they said people do not get excited until the trees are red.

Interestingly, resource managers and planners generally did *not* feel the recent insect outbreaks would substantially reduce near-term timber harvest levels below what would have occurred in the absence of these events. This is because the standing volume of timber and net growth rates are not the major limiting factors with respect to harvest levels on either the Payette or Boise National Forests. Recent developments related to a variety of other issues were felt to be far more limiting than the quantity of timber available for harvest. These include land and harvest component allocations made in the forest plan, threatened and endangered status of various salmon species that spawn in the Salmon River drainage, the unanticipated complexity of documentation related to compliance with the National Environmental Policy Act, and administrative regulations and changing public attitudes.

Forest managers in the region indicated three potential major economic losses or costs associated with the forest conditions in the area: (1) increased costs of forest management activities, particularly in the form of significantly higher fire control costs, (2) loss through decline in the value of various timber products resulting from the deterioration of wood quality in dead trees, and (3) loss through reduced net growth and a reduction in the current inventory of merchantable timber.

It is clear that there has been a substantial loss of potentially merchantable timber and there is the potential for even greater mortality through future insect outbreaks and fire. It is unclear, given the constraints under which the national forests operate, if there has been any appreciable change in the volume of timber products actually available within southern Idaho because of the recent insect and fire situation. Salvage timber sales do not add to the amount of timber removed from the national forests when harvested from suited land where dead timber replaces green timber in the allowable sale quantity of the national forests. If salvage sales are to capture the economic value, they must be made expeditiously.

Salvage sales will remain controversial with some resource managers, scientists, environmentalists, and members of the public at large. The sharply different views held by resource managers and some members of the environmental community regarding salvage sales and thinnings designed to improve forest health point to the need to develop a more participatory planning process. The proposed planning process should recognize the need for and value of collaborative efforts designed to increase the general public's understanding of and involvement in resource management decisions. Simultaneously, resource managers should be willing to learn from the public as well. After all, everyone shares a common goal–a healthy forest.

REFERENCES

Aplet, G. 1992. Forest health: ecological crisis or timber driven hype? *Forest Watch* 13(2):19-22. Cascade Holistic Economic Consultants. Portland, OR.

Boise National Forest. 1990. Land and resource management plan. Boise, ID.

———1992a. Forest Health Strategy, pamphlet. Boise, ID.

———1992b. South Whitehawk salvage sale environmental assessment. Boise, ID.

———1992c. Cayuse Point salvage sale timber sale contract. Boise, ID.

———1992d. Position paper on the rehabilitation and recovery of insect damaged resources. Boise, ID. 4 p. + attachments.

Boise Payette Lumber Company. 1947. Report to the shareholders. Boise, ID. 12p.

Daniel, T.W., J.A. Helms and F.S. Baker. 1979. *Principles of Silviculture*. 2nd edition, McGraw-Hill. New York. 500p.

Dunning, D. 1928. A tree classification system for forests of the Sierra Nevada. *Journal of Agriculture Research* 36: 755-771.

Garber, A. 1992. Forest dilemma: log it or leave it. *Idaho Statesman* (February 3). Boise, ID.

Gray, G. W. 1992. Forest health emergency threatens western forests. *Resource Hotline* 8(9):1-4. American Forests. Washington, DC.

Gehrke, C. 1993. Have past management practices affected forest health? In: Proc. Forest Health in the Inland West. USDA Forest Service, Boise National Forest, Dept. of Forest Resources, University of Idaho. Moscow.

Idaho Department of Lands and USDA Forest Service. 1991. Idaho forest pest conditions. Report No. 91-2. Boise, ID.

Idaho Forest Products Commission. 1993. Keeping the [forest] miracle alive. Pamphlet. Boise, ID. 10 p.

Keen, F.P. 1936. Relative susceptibility of ponderosa pines to bark-beetle attack. *Journal of Forestry* 34(10): 919-927.

Keene, R. 1993. Salvage logging: health or hoax? *Inner Voice*, (March/April) 5(2):1,4. Association of Forest Service Employees for Environmental Ethics. Eugene, OR.

Lawson, T. 1992. The forest doctor: spin doctoring salvage sales. *Forest Watch* 13(10):4. Cascade Holistic Economic Consultants. Portland, OR.

McGuire, J. R. 1958. Definitions. In: pp. 629-637. Timber resources for America's Future. USDA Forest Service Forest Resource Report No. 14. Washington, DC.

McLean, H. E. 1993. Lookout: the Boise quickstep. *American Forests* 99(1/2): 11-14.

Mutch, R. W. 1993. Restoring fire to the West. *Inner Voice* 5(2):10-11. Association of Forest Service Employees for Environmental Ethics. Eugene, OR.

O'Laughlin, J. 1994. Assessing forest health conditions in Idaho with forest inventory data. In: pp. 221-247. Sampson, R.N. and D.L. Adams (Eds.). Assessing Forest Ecosystem Health in the Inland West. Proceedings of the American Forests scientific workshop, November 14-20, 1993, Sun Valley, ID. The Haworth Press, Inc. New York.

O'Laughlin, J., J. G. MacCracken, D. L. Adams, S. C. Bunting, K. A. Blatner and C. E. Keegan III. 1993. Forest health conditions in Idaho, Report No. 11, Idaho Forest, Wildlife and Range Policy Analysis Group, University of Idaho. Moscow.

Payette National Forest. 1986. Land and resource management plan. McCall, ID.

Reineke, L.H. 1933. Perfecting a stand-density index for even-aged forests. *Journal of Agricultural Research* 46: 627-638.

Swisher, L. 1992. Lawmakers debate "forest health." *Lewiston Morning Tribune* (August 24):5A. Lewiston, ID.

Swisher, L. 1993. Forest health wanted by all; how to get there is the issue. *Capital Press.* (Jan. 1):9B. Salem, OR.

USDA Forest Service. 1992. Land areas of the national forest system. FS-383. Washington, DC.

Silviculture, Fire
and Ecosystem Management

Russell T. Graham

ABSTRACT. Fire and timber harvesting are two major factors affecting the development of forests in the Inland West. Prior to 1900, wildfires burned at various intervals and intensities, regenerating and tending forests creating a mosaic of conditions. After 1900, fire exclusion and timber harvesting created forests that are overstocked with trees and prone to epidemics of insects and diseases and forest replacing fires. The tools and methods developed by silviculturists for producing high value timber products can be used to manage forests for sustainability. If forests of the Inland West are going to be sustained, some type of active management other than fire exclusion is going to be needed. These management activities can best be determined by analyzing large tracts of forests in both temporal and spatial scales. These ecosystem analyses can guide the development of silvicultural systems for the long-term maintenance of ecosystems.

INTRODUCTION

Two major factors affecting the development of forests and landscapes in the Inland West are fire and timber harvesting. Prior to 1900, wildfires both intentional and natural, burned at a variety of intervals and intensities (Habeck and Mutch 1973; Wellner 1970). These fires ranged from fre-

---

Russell T. Graham is Research Silviculturist, U.S. Department of Agriculture Forest Service, Intermountain Research Station, Moscow, ID 83843.

[Haworth co-indexing entry note]: "Silviculture, Fire and Ecosystem Management." Graham, Russell T. Co-published simultaneously in the *Journal of Sustainable Forestry* (The Haworth Press, Inc.) Vol. 2, No. 3/4, 1994, pp. 339-351; and: *Assessing Forest Ecosystem Health in the Inland West* (eds: R. Neil Sampson and David L. Adams) The Haworth Press, Inc., 1994, pp. 339-351. Multiple copies of this article/chapter may be purchased from The Haworth Document Delivery Center [1-800-3-HAWORTH; 9:00 a.m. - 5:00 p.m. (EST)].

339

quent surface fires in ponderosa pine (*Pinus ponderosa*) forests to infrequent stand replacing fires in the cedar/hemlock (*Thuja plicata/Tsuga heterophylla*) zone (Arno 1980). Fires influence stand regeneration, composition, and structure. In fact, most forests of the Inland West are dependent on fire for their existence (Habeck and Mutch 1973; Agee 1993). Since 1900, fire control practices have excluded fire from many forests, making them more prone to epidemic outbreaks of insects, diseases, and highly susceptible to catastrophic fire (Agee 1993).

At approximately the time when fire exclusion began in Inland forests, timber harvesting started to increase. At first the valley bottoms along railroads and major rivers were harvested to supply products for expanding communities (Hutchison and Winters 1942). Forests around mining centers were often heavily harvested sometimes to alpine tundra (Pfister et al. 1977). Railroads often selectively harvested many stands for ties and fuel wood (Steele et al. 1983). The effects of this timber harvesting on forest development are still evident today.

The development of scientifically sound silvicultural practices for the production of wood crops also started in the early 1900s. In northern Idaho, the Priest River Experimental Station was established in 1911. It emphasized silvicultural treatments for regenerating and tending Northern Rocky Mountain forests (Wellner 1976; Haig et al. 1941). In 1908 the Fort Valley Experimental Forest in northern Arizona was established to provide similar information for managing ponderosa pine forests (Pearson 1950).

The information developed at these research centers included regeneration methods (clearcutting, seed tree, shelterwood, selection), site preparation (mechanical, fire, chemical), intermediate treatments (weedings, cleanings, thinnings, liberation cuttings) and all of their variants. Studies at these two locations and subsequently at additional sites provided a large amount of information for producing forest crops throughout the Inland West. This knowledge can also be applied to the manipulation of vegetation to create a variety of stands and landscapes for other uses such as wildlife habitat, water quality, esthetics, and biodiversity.

SILVICULTURE AND TIMBER PRODUCTION

During the middle to late 1800s, high valued ponderosa pine was the primary species harvested in the Inland West (Hutchison and Winters 1942; Pearson 1950). This resource helped the development of the mining and farming communities located along major river valleys. In the Northern Rocky Mountains south of the Canadian border, the amount of ponderosa pine harvested peaked in 1910 and started to decrease when much of the accessible timber had been cut. Also at this time, the amount of western white pine (*Pinus monticola*) harvested started to increase, with much of the volume going for toothpick and match stick blocks (Anderson 1935). These early harvests were largely uncontrolled until the establishment of the U.S. Department of Agriculture, Forest Service in 1905.

By the early 1900s, management objectives for timber production were established for lands administered by the Forest Service. These objectives directed the Forest Service not to place inferior species (e.g., lodgepole pine (*Pinus contorta*), grand fir (*Abies grandis*), western hemlock) on the market (USDA 1926). Therefore, management plans and silvicultural practices were developed that discriminated against these species. For example, techniques were developed for killing western hemlocks to release western white pines and prepare sites for regeneration (Foiles 1950). Tree classification systems were developed for ranking individual trees as to their risk of dying (Keen 1943, Wellner 1952). These silvicultural practices were designed to harvest a portion of forest stands, leaving a viable future forest. However, many of these partial cuttings decreased the health and vigor of forest stands (Graham 1980). Because of the management objectives and the silvicultural techniques used to achieve the objectives, a disproportionate amount of western white pine and ponderosa pine were harvested (Figure 1) (Hutchison and Winters et al. 1942).

Partial cutting of many forest stands in the Inland West, removing the high quality western white pine, western larch (*Larix occidentalis*), and ponderosa pine, tended to develop forests containing high amounts of shade tolerant grand fir, white fir (*Abies concolor*), and Douglas-fir (*Pseudotsuga menziesii*). These forests are highly susceptible to epidemic outbreaks of insects (tussock moth (*Orgyia pseudotsugata*), spruce budworm (*Choristoneura occidentalis*), mountain pine beetle (*Dendroctonus ponderosae*) and diseases (*Armillaria* spp. *Phellinus weirii*) (Hessburg et al. 1993). Because western white pine, ponderosa pine, and western larch are all relatively shade intolerant, partial cutting of mixed stands containing these species did not create conditions conducive to their regeneration, thereby decreasing the proportion of these species present in Inland forests. In addition, western white pine blister rust (*Cronartium ribicola*) (an introduced disease) further decreased the amount of western white pine. Therefore, by the 1950s it was recognized that more intensive silvicultural practices were going to be needed if the management objectives of producing high quality timber products were going to be achieved.

To effectively accomplish these tasks, a tremendous amount of knowledge was developed on the characteristics (regeneration, growth, longevity, genetics, structure, etc.) of forest vegetation and the response of forest

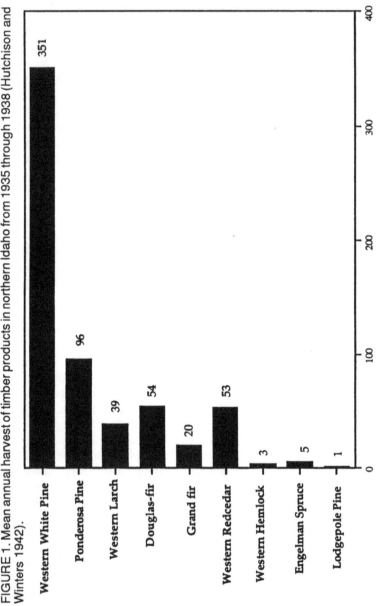

FIGURE 1. Mean annual harvest of timber products in northern Idaho from 1935 through 1938 (Hutchison and Winters 1942).

vegetation to disturbance (Haig et al. 1941; Pearson 1950; Schmidt 1988; Schubert and Adams 1971). High yield forestry concepts were developed and used. Clearcutting was often the regeneration method of choice because it was economical, efficient, and usually could regenerate the high value western white pine, western larch, and ponderosa pine. Site preparation methods, nursery practices, and improved planting stock were developed and used. In addition, cleaning, weeding, and thinning schedules were developed for maximizing mean annual increments. Using these approaches, rotation ages as low as 70 years for some forest types in the Inland West were prescribed, far younger than the 200 year-old plus stands that were harvested. This body of knowledge was applied very effectively in tending and producing forest crops at the stand level and for maximizing efficiency while attempting to mitigate the effects of timber management on other uses such as wildlife and scenery (Kennedy and Quigley 1993).

As Kessler (1993) stated "In this century, the profession's main service to society has been to produce valuable commodities from forest lands. Foresters have served society well in this regard, using science and improved practices to make the land produce more efficiently and economically." But, the production of commodities is not the only objective being addressed by today's forest managers. Grizzly bears (*Ursus horribillis*), northern spotted owls (*Strix occidentalis caurina*), chinook salmon (*Oncorhynchus tshawytscha*), northern goshawks (*Accipiter gentilis*), and old-growth forests are all emerging issues and often dominant in forest management discussions. Therefore, silviculturists are constantly being challenged to meet the management objectives of society, yet maintain the health and vigor of forests. Often they succeed but sometimes they fail.

FIRE IN THE INLAND WEST

Wildfire, in contrast to silviculture, does not alter forest structure and composition according to management objectives. Fire is opportunistic and unpredictable; it burned at a variety of frequencies and intensities throughout most of the Inland West forests prior to 1900 (Habeck and Mutch 1973; Arno 1980; Wellner 1970). These fires tended to regulate forest health by regenerating and maintaining forest structures and functions (Habeck and Mutch 1973).

Fire exclusion, starting in 1900 and becoming progressively more effective, changed the character of these forests. Meadows and open ponderosa pine and Douglas-fir forests were invaded with trees (Habeck 1990). The exclusion of fire has led to the development of dense understories of

Douglas-fir, white fir, and grand fir in stands that were once open-grown ponderosa pine (Agee 1993). Surface fires historically cleaned these forests, perpetuating ponderosa pine. In the early 1900s a combination of livestock removal, favorable weather conditions, and seed crops, led to regeneration of large amounts of ponderosa pine (Pearson 1950). Today, because of the multiple canopy layers and high tree densities, the wildfires that do occur tend to be large and intense, potentially causing wide-scale damage to watersheds and soils (Harvey et al. 1993).

Historical fires varied in intensity and size, and left a mosaic on the landscape (Arno 1980; Habeck and Mutch 1973; Agee 1993). These fires prepared the forest floor for tree regeneration and, in the case of lodgepole pine, opened serotinus cones, releasing seed (Lotan and Perry 1983). Burned surfaces and exposed mineral soil were ideal sites for tree regeneration (Haig et al. 1941). Openings caused by fires also allowed the shade intolerant species such as western white pine, western larch, and ponderosa pine to become established.

Understory vegetation such as buck brush (*Ceanothus* spp.) regenerates aggressively when fires scarify buried seed, and willows (*Salix* spp.) produce new sprouts after burning (Noste and Bushey 1987). Likewise, pinegrass (*Calamagrotis rubescens*) produces luxuriant growth after the forest floor is burned. Many of the positive responses plants have after a fire can be related to the flush of nutrients released from the organic layers on the soil surface (Hungerford et al. 1991; Harvey et al. 1987). The large amounts of coarse woody debris left after a fire can contribute to the organic component of the forest floor for years to come. These materials are also sites for non symbiotic nitrogen fixation and provide substrata for ectomycorrhizae (Harvey et al. 1987). Ectomycorrhizae are not only important for plant growth but their fruiting bodies are important in the food chains of many small mammals (Maser et al. 1986). Therefore, fire is a good regulator of forest health because it regenerates stands, maintains site-adapted species, thins stands, and recycles nutrients and carbon.

Timber harvesting and fire exclusion have greatly influenced the development of the Inland West. In turn, because of the conditions created, larger and more devastating fires are occurring. This situation, along with the apparent loss of habitat for many wildlife species, is causing changes in philosophies of how forests should be managed. The most rational approach to solving these problems is to take a new holistic approach to forest management and to the management of ecosystems.

ECOSYSTEM MANAGEMENT

Ecosystem management, although evolving, is the blending of economic, social, and ecological needs when managing forests (Jensen and Everett 1993). Most definitions of the concept rely on the analysis of large tracts of forests that may cross ownership and administrative boundaries (Bourgeron and Jensen 1993). These large tracts should be analyzed in a hierarchy of scales in both time and space (from past to future and from global to site). This approach can be used to show how the processes and functions within the analysis area affect and are affected by the processes and functions at the other scales. Depending on the scale, this analysis can be both very coarse (e.g., changes in dominant tree species over large tracts) and fine (e.g., changes in intermittent stream water quality) depending on the ecosystem element, process, or function that is of concern.

Reference conditions at all temporal and spatial scales can be developed using the concept of historical variability that can be derived from historical records, fire histories, and other historical data (Swanson et al. 1993). But the development of reference conditions should not preclude the use of process models and other sources of information. The elements analyzed using these techniques would vary but could include such items as stream stability, fire frequency, forest seral stage, and patch size to name a few.

Reference conditions can also be compared to the existing conditions of a forest. This comparison can be used to determine if any forest treatment would be necessary to direct an ecosystem toward a more stable condition (e.g., one that might be within the range of historical variability). Also, these reference conditions can be used to evaluate management alternatives (e.g., high yield forestry or a campground) as to the effect they might have on forest sustainability. Because of wildfire exclusion, timber harvesting, grazing, and urban encroachment, it is likely that some type of active forest treatment other than fire control is going to be needed in many western forest ecosystems. The practice of silviculture is the source of these treatments.

SILVICULTURE, FIRE
AND ECOSYSTEM MANAGEMENT

The exclusion of fire in the Inland West is a major cause of unstable forest conditions. Reintroducing fire will be difficult because of high fuel loadings, dense multistoried stands, and the large tracts that are in need of prescribed fire. Fires burning in these stands would likely do resource damage (soil sterilization, killing of large trees) and would be difficult to control. In addition, the amount of smoke released into the air is restricted in many areas throughout the West. Therefore, it is not likely that fire can be used exclusively to treat these large forested areas (Agee 1993). The alternative is to use silvicultural practices that include prescribed fire to produce sustainable forests.

Within the practice of silviculture, the knowledge and tools exist to complement, if not replace, many processes associated with fire in many forest ecosystems. Silvicultural techniques can remove high forest cover, thin trees, prepare the forest floor for tree regeneration, stimulate the growth of lower vegetation, decrease the incidence of disease and insects, maintain site adapted species, and recycle nutrients. Decomposition of organic matter can be encouraged by using practices that increase soil heating, maintain soil moisture, and maintain soil aeration thereby recycling nutrients and carbon. One effect fire has on forest soils that would be difficult to replace is the long-lasting increase in pH (Jurgensen et al. 1981). This fire effect could be very important in forests subjected to acid deposition or naturally developing acidic soils. Therefore, the best alternative for implementing ecosystem management is to use a combination of silvicultural tools that include prescribed fire.

IMPLEMENTATION

Effective fire exclusion for the past 60 years has led to the development of large amounts of biomass on the forest floor and dense stands of small and midsized trees throughout the West. In addition, many of these forests have been harvested, removing many of the large trees that were fire, disease, and insect resistant. The magnitude of the task of returning these forests to a condition similar to those before fire exclusion is horrendous, but the alternative of not trying will only continue to perpetuate forests prone to insect, disease, and forest replacing fires. This is evident by the major wildfires that have occurred throughout the West in recent years.

Forests can be managed for sustainability and function using silviculture. Innovative uses of silvicultural practices such as species selection, thinning, and biomass removal can go a long way in producing forests that are vigorous and healthy. For example, the technology is available to produce cable yarding equipment that can work on steep slopes and at great distances to remove small diameter material. Bunching of small material with cable machines and using helicopters to remove biomass is also possible. Using these techniques could remove forest biomass with minimum disturbance to the soil resource, the foundation of healthy forest ecosystems (Graham 1981).

But the cost of developing silvicultural systems for sustaining healthy forest ecosystems and the technology needed to implement these prescriptions will not be inexpensive. The cost of such forest management practices can not be supported by the amount of products removed at today's stumpage prices. But as the price of lumber increases and the technologies for utilizing small material become more efficient, the utilization of these materials will become more cost effective. Even if the value of the material does not pay for its removal, regulation of the species composition and density of Inland West forests through silviculture will be necessary. The alternative–of no management–is unacceptable.

CONCLUSIONS

Fire exclusion and silviculture, through the practices used for producing high yields of forest products, have been the main factors in the development of the current forests in the Inland West. These factors, interacting with the dynamics of forest succession and growth, have produced forest landscapes that are overly dense and highly susceptible to epidemics of insects and diseases and the occurrence of catastrophic fires. These conditions are not acceptable or sustainable.

It appears that the most logical approach to changing these conditions is to analyze and manage landscapes. This ecosystem approach still must rely on the establishment of forest management objectives that society can agree on. Using these management objectives, reference conditions, and the tools and methods developed in the practice of silviculture, silvicultural systems for maintaining ecosystem processes and functions can be developed (Figure 2). By using innovative techniques of biomass removal and carefully applying fire, it may be possible to use fire as a tool even in areas where smoke emissions are restricted. Silviculture is the art and science of managing forests to meet management objectives (Smith 1962). If society desires forests to be managed for sustainability, silvicultural prescriptions can be developed using many of the same tools that successfully produced timber crops.

FIGURE 2. An ecosystem approach to managing Inland West ecosystems should include the protection of the soil and organic resource. If these ecosystem elements and their inherent processes (nitrogen fixation, ectomycorrhize activity) are in a healthy and vigorous state, the other elements of the ecosystem will also most likely be in a healthy state.

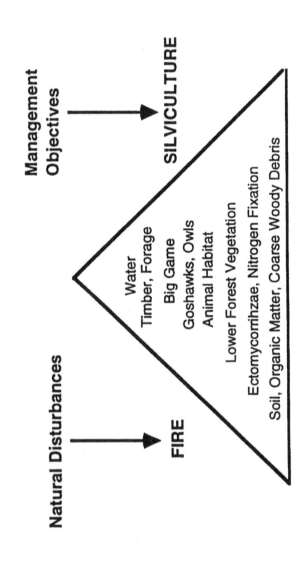

Management Objectives → SILVICULTURE

Natural Disturbances → FIRE

Water
Timber, Forage
Big Game
Goshawks, Owls
Animal Habitat

Lower Forest Vegetation

Ectomycorrhizae, Nitrogen Fixation

Soil, Organic Matter, Coarse Woody Debris

REFERENCES

Agee, J. K. 1993. Fire and weather disturbances in terrestrial ecosystems of the eastern Cascades. In: pp. 360-414. Hessburg, P. F. (Comp.). Eastside forest ecosystem health assessment, Volume III–Assessment. U.S. Dept. Agric., For. Serv., Pacific Northwest Research Station. Wenatchee, WA.

Anderson, I. V. 1935. Match plank and commercial lumber from western white pine logs. App. For. Note. Note 72. U.S. Dept. Agric., For. Serv., Northern Rocky Mountain Forest and Range Experiment Station. Missoula, MT. 3 p.

Arno, S. F. 1980. Forest fire history in the northern Rockies. *Journal of Forestry.* 78:460-465.

Bourgeron, P. S. and M. E. Jensen. 1993. An overview of ecological principles for ecosystem management. In: pp. 49-60. Jensen, M. E. and P. S. Bourgeron (Eds.). Eastside forest ecosystem health assessment, Volume III–Ecosystem management: ecosystem principles and applications. U.S. Dept. Agric., For. Serv. Missoula, MT.

Foiles, M. W. 1950. Recommendations for poisoning western hemlock. Res. Note. 77. Missoula, MT: U.S. Dept. Agric., For. Serv., Northern Rocky Mountain Forest and Range Experiment Station. 1 p.

Graham, R. T. 1980. White pine vigor–A new look. Res. Pap. INT-254. U.S. Dept. Agric., For. Serv., Intermountain Forest and Range Experiment Station. Ogden, UT. 15 p.

Graham, R. T., D. Minore, A. E. Harvey, M. F. Jurgensen and D. S. Page-Dumroese. 1991. Soil management as an integral part of silvicultural systems. In: pp. 59-64. Harvey, A. E. and L. F. Neuenschwander (Comps.). Proceedings: Management and productivity of western-montane forest soils. Gen. Tech. Rep. INT-280. U.S. Dept. Agric., For. Serv., Intermountain Research Station. Ogden, UT.

Habeck, J. R. 1990. Old-growth ponderosa pine-western larch forests in western Montana: ecology and management. *Northwest Environmental Journal.* 6:271-292.

Habeck, J. R. and R. W. Mutch. 1973. Fire-dependent forests in the northern Rocky Mountains. *Quatanery Research.* 3:408-424.

Haig, I. T., K. P. Davis and R. H. Weidman. 1941. Natural regeneration in the western white pine type. Tech. Bulletin. No. 767. U.S. Dept. Agric. Washington, DC. 99 p.

Harvey, A. E., M. J. Geist, G. I. McDonald, M. F. Jurgensen, P. H. Cochran, D. Zabowski and R. T. Meurisse. 1993. Biotic and abiotic processes in eastside ecosystems: The effects of management on soil properties, processes, and productivity. In: pp. 102-173. Hessburg, P. F. (Comp.). Eastside forest ecosystem health assessment, Volume III–Assessment. U.S. Dept. Agric., For. Serv., Pacific Northwest Research Station. Wenatchee, WA.

Harvey, A. E., M. F. Jurgensen, M. J. Larsen and R. T. Graham. 1987. Decaying organic materials and soil quality in the Inland Northwest: A management opportunity. Gen. Tech. Rep. INT-225. U.S. Dept. Agric., For. Serv., Intermountain Research Station. Ogden, UT. 15 p.

Hessburg, P. F., R. G. Mitchell and G. M. Filip. 1993. Historical and current roles of insects and pathogens in eastern Oregon and Washington forested landscapes. In: pp. 485-535. Hessburg, P. F. (Comp.). Eastside forest ecosystem health assessment, Volume III–Assessment. U.S. Dept. Agric., For. Serv., Forest Service, Pacific Northwest Research Station. Wenatchee, WA.

Hungerford, R. D., M. G. Harrington, W. H. Frandsen, K. C. Ryan, K. C. and G. J. Niehoff. 1991. Influence of fire on factors that affect site productivity. In: pp. 32-50. Harvey, A. E. and Neuenschwander, F. L. (Comps.). Proceedings–Management and productivity of western-montane forest soils. Gen. Tech. Rep. INT-280. U.S. Dept. Agric., For. Serv., Intermountain Research Station. Ogden, UT.

Hutchison, S. B. and R. K. Winters. 1942. Northern Idaho forest resources and industries. Miscellaneous Publication. 508. U.S. Dept. Agric., For. Serv. Washington, DC. 75 p.

Jensen, M. E. and R. Everett. 1993. An overview of ecosystem management principles. In: pp. 7-16. Jensen, M. E. and P. S. Bourgeron (Eds.). Eastside forest ecosystem health assessment, Volume II–Ecosystem management: ecosystem principles and applications. U.S. Dept. Agric., For. Serv. Missoula, MT.

Jurgensen, M. F., A. E. Harvey and M. J. Larsen. 1981. Effects of prescribed fire on soil nitrogen levels in a cutover Douglas-fir/western larch forest. Res. Paper. INT-275. U.S. Dept. Agric., For. Serv., Intermountain Research Station. Ogden, UT. 6 p.

Keen, F. P. 1943. Ponderosa pine tree classes redefined. *Journal of Forestry* 41: 249-253.

Kennedy, J. J. and T. M. Quigley. 1993. Evolution of Forest Service organizational culture and adaptation issues in embracing ecosystem management. In: pp. 17-28. Jensen, M. E. and Bourgeron, P. S. (Eds.). Eastside forest ecosystem health assessment, Volume II–Ecosystem management: ecosystem principles and applications. U.S. Dept. Agric., For. Serv. Missoula, MT.

Kessler, W. B. 1993. Change, the essence of life. *Journal of Forestry*. 91:68.

Lotan, J. E. and D. A. Perry. 1983. Ecology and regeneration of lodgepole pine. Agric. Handb. No. 606. U.S. Dept. of Agric. Washington, DC. 51 p.

Maser, C., Z. Maser, J. W. Witt and G. Hunt. 1986. The northern flying squirrel: A mycophagist in southwestern Oregon. *Canadian Journal of Zoology* 64: 2086-2089.

Noste, N. V. and C. L. Bushey. 1987. Fire response of shrubs of dry forest habitat types of Montana and Idaho. Gen. Tech. Rep. INT-239. U.S. Dept. of Agric., For. Serv., Intermountain Research Station. Ogden, UT. 22 pp.

Pearson, G. A. 1950. Management of ponderosa pine in the southwest as developed by research and experimental practices. Monograph. 6. U.S. Dept. of Agric., For. Serv. Washington, DC. 218 p.

Pfister, R. D., B. L. Kovalchik, S. F. Arno and R. C. Presby. 1977. Forest habitat types of Montana. Gen. Tech. Rep. INT-34. U.S. Dept. of Agric., For. serv., Intermountain Forest and Range Experiment Station. Ogden, UT. 174 p.

Schmidt, W. (Ed.). 1988. Future forests of the mountain west: A stand culture symposium. Gen. Tech. Rep. INT-243. U.S. Department of Agriculture, Forest Service, Intermountain Research Station. Ogden, UT. 402 p.

Schubert, G. H. and R. S. Adams. 1971. Reforestation practices for conifers in California. State of California, The Resources Agency, Department of Conservation, Division of Forestry. Sacramento. 359 p.

Smith, D. M. 1962. The practice of silviculture. John Wiley. New York. 578 p.

Steele, R., S. V. Cooper, D. M. Ondov, D. W. Roberts and R. D. Pfister. 1983. Forest habitat types of eastern Idaho-western Wyoming. Gen. Tech. Rep. INT-144. U.S. Dept. of Agric., For. Serv., Intermountain Forest and Range Experiment Station. Ogden, UT. 122 p.

Swanson, F. J., J. A. Jones, D. A. Wallin and J. Cissel. 1993. Natural variability–implications for ecosystem management. In: pp. 85-112. Jensen, M. E. and Bourgeron, P. S. (Eds.). Eastside forest ecosystem health assessment, Volume II–Ecosystem management: principles and applications. U.S. Department of Agriculture, Forest Service. Missoula, MT.

USDA. 1926. The national forest manual: Regulations and instructions. U.S. Dept. of Agric., For. Serv. Washington, DC. 9 Sections.

Wellner, C. A. 1952. A vigor classification of tree-vigor for western white pine trees in the Inland Empire. Res. Note. 110. U.S. Dept. of Agric., For. Serv., Northern Rocky Mountain Forest and Range Experiment Station. Missoula, MT. 6 p.

Wellner, C. A. 1970. Fire history in the Northern Rocky mountains. In: pp. 42-64. The role of fire in the Intermountain West, Symposium proceedings. Intermountain Fire Research Council. Missoula, MT.

Wellner, C. A. 1976. Frontiers of forestry research-Priest River Experimental Forest 1911-1976. U.S. Dept. of Agric., For. Serv., Intermountain Forest And Range Experiment Station. Ogden, UT. 148 p.

Maintaining and Creating Old Growth Structural Features in Previously Disturbed Stands Typical of the Eastern Washington Cascades

C. D. Oliver
C. Harrington
M. Bickford
R. Gara
W. Knapp
G. Lightner
L. Hicks

C. D. Oliver is Professor of Silviculture, University of Washington College of Forest Resources, AR-10, Seattle, WA 98195.

C. Harrington is Research Silviculturist, USDA Forest Service, Pacific Northwest Research Station, Olympia, WA 98502.

M. Bickford is Forest Silviculturist, USDA Forest Service, Wenatchee National Forest, Wenatchee, WA 98801.

R. Gara is Professor of Entomology, University of Washington College of Forest Resources, AR-10, Seattle, WA 98195.

W. Knapp is Silviculturist, A.G. Crook Company, Beaverton, OR 97006.

G. Lightner is Silviculture Specialist, USDA Forest Service, Region 6, Portland, OR 97208.

L. Hicks is Manager of Fish and Wildlife Resources, Plum Creek Timber Company, 2300 F.I.C., 999 Third Avenue, Seattle, WA 98104.

This paper is based on a report from the Washington State Working Subgroup of the Silviculture Subcommittee of the Spotted Owl Recovery Team, U.S. Department of Interior Fish and Wildlife Service; December 9, 1991.

[Haworth co-indexing entry note]: "Maintaining and Creating Old Growth Structural Features in Previously Disturbed Stands Typical of the Eastern Washington Cascades." Oliver, C.D. et al. Co-published simultaneously in the *Journal of Sustainable Forestry* (The Haworth Press, Inc.) Vol. 2, No. 3/4, 1994, pp. 353-387; and: *Assessing Forest Ecosystem Health in the Inland West* (eds: R. Neil Sampson and David L. Adams) The Haworth Press, Inc., 1994, pp. 353-387. Multiple copies of this article/chapter may be purchased from The Haworth Document Delivery Center [1-800-3-HAWORTH; 9:00 a.m. - 5:00 p.m. (EST)].

353

ABSTRACT. Changes in old growth structural features as well as susceptibilities to disturbances were projected in stands typical of the eastern Washington Cascades. Projected changes with and without silvicultural operations were made. Doing no silvicultural activities in these stands will not rapidly increase old growth structural features and will allow the stands to become very susceptible to insects and wind breakage, followed by fires. Specifically designed silvicultural operations can maintain or rapidly increase old growth structural features and reduce susceptibilities to most disturbances. Unless some trees killed in the silvicultural operations are removed, the treated stands will become very susceptible to fires. Removing some of the thinned trees can also offset the costs of doing the operations. A landscape approach of treating various stands with different silvicultural regimes will probably best maintain a dynamic balance of structural features, a reduced susceptibility to various disturbances, and a steady flow of wood for manufacturing.

INTRODUCTION

Uneven-age harvesting and fire exclusion in mixed-conifer stands have been predominant practices for many decades at mid elevations (the Abies grandis and Pseudotsuga menziesii zones of Franklin and Dyrness 1973) in the eastern Washington Cascade Range (Oliver et al. 1993). The practices have generally increased the habitat for some old-growth related species, such as the spotted owl. Past logging practices (primarily high-grading) have increased the occurrence of hollow, rotted trees and diseased (mistletoe) limbs which the spotted owl favors for nesting. Stand-replacing fires in the late 1800s and early 1900s, followed by fire exclusion, have resulted in dense stands of a single cohort (age class) with trees of small diameters.

An initial reaction to the decline of old growth-related species populations is to stop all forestry-related activities in large forested areas, in case these activities reduce habitat. Stopping all activities carries consequences in itself, including the possibility that the stands will grow susceptible to destruction by winds and heavy snows, fires, insects, or a combination of these. Many eastern Washington forests consist of partially harvested, multiple cohort (uneven-age) stands and dense single cohort stands which grew after fires. It is unknown whether old growth structures can be maintained or achieved rapidly in these forests if they are left alone.

Silvicultural operations such as thinning are sometimes used to increase timber value; however, they may also change the structural features of stands for the benefit of wildlife. Silvicultural operations may also reduce risks of wind or snow breakage, fires, insects, and diseases. Removal and utilization of some thinned timber may offset the costs of manipulating stands to reduce these risks.

Objectives

This study addressed typical, multiple cohort (uneven-age) stands and dense, single cohort stands in the eastern Cascade Range of Washington state, U.S.A., to determine the following:

- Will avoiding future forestry activities maintain and enhance old growth structural features in previously selectively harvested stands?
- Will avoiding future forestry activities achieve old growth structural features in dense, single cohort forests?
- Can silvicultural treatments be done to increase old growth structural features more rapidly than would occur through growth without manipulations in these two stand types?
- How susceptible will these stands be to destruction by insects, winds, snow, and fire with and without silvicultural treatments designed to achieve old growth structures?
- What will be the costs of silvicultural treatments designed to achieve old growth structures?
- How can these susceptibilities and costs be minimized?

This study projected changes in certain structural features important for old growth species both with and without silvicultural treatments in two representative stand types of the eastern Washington Cascades. The PROGNOSIS computerized stand projection model along with information from other sources not available with PROGNOSIS were used to project changes in stand structures (East Cascades variant of Prognosis model of Stage [1973]; this variant was authored by R.Johnson and G.Lightner).

The first part of this study was concerned with creating and maintaining viable old growth structural features, not with obtaining revenue from the forest; therefore, this analysis first assumed that no wood would be extracted, even when silvicultural treatments killed trees and left them standing or felled them. The second part considered the susceptibilities to disturbances with and without active silvicultural treatments when the killed trees are left in the woods following the treatments. The third part considered the costs/returns and susceptibilities to disturbances when the merchantable killed trees were left in the woods versus removed in commercial operations.

Stand Structural Features Used by "Old Growth" Species

Structural features of forests, and associated habitats and plant and animal populations, constantly fluctuate with tree growth and natural and human disturbances (Oliver and Larson 1990; Hunter 1990; Sprugel 1991). Changing forest structures after disturbances have been described in several ways, but generally develop through several structural stages–first, open areas ("stand initiation stage," Oliver 1981); then, very dense stands of trees with small diameters ("stem exclusion stage"); followed by somewhat more open stands with an understory ("understory reinitiation stage"); and, barring disturbances, eventually "old growth" structures. Minor disturbances which kill only some trees can create park-like, open stands soon after the disturbance, followed by stands with dense understories, and then "old growth" structures (Johnson et al. 1993).

These structural stages are not obligatory or predetermined, but are a natural result of interactions among trees (Oliver and Larson 1990). Certain plant and animal species require specific features commonly found in one or a few of these structures. Presently, endangered species such as the spotted owl are believed to require certain structural features found in old growth stands, although other species in danger of becoming extinct require other structural conditions–such as open areas which occur immediately after a disturbance. Animal species which use each structure as habitat have existed in a dynamic state, expanding and declining in number and moving from one place to another as the stand structure changed.

Various structural features of "old growth" stands have been described by Franklin et al. (1986). Thomas (1979) and Hunter (1990) have described structural features of old forests important to various animals. Some structural features of stands important for species associated with old growth forests are:

- Areas of several canopy layers and long crown lengths.
- Some or all of the forest floor covered with ground vegetation. Ground vegetation is loosely defined, but generally is below about 4.5 to 6 feet.
- Some areas of narrow tree spacings (high tree numbers) and other areas of wide tree spacings (low tree numbers).
- A high combination of tree diameters and tree numbers (often expressed as basal area).
- A relatively high percentage of the overall area covered by foliage. Foliage coverage is often expressed as the proportion of canopy closure (percent of total area covered by foliage). Foliage coverage is also somewhat related to basal area.

- Large tree diameters for developing hollow trees for nests and prey habitat and for forming future down logs, and tall tree heights for providing more vertical space within the forest.
- Snags (dead standing trees) and large, hollow, living trees (a.k.a. "green snags"). A minimum snag size may be necessary for nesting sites in some areas.
- Down logs similarly provide habitat and food fungal substrata. Large logs are again more beneficial than small logs.

METHODS

Representative Stands

Two representative stand conditions–one typical of eastern Washington multiple cohort (uneven-age) stands and one typical of dense, single cohort (even-age) stands–were used to project changes in structures and likelihood of destruction by different disturbance agents with and without silvicultural treatments (Table 1).

Projecting Structural Features

Structural features were described and projected for the representative stands of Table 1. No model presently projects the full range of structural features; consequently, several models as well as estimates based on professional experience (where no model existed) were combined (on a computer spread sheet).

Projected changes in structural features were made for 60 years–from 1990 through 2050. To gain initial (1990) values of snags, down logs, and crown depths, projections were begun in 1950.

The PROGNOSIS model (Stage 1973) was used to project tree height and diameter distributions, mortality, and live crown ratios. Crown radii were calculated using equations from Moeur (1985). Percent crown closure was calculated by summing the crown projection areas for each tree based on its crown radius. Understory vegetation presence and height were estimated to be inversely proportional to amount of overstory closure, based on professional experience.

Numbers of hard and soft snags by diameter class were calculated from data on mortality and thinned (harvested) trees in PROGNOSIS using the Snag Recruitment Simulator (SRS, East-side version, Marcot 1990). All thinned (harvested) trees were assumed to be killed (i.e., by girdling or

TABLE 1. Characteristics of stands used in this report. Stand characteristics were chosen as typical of stands at mid elevations (*Abies grandis* and *Pseudotsuga menziesii* zones of Franklin and Dyrness 1973) in eastern Washington.

Multiple Canopy Stand:	Dense, Single Cohort Stand:
Upper canopy stratum consists of ponderosa pines (*Pinus ponderosa* Dougl. ex Laws.) 170 years old in 1990. These trees began in 1820 after a natural, stand replacing disturbance with 1000 trees per acre. Lower stratum consists of Douglas-firs (*Pseudotsuga menziesii* [Mirb.] Franco)and grand firs (*Abies grandis* [Dougl. ex D.Don] Lindl.) 70 years old. These trees began after a ground fire with 750 trees/acre.	Stand contains a mixture of Douglas-firs, grand firs, and ponderosa pines 70years old in 1990. The trees began in 1920 after fires, with 600 Douglas-firs/acre, 200 grand firs/acre, and 200 ponderosa pines/acre.

other means) and left standing to form snags. Down log numbers by size classes were calculated from differences in snag recruitment and snag number. Log lengths were assumed to be 30 percent of total tree heights (at time of death). Because of the dry climate, logs were assumed not to decay appreciably during the 60 years of this study. The projections illustrate trends and relative values of structural features more accurately than they show precise values.

Projecting Susceptibility to Disturbances

A stand's susceptibility to disturbances is related to specific structural features and to external conditions such as insect populations and weather patterns. Measurable relationships between stand structural features and likelihood of destruction by various agents were developed based on these known biological relationships between stand structures and disturbances. Susceptibilities to different disturbances were converted to a proportion, with a "susceptibility threshold" of 100 percent indicating very high vulnerability of a stand to the disturbance agent. These relationships show trends of stand susceptibility with changes in certain structural features rather than predict precise values. Additional research should increase the accuracy of specific predictions, but the general trends would be expected to remain as illustrated.

Wind or Snow Breakage–Tree susceptibility to wind or snow breakage was calculated by its height/diameter ratio (Halle, Oldeman, and Tomlinson 1978). Although not calibrated for eastern Washington species, a general rule is that trees are susceptible to breakage when the ratio of height to lower stem diameter (measured at 4.5 feet) is over 100 (assumes that height and diameter are expressed in same units). A value of 100 was assumed to give a susceptibility threshold of 100 percent.

Susceptibility to Fire–Fuel loads were calculated for dead trees from weight of dry crowns and limbs (material below 6 inches diameter) by species and diameter whenever trees died using values from Brown et al. (1977). Fuels were assumed to remain for 50 years on the dry sites of this study. Thirty tons of fuel per acre was considered a susceptibility threshold of 100 percent. This study probably underestimated fuel loads because it did not consider dead limbs or foliage shed from living trees, and it did not consider dead materials on the ground from previous stands or before 1950 (when calculations started in this analysis). When trees were removed in silvicultural operations, fuel loads were reduced because the flammable materials were reduced, crushed to the ground, or otherwise mitigated.

Insect Susceptibility–In the study stands, major insect concerns are western spruce budworm (Choristoneura occidentalis Freeman) and mountain pine beetle (Dendroctonus ponderosae Hopkins). Susceptibility to attack by western spruce budworm is related to the amount of Douglas-fir and grand fir in the stand, the stand density, and the presence of multiple canopy strata with Douglas-fir (Dfir) and/or grand fir (Gfir) in an upper stratum and grand fir in the lower stratum. A susceptibility value was developed from the conceptual relationship by Dr. Gara and Dr. Oliver as follows:

$$1.5 * \left[\left(\frac{BA_{GF} + DF}{BA_{total\ stand}} \times 100 \right) - 60 \right]$$

$$+ \ [30 * (BA_{total} - 100)]$$

$+ 40$ (if stand has *DF* and *GF* in upper stratum and *GF* in lower stratum)

BA=Basal area, *DF*=Douglis-fir, *GF*=grand fir

Susceptibility to attack by mountain pine beetle is believed to increase with increasing total stand density, closer spacing of ponderosa pines, and larger diameters of ponderosa pines. A susceptibility value was similarly developed by Dr. Gara and Dr. Oliver from the conceptual relationship as follows:

$$30 * (BA_{total} - 100)$$

$$+ \ 0.6 * (TPA_{Ponderosa\ pine})$$

$$+ \ 1.7 * (DBH_{Ponderosa\ pine})$$

$$\left(BA \text{ in } Ft^2\big/_{acre} \ DBH \text{ in inches} \right)$$

BA=Basal area, *TPA*=Trees per acre, *DBH*=Diameter at breast height

Silvicultural Regimes

Two active silvicultural regimes were used for manipulating each stand (Table 2). The choice of not managing each stand is considered a third (No Activity) regime. Most regimes in this study were single cohort (even-age) regimes involving thinning treatments, although one regime–Multiple Canopy–was a multi-cohort (uneven-age) regime and involved regeneration as well as selection cutting treatments. Specifically chosen regimes likely to achieve old growth structural features were utilized. Other, more creative, regimes may actually achieve old growth features more effectively.

Each regime was projected until 2050 A.D., with trees killed and left in the stand and again with merchantable trees removed using standard thinning and harvesting treatments. When not removed, killed trees were left standing and first became hard snags, then soft snags, and finally downed logs.

The silvicultural treatments used are common; consequently, it is certain that they can be accomplished. On the other hand, their timing, sequence, and the resultant changes in stand structures are not common to present forest management.

Costs and Returns from Silvicultural Treatments

Costs per acre for each silvicultural regime are shown in Table 3. These costs do not include treatments to reduce susceptibilities to disturbance agents and to control fires once started. Costs of killing trees by girdling to achieve the various structures are estimated to be $1.00 to $3.00 per tree, depending on tree size.

Returns to the landowner when merchantable trees were removed are estimated. These values vary greatly with numbers and sizes of logs removed, regional log prices, and time. For this paper, a stumpage value (return to the landowner) of $250.00 per thousand board feet is assumed for trees over 12 inches (at 4.5 feet). A value of $25.00 per green ton minus $825.00 per acre harvesting cost was assumed to be returned to the landowner for trees less than 12 inches in diameter. Costs and return values are for comparison purposes, and are not discounted to a single time.

RESULTS

Changes in Stand Structures Without Timber Removal

Barring disturbances (discussed later), relatively few structural changes will occur in either the multi-canopy stands or the dense, single-cohort stands between 1990 and 2050 without silvicultural manipulations. Unless altered by silvicultural treatments or destroyed by winds, snows, fires, or insects, the multiple canopy stands will continue to have relatively deep foliage, two canopy layers, trees close together, and trees and snags of large diameters (Figures 1 through 5). The two canopies will become less

TABLE 2. Representative silvicultural regimes for manipulating each stand (Table 1) in this study. Detailed description of regimes can be found in Oliver et al. (1991).

MULTIPLE CANOPY STANDS:	DENSE, SINGLE COHORT STANDS:
No Activities Silvicultural Regime: No Management Activities are done in stand. Low Density Regime: Living trees in both strata are kept in a wide spacing by periodically killing many trees in both strata, beginning in 1992. A third, forest floor stratum is allowed to develop naturally. Specific Activities: 1992: Thin 70% of Douglas-fir and 50% of grand fir. 2021: Thin all Douglas-firs and 80% of grand fir (primarily regeneration).	No Activities Silvicultural Regime: No Management Activities are done in stand. Multiple Canopy Strata Regime: All but a few overstory trees (6 trees/acre) are killed in 1992; and trees of many species are allowed to develop naturally in the understory. Second cohort (age class) is allowed to develop rapidly by later killing some (but not all) of the oldest trees. Second cohort is kept at a wide spacing by killing some trees. A forest floor stratum is allowed to develop naturally. Specific Activities: 1992: Thin all Douglas-firs and grand firs and 85% of ponderosa pines.

High Density Regime:
Living overstory trees in both strata are kept at a narrow spacing (but not as narrow as the No Activities Regime by periodically killing some trees beginning in 1992. A third, forest floor stratum is allowed to develop naturally.

Specific Activities:
1992: Thin 40% of ponderosa pines.
2021: Thin 85% of Douglas-firs and 55% of grand firs (both regeneration and older trees).

2011: Thin 95% of Douglas-firs and all grand fir (both regeneration and older trees).
2041: Thin 50% of Douglas-firs and all grand fir (both regeneration and older trees).

High Density Regime:
Living overstory trees are kept at a narrow spacing (but not as narrow as No Activities Regime) by periodically killing some trees beginning in 1992. Understory is allowed to develop naturally.

Specific Activities:
1992: Thin 50% of Douglas-firs and 30% of grand firs
2011: Thin 70% of Douglas-firs and 90% of grand firs (both regeneration and older trees).
2041: Thin 95% of Douglas-firs, 90% of grand firs (both regeneration and older trees), and 5% of ponderosa pines.

TABLE 3. Costs/acre or return/acre over 60 years for each silvicultural regime and each stand. COSTS assume no timber removal is done when trees are artificially killed and no risk reduction measures are taken. RETURNS assume all timber killed is removed (none is left to create snags and logs). Where some trees are left and others removed and sold, costs or returns would be between values shown here. (Costs and returns were not discounted to a single time.)

	COST/ACRE No Removal	RETURN/ACRE Full Removal
Multiple Canopy Stands		
No activities regime:	$ 0	$ 0
Low density regime:	$1,030	$14,042
High denisty regime:	$ 742	$ 1,215
Dense, single cohort stands		
No activities regime:	$ 0	$ 0
Multiple strate regime:	$2,822	$ 1,777
High density regime:	$1,515	$ 1,862

FIGURE 1. 60-year, stand silhouette projections showing structural features typical of multiple canopy stands in eastern Washington under different silvicultural regimes. (Scale on left (in feet) represents both horizontal and vertical. Black crown = Douglas-fir; white crown = grand fir; spotted crown = ponderosa pine.)

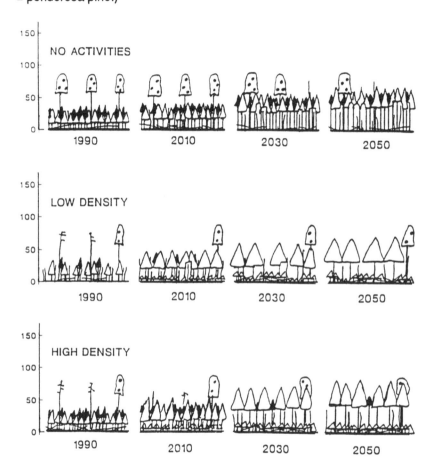

distinct after about 40 years (Figure 1). Large snags and down logs will begin appearing in two to four decades (Figure 2). There will continue to be little understory, however (Figure 1).

Dense, single cohort stands not altered by silvicultural treatments or destroyed by disturbances will continue to have similar features for the next 60 years (Figures 6-9). They will have a closed canopy of closely

FIGURE 2. Projections of species diameters, and percent crown closure, hard snag numbers, and down log lengths for multiple canopy stands (Figure 1). (Diameters show largest tree of each species and stand. Snags and logs are in 3 diameter classes. N = No Activity Regime; L = Low Density Regime; H = High Density Regime. Snag & down log scales are logarithmic.)

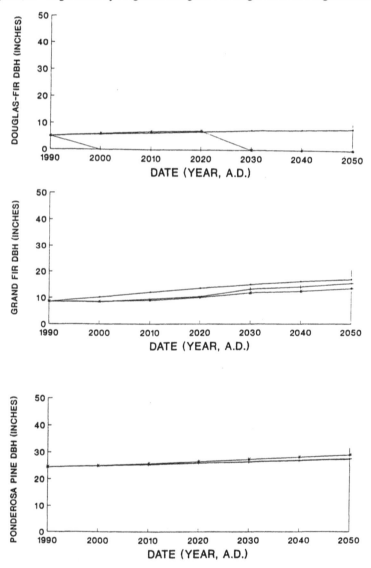

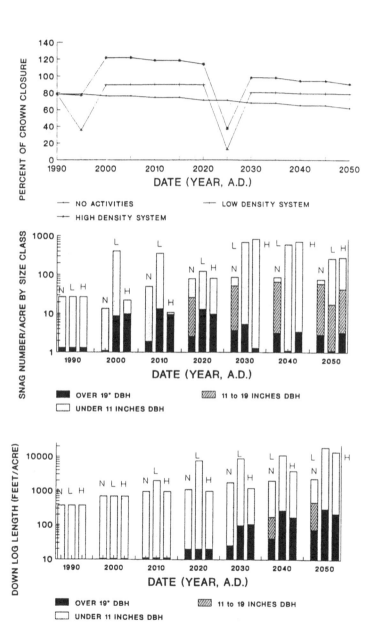

FIGURE 3. Projections of susceptibility of multiple canopy stands following different silvicultural regimes to destruction by common wind/snow and insect disturbance agents. Susceptibility of 100 = extreme likelihood of disturbance. Susceptibility values over 200 were assigned a value of 200. Wind and snow susceptibility listed by species. [ ] = No Activities Regime; + = High Density Regime; * = Low Density Regime.

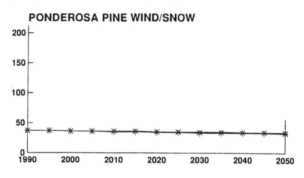

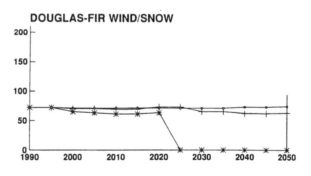

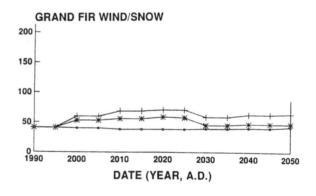

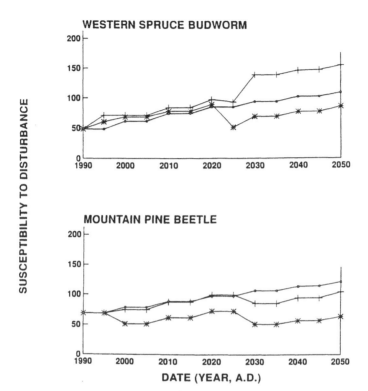

spaced, small trees and some small snags and down logs (Figure 7). There will continue to be only one canopy layer and a shallow foliage layer with no ground vegetation (Figure 6).

The active silvicultural manipulations will dramatically change the old growth structural features in both the multiple canopy and dense, single cohort stands. Some old growth features will immediately increase, while others will initially decrease and later increase. Changes in the various structural features will be as follows:

Canopy Layers–The number of canopy layers will increase with the Multiple Strata regime in the dense, single cohort stands (Figures 6 and 9); however, the active regimes may reduce the distinctiveness of different canopy layers in stands already containing multiple canopy strata (Figure 1).

Foliage Depth–Foliage depth will increase in all regimes as the number of living trees is reduced (Figures 1 and 6).

Crown Closure–Reducing the number of living trees through silvicultural treatments will temporarily decrease crown closure (Figures 2 and

FIGURE 4. Projections of susceptibility of multiple canopy stands (top) and dense, single cohort stands (bottom) to fires with no silvicultural activities ([ ]) and with various regimes where the thinned trees are left as snags or logs are similar to effects of insects or wind/snow in leaving stands very suscepti- ble to fires. Susceptibility values are similar to Figure 3. + = High Density Regime; * = Low Density Regime (top) or Multiple Strata Regime (bottom); Black boxes = Thinning regimes where killed trees are removed.

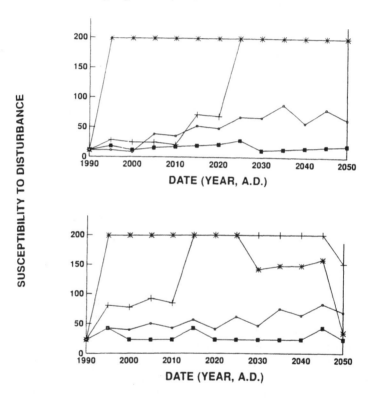

7)–most dramatically when dense, single cohort stands are converted to multiple canopy stands. Percent canopy closure will increase to very high levels where light thinning treatments (high density regime) left vigorous trees in several canopy strata (Figure 2).

   Ground Vegetation–Few silvicultural regimes will maintain understory vegetation because the shade tolerant grand firs (and Douglas-firs to a lesser extent) will regenerate abundantly whenever there is a thinning or similar disturbance (Figures 1 and 6).

Tree Spacing–Tree spacing (Figures 1 and 6) will generally increase with time and with lower density silvicultural regimes, except where the Multiple Canopy regime allows abundant regeneration to develop.

Tree Size–Tree sizes will also increase with time, but will increase more rapidly with lower density silvicultural regimes (Figures 2 and 7).

Snags and Down Logs–There will be many snags and down logs (Figures 2 and 7); however, down logs will be very small even under low density silvicultural regimes during the first few decades–until the snags fall over. Active felling of dead trees could create large down logs more rapidly. The down logs will last many decades.

Changes in Susceptibilities to Disturbances Without Tree Removal

Without silvicultural operations, the stands will become increasingly susceptible to all disturbances (Figures 3, 4, and 8). Douglas-firs and grand firs in the dense single cohort stands are already susceptible to wind/snow breakage, and this susceptibility will increase. Both stand types will become increasingly susceptible to the western spruce budworm and the mountain pine beetle. Even without wind or insect disturbances adding to the dead trees, both stand types will become increasingly susceptible to fires (Figure 4).

The different silvicultural regimes will have varied effects on the multiple canopy stands' susceptibilities to wind/snow and insect disturbances (Figure 3). The stands are generally not susceptible to wind and snow breakage, and the silvicultural regimes will not dramatically increase their susceptibility. The light thinning regime will reduce the stand's susceptibility to mountain pine beetle but will eventually increase the stand's susceptibility to western spruce budworm. The heavy thinning regime will reduce the stand's susceptibility to both the budworm and the beetle.

The dense, single cohort stands will become less susceptible to wind/snow and insect disturbance agents following all active silvicultural regimes (Figure 8). Thinning the dense, single cohort stands will make the Douglas-firs much less likely to be destroyed by winds or snow or the whole stand much less likely to be destroyed by the mountain pine beetle. If Douglas-firs or grand firs had been left in the overstory in the "Multiple Strata" regime (instead of ponderosa pines), the stand's susceptibility to western spruce budworm would have been much greater. Thinning will increase the probability of wind/snow breakage in grand firs; however, the broken grand firs will be very small and will not have a marked effect on the stand's structure. Thinning will initially reduce stand susceptibility to western spruce budworm by reducing basal area. If thinning results in an

FIGURE 5. Projections of some "old growth" structural features in multiple canopy strata stands following different silvicultural regimes. Darkened area shows presence of feature.

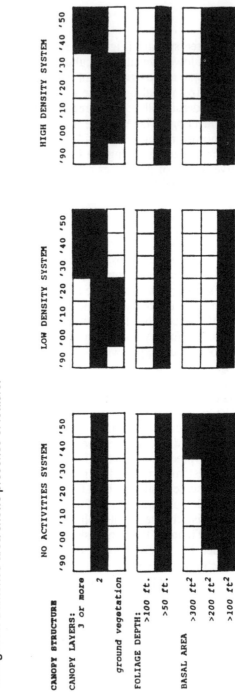

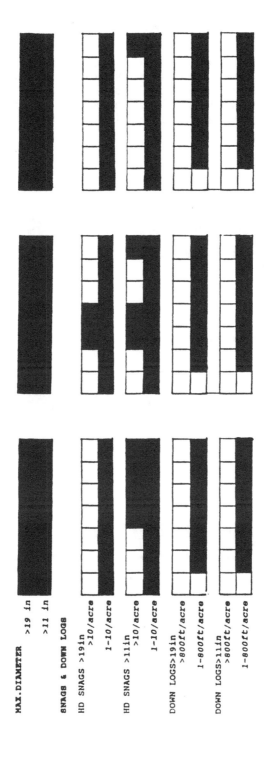

MAX. DIAMETER   >19 in

            >11 in

SNAGS & DOWN LOGS

HD SNAGS >19in
        >10/acre
        1-10/acre

HD SNAGS >11in
        >10/acre
        1-10/acre

DOWN LOGS>19in
        >800ft/acre
        1-800ft/acre

DOWN LOGS>11in
        >800ft/acre
        1-800ft/acre

FIGURE 6. 60-year, stand silhouette projections showing structural features typical of dense, single cohort stands (referred to here as "pole" stands) under different silvicultural regimes. Low Density = Multiple Strata Regime (Scale on left (in feet) represents both horizontal and vertical); Black crown = Douglas-fir; White crown = grand fir; Spotted crown = ponderosa pine.

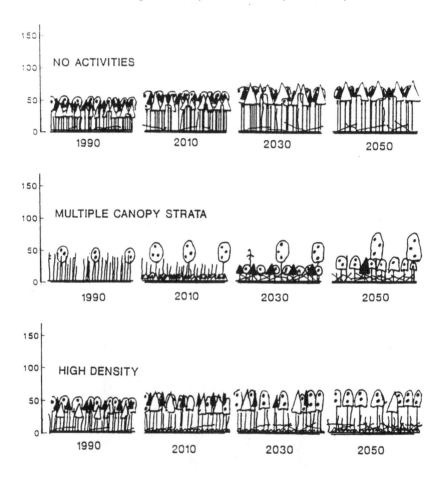

increase in grand firs in the understory, however, stand susceptibility to budworm will increase over time. This trend could be reduced if thinning was heavy enough to favor growth of nonsusceptible species (e.g., ponderosa pine). Creating a multiple canopy structure in dense young stands will substantially reduce the susceptibility of the stand to wind/snow and insects. After four to five decades, the grand firs will become susceptible to wind/snow breakage; however, these trees will be very small and will have little impact on stand structure.

Where silvicultural activities kill trees but do not remove any of them (to create more snags and logs for "old growth" habitat), both multiple canopy and dense single cohort stands will become extremely susceptible to fires. This extreme susceptibility would also occur where trees were killed by the common wind/snow and insect disturbances (Figure 3). Removing most or all of the killed trees during the silvicultural operations or performing other silvicultural treatments to remove fuels can dramatically reduce the stand's susceptibility to fires (Figure 4).

Effects of Timber Removal
During Silvicultural Treatments

Stand Structures–If all merchantable trees are removed in thinnings, the absence of snags and downed logs will be the most noticeably changed structural features. If twigs, foliage, and buds are not removed, the nutrient losses will be minimal. Only young, small snags and very old snags will then be present, since most large trees will be removed before they die. Old down logs will eventually disappear, and few new down logs will be recruited. To provide these features in the future, some trees would need to be designated and left for snag and down log recruitment.

It is highly unlikely that all trees killed in the silvicultural treatments will need to remain in the forest for snags and for down logs. For example, some treatments will create nearly 1,000 snags per acre. At the end of 60 years, there will be over 10,000 linear feet of down logs per acre with some regimes. Evenly distributed, these forests would contain a down log every 4 feet across the forest floor. Since most of a tree's timber value is in the lower one third of the bole, removing this portion and leaving the top as a down log may be a cost-effective way of producing down logs, provided the top is large enough to serve structural purposes. Removing some of the trees during thinnings and lopping the tree tops and branches to ensure contact with the soil and rapid decomposition will markedly reduce fire susceptibility.

Costs–If no silvicultural activities are done in the stands, the primary direct costs will be fighting and recovering from the catastrophic fires which will occur with or without preceding wind or snow breakages and insect disturbances. If active silvicultural treatments are imposed and no timber is removed to offset their costs, the different regimes will require between $742.00 and $1,515.00 per acre (not discounted) over the next 60 years to implement (Table 3). Where thousands of acres are to be sustained for old growth and no timber is removed, millions of dollars will be neces-

FIGURE 7. Projections of species diameters, and percent crown closure, hard snag numbers, and down log lengths for dense, single cohort stands (Figure 6). (Diameters show largest tree of each species and stand. Snags and logs are in 3 diameter classes. N = No Activity Regime; C = Multiple Strata Regime; T = High Density Regime. Snag & down log scales are logarithmic.

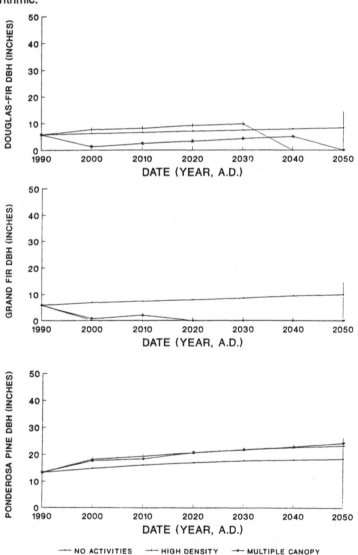

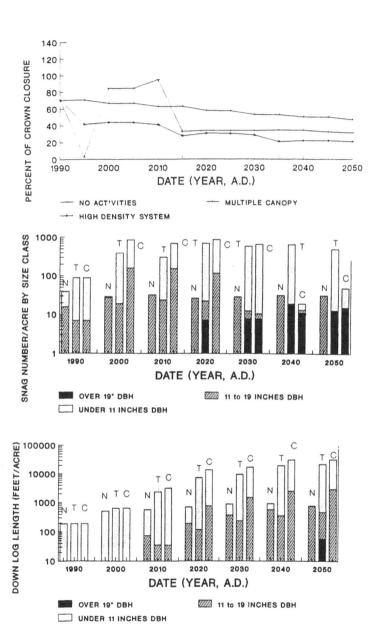

FIGURE 8. Projections of susceptibility of dense, single cohort stands following different silvicultural regimes to destruction by common wind/snow and insect disturbance agents. Susceptibility of 100 = extreme likelihood of disturbance. Susceptibility values over 200 were assigned a value of 200. Wind and snow susceptibility listed by species. [ ] = No activities regime; + = High Density Regime; * = Multiple Strata Regime.

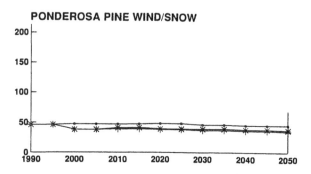

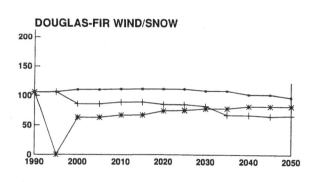

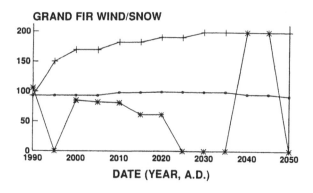

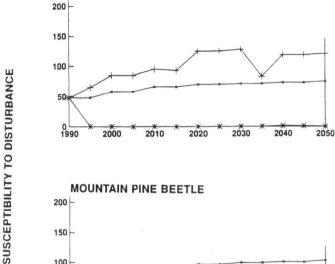

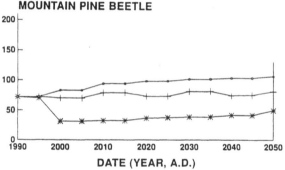

sary to create and maintain suitable habitat. These costs do not include the direct costs of fighting and recovering from catastrophic fires and the indirect costs of losses of habitat and water quality following fires.

If all merchantable timber is removed in the silvicultural treatments, the different regimes will yield a return of between $1,215 and $14,042 per acre (not discounted) over the next 60 years (Table 3 and Figure 10). This return does not include the direct and indirect savings created by avoiding catastrophic wildfires and unemployment.

Removing trees as forest products can reduce costs of implementing the various systems–or create a monetary gain–as well as reducing the stand's susceptibility to fires (Figure 4). Removing some trees for forest products and leaving others for wildlife habitat could bring enough revenue to cover the costs of creating habitat while reducing fire danger. Wildlife and ecological expertise will be necessary to determine the appropriate numbers of snags and down logs to leave.

FIGURE 9. Projections of some "old growth" structural features in dense, single cohort stands following different silvicultural regimes. Darkened area shows presence of feature.

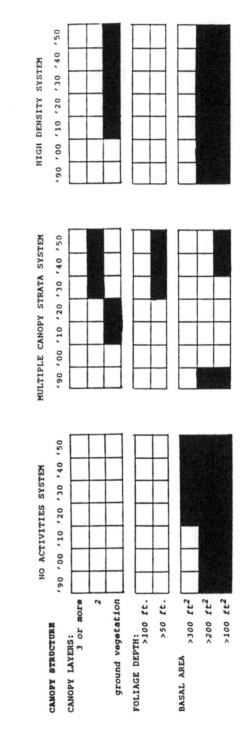

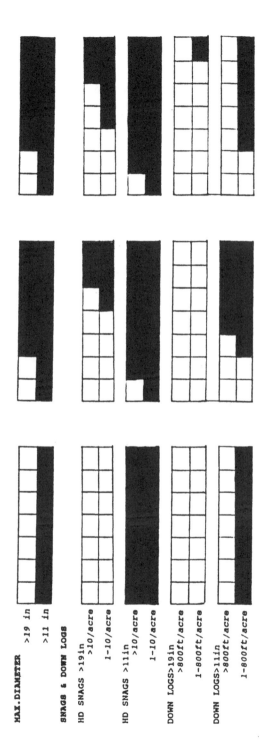

MAX.DIAMETER   >19 in
                >11 in

SNAGS & DOWN LOGS

HD SNAGS >19in
         >10/acre
         1-10/acre

HD SNAGS >11in
         >10/acre
         1-10/acre

DOWN LOGS>19in
        >800ft/acre
        1-800ft/acre

DOWN LOGS>11in
        >800ft/acre
        1-800ft/acre

FIGURE 10. Living (line) and dead volume (block) by time for multiple canopy stands and dense stands, single cohort and silvicultural regimes in eastern Washington. Removing some volume would reduce risk of fire and provide a montary return or offset the costs of silvicultural activities to create old growth structures.

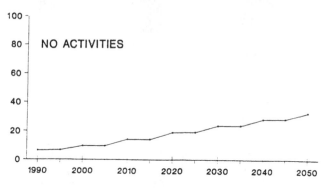

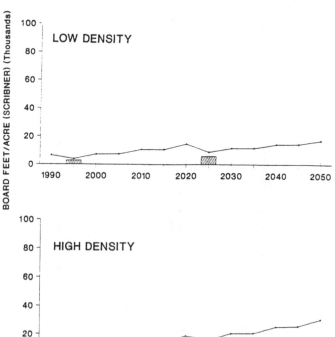

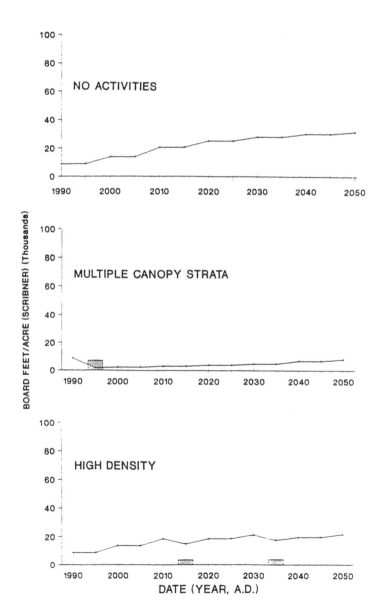

DISCUSSION

Effects of Silvicultural Activities
With and Without Removal of Timber

Avoiding all treatments in the multiple canopy stands will maintain some "old growth" structural features, but will not allow others to be achieved for many decades. In addition, most stands will probably burn–with or without preceding western spruce budworm or mountain pine beetle outbreaks.

Avoiding all silvicultural treatments in the dense, single cohort stands will ensure they will not be suitable habitat for old growth species for the next 60 years. Some stands will become susceptible to wind and snow breakage, which can lead to insect outbreaks and fire. Other stands will first become susceptible to the western pine beetle and, afterwards, fire. Still others may burn without previous wind, snow, or insect disturbances.

Appropriate silvicultural treatments such as thinnings or "uneven-age" harvesting will create many of the "old growth" structural features more rapidly than by avoiding management. These treatments can reduce the susceptibility of the stands to wind, snow, and insect disturbances; however, the stands will become extremely susceptible to fires if some wood is not removed (and the slash treated appropriately).

Silvicultural treatments which remove some wood for forest products will be more effective in both providing greater amounts of "old growth" structures and reducing risks of stand destruction than either avoiding all management or doing silvicultural treatments without removing timber. Removing timber for wood processing can also make creation of "old growth' structures financially profitable, or at least reduce the costs of producing the structures (Lippke and Oliver 1993).

Other silvicultural treatments are also possible, but were not simulated here. For example, tree tops can be removed with dynamite or a chain saw to create snags which rot differently from girdled trees. Hollow, living trees can be created by scarring living trees. Understory density and species can be manipulated by scarifying the forest floor, although little understory vegetation will grow if the overstory is extremely dense. Some forms of "nest platforms" can be constructed in trees if such structures are helpful for endangered species (e.g., the spotted owl). Other silvicultural regimes can also make the stands less susceptible to various disturbance agents at various times.

The actual silvicultural treatments can be done during seasons when the old growth species would be least disturbed. Each treatment described in this paper will occur about once every twenty years in a given stand and last for one day to two weeks, depending on the technique. Each stand, therefore, will be disturbed for 0.013 to 0.19 percent of the time by these silvicultural treatments.

Landscape Management

There is no single silvicultural regime (including avoiding all activities) which will create and maintain a structure ideally suited for old growth species throughout time. The structural features will change under each silvicultural regime, with some useful features becoming present as others decline. In fact, it is probably not biologically possible to have and maintain (through natural or artificial means) a stand which will be perpetually ideal "old growth" habitat. Structural features useful for old growth species can probably best be achieved in a forest by maintaining a mixture of stands across a landscape, with the structural features developing in different stands at different times (Oliver 1992). This mixture will also keep large areas from developing susceptibility to a disturbance agent and being destroyed at one time. It will probably be necessary to do stand-replacement harvests (varieties of clearcut, shelterwood, and seed tree harvests) to some stands to maintain open habitats (Young 1992) where natural disturbances do not create sufficient amounts across the landscape.

Each active silvicultural regime will create some structural features of old growth at some times. Doing no silvicultural treatments will generally create the fewest old growth features where the features do not presently exist and may leave stands susceptible to various disturbances in the long term. Which regime is used on a given stand will depend on the present mixture of stand structures across the landscape, how this mixture will change with time, and what is the desired distribution of structures across the landscape. It will probably be appropriate to prescribe different silvicultural regimes for different stands within a landscape so that different stands produce desired structural features and avoid becoming susceptible to disturbances at different times. This "landscape management" will keep various "old growth" and other structural features available within the landscape as a whole at all times (Boyce 1985).

Practical Considerations

Maintaining a variety of stand structures across a landscape while removing timber, with each stand developing to certain structural features at different times, will also have practical advantages. Silvicultural treatments required will be diversified and staggered, so a relatively constant work force and budget can be maintained. Also, any products removed will flow at a relatively steady rate to provide steady manufacturing employment. The wood produced will provide an environmentally sound substitute for more polluting steel, aluminum, brick, or concrete (Kershaw et al. 1992) or for wood produced from regions of the world with fewer environmental safeguards. Finally, the diversity of stand structures and times of treatments will reduce the risk of catastrophic disturbances caused by synchronous buildups of fuels or insects across a large landscape.

REFERENCES

Boyce, S.G. 1985. Forestry decisions. USDA Forest Service General Technical Report SE-35. 318 p.

Brown, J.K., J.A. Kendall Snell and D.L. Bunnell. 1977. Handbook for predicting slash weight of western conifers. USDA Forest Service General Technical Report INT-37. 35 p.

Franklin, J.F. and C.T. Dyrness. 1973. Natural vegetation of Oregon and Washington. USDA Forest Service General Technical Report PNW-8. 417 p.

Franklin, J.F., F. Hall, W. Laudenslayer, C. Maser, J. Nunan, J. Poppino, C.J. Ralph and T. Spies. 1986. Interim definitions for old-growth Douglas-fir and mixed-conifer forests in the Pacific Northwest and California. USDA Forest Service Research Note PNW-447. 7 p.

Halle, F., R.A.A. Oldeman and P.B. Tomlinson. 1978. Tropical Trees and Forests: An Architectural Analysis. Springer-Verlag. New York.

Hunter, M.L., Jr. 1990. Wildlife, forests, and forestry. Regents/Prentice Hall. Englewood Cliffs, NJ.

Johnson, C.G., R.R. Clausnitzer, P.J. Mehringer and C.D. Oliver. 1993. Biotic and abiotic processes of eastside ecosystems: the effects of management on plant and community ecology, and on stand and landscape vegetation dynamics. In: Hessburg, P. F. (Ed.). Eastside Forest Ecosystem Health Assessment. Volume III. Assessment. USDA Forest Service, National Forest System, Forest Service Research. Wenatchee, WA.

Kershaw, J.A., Jr., C.D. Oliver and T.M. Hinckley. 1992. Effect of harvest of old growth Douglas-fir stands and subsequent management on carbon dioxide levels in the atmosphere. *Journal of Sustainable Forestry* 1(1):61-77.

Lippke, B. and C.D. Oliver. 1993. Managing for multiple values. *Journal of Forestry* 91:14-18.

Marcot, B. 1990. SNAGS snag recruitment simulator. USDA Forest Service, Pacific Northwest Forest and Range Experiment Station. Portland, OR.

Moeur, M. 1985. COVER: a user's guide to the CANOPY and SHRUB extension of the Stand Prognosis Model. USDA Forest Service. General Technical Report INT-190. 49 p.

Oliver, C.D. 1981. Forest development in North America following major disturbances. *Forest Ecology and Management* 3:153-168.

Oliver, C.D. 1992. Enhancing biodiversity and economic productivity through a systems approach to silviculture. The Silviculture Conference. Forestry Canada. Ottawa, Ontario. pp. 287-293.

Oliver, C.D. and B.C. Larson. 1990. Forest Stand Dynamics. McGraw-Hill. New York.

Oliver, C.D., L.L. Irwin and W.H. Knapp. 1992. Eastside forest management practices: historical overviews, extent of their application, and their effects on sustainability of ecosystems. In: Hessburg, P.F. (Ed.). Eastside Forest Ecosystem Health Assessment. Volume III. Assessment. USDA Forest Service, National Forest System, Forest Service Research. Wenatchee, WA.

Sprugel, D.G. 1991. Disturbance, equilibrium, and environmental variability: What is "natural" vegetation in a changing environment? *Biological Conservation* 58:1-18.

Stage, A.L. 1973. Prognosis model for stand development. USDA Forest Service Research Paper INT-137. 32 p. (Version 6.0 East Cascades)

Thomas, J.W. (Ed.). 1979. Wildlife habitats in managed forests: The Blue Mountains of Oregon and Washington. USDA Forest Service, Agricultural Handbook No 553. 512 p.

Young, M.R. 1992. Conserving insect communities in mixed woodlands. In: pp. 277-296. Cannell, M.G.R., D.C. Malcolm and P.A. Robertson (Eds.). The Ecology of Mixed-Species Stands of Trees. Blackwell Scientific Publications. Oxford, England.

Forest Health
and Wildlife Habitat Management
on the Boise National Forest, Idaho

John R. Erickson
Dale E. Toweill

ABSTRACT. The National Forest Management Act (NFMA) requires the Forest Service to provide for diversity of plant and animal communities and maintain viable wildlife populations. Changes in forest stand structure, species composition and disturbance patterns within ponderosa pine (*Pinus ponderosa*)-Douglas-fir (*Pseudotsuga menziesii*) habitat types on the Boise National Forest make it difficult to meet NFMA direction. Three management strategies, including "no action," were evaluated in terms of the risk of wildfire's effects on plant community diversity and distribution, dispersal, and local population viability for the pileated woodpecker (*Dryocopus pileatus*) and flammulated owl (*Otus flammeolus*). The no action alternative appeared to have the greatest long-term risk to plant community diversity and wildlife species distribution and dispersal. Landscape analysis that considers the capabilities and risks associated with different management strategies is recommended to meet NFMA direction while responding to diverse public expectations of the Forest.

John R. Erickson is Forest Wildlife Biologist, Boise National Forest, Boise, ID 83702.

Dale E. Toweill is Wildlife Program Coordinator, Natural Resources Policy Bureau, Idaho Department of Fish and Game, Boise, ID 83707.

[Haworth co-indexing entry note]: "Forest Health and Wildlife Habitat Management on the Boise National Forest, Idaho." Erickson, John R. and Dale E. Toweill. Co-published simultaneously in the *Journal of Sustainable Forestry* (The Haworth Press, Inc.) Vol. 2, No. 3/4, 1994, pp. 389-409; and: *Assessing Forest Ecosystem Health in the Inland West* (eds: R. Neil Sampson and David L. Adams) The Haworth Press, Inc., 1994, pp. 389-409. Multiple copies of this article/chapter may be purchased from The Haworth Document Delivery Center [1-800-3-HAWORTH; 9:00 a.m. - 5:00 p.m. (EST)].

INTRODUCTION

Douglas-fir habitat types comprise approximately 40-50% of the forested area on the Boise National Forest. These habitat types experienced dramatic changes in stand structure (tree density, overstory layers, understory fuels and diversity) over the last 70-100 years. These changes affect the Boise National Forest's ability to meet NFMA requirements for diversity and species' population viability.

The National Forest Management Act (NFMA) requires forests to "provide for diversity of plant and animal communities and tree species consistent with the overall multiple-use objectives of the planning area." In accomplishing this, NFMA requires an evaluation of diversity "in terms of its prior and present condition" and "consider how diversity will be affected by various mixes of resource outputs and uses." The Act also requires forests to "maintain viable populations of all native and desired non-native wildlife vertebrate species in the planning area" (36 CFR 219.26). A viable population is one that "has the estimated numbers and distribution of reproductive individuals to insure its continued existence is well distributed in the planning area" (36 CFR 219.19).

Changes in stand structure, species composition and disturbance patterns, particularly the trend towards extensive stand replacement fire, can affect plant and animal community diversity and local wildlife population distribution, dispersal, and viability. Changes in disturbance frequency and intensity can have substantial effects on habitat distribution and patch size (Turner et al. 1989). The pattern of successional stages across the landscape may enhance or retard the spread of disturbances, and disturbances may generate new landscape patterns (Pickett and White 1985, Franklin and Forman 1987, Krummel et al. 1987, Remilard et al. 1987, Turner and Bratton 1987). In order to meet NFMA direction for diversity and population viability, land managers should analyze management activities in relation to past, current, and predicted future disturbance patterns.

PRESETTLEMENT FIRE PATTERNS

Prior to settlement, drought, combined with high winds and lightning, resulted in stand replacement fires within the ponderosa pine-Douglas-fir habitat types. However, these events were thought to be infrequent and not the dominant factor determining the stand structure and species composition or the pattern of successional stages across the landscape. Studies of fire scars on trees in the Boise National Forest (Steele et al. 1986) indicate that the current trend of stand replacement fire is not typical of presettlement disturbance patterns. Fire scars on trees indicate that, prior to 1900, the dominant fire regime was frequent low-intensity understory fires (Steele et al. 1986). Fire frequencies ranged from a mean of 9.8 to 21.7 years (Table 1, Figure 1).

This presettlement disturbance pattern had direct effects on stand structure and composition. The estimated structure of presettlement stands within the Douglas-fir habitat type was a single open canopy layer (20-50 % canopy closure), low tree densities consisting of large mature and old-growth ponderosa pine, and infrequent large snags. Stands were generally open and park-like without laddering fuels that could carry the fire into the canopy. Current stand data provide evidence of the structure and distribution of these remnant mature and old growth ponderosa stands (stands with 10-20 ponderosa pine trees/acre at > 20" diameter at breast height (dbh) now dominated by immature Douglas-fir 8-14" dbh) on the Boise National Forest.

The presettlement vegetation pattern influenced wildlife species diversity and distribution. Species such as the flammulated owl and white-headed woodpecker (*Picoides albolarvatus*) benefitted from mature, open stands of ponderosa pine with open understories. Because these habitat types were not as vulnerable to stand replacement fire, dramatic shifts in the amount and distribution of wildlife habitat were thought to be uncommon in these habitat types.

Areas protected from frequent understory fire (e.g., riparian areas and natural fuel breaks) developed into late successional stages. Fires were generally infrequent (250-350 + years). Habitat distribution was generally patchy and linear. Riparian areas provided closed forest canopy corridors

TABLE 1. Mean fire interval of sampled stands with Douglas-fir habitat types arranged on a dry-to-moist gradient (Steele et al. 1986).

Stand Number	Elevation (Ft.)	Habitat Type (potential climax)	Mean fire Interval (1700–1895)
5	5,625	Ponderosa pine/mountain snowberry	11.4
3	5,000	Douglas-fir/elk sedge	9.8
1	4,600	Douglas-fir/white spirea	10.3
4	4,975	Douglas-fir/white spirea	18.1
6	4,850	Douglas-fir/nine bark (dry)	12.8
2	5,820	Douglas-fir/nine bark (typical)	15.9
7	5,600	Douglas-fir/nine bark (moist)	21.7

FIGURE 1. Fire frequency, based on sample sites from Steel et al. (1986), in the Boise basin, 1700-Present.

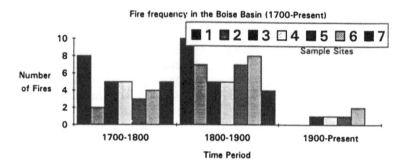

through more extensive open canopy ponderosa uplands and provided habitat for species tied to more closed forest characteristics (pileated woodpecker and marten (*Martes americana*). Most of these areas probably did not experience dramatic shifts in successional stages. As a result, species distribution, dispersal and viability associated with these habitats remained relatively stable.

CURRENT SITUATION

The lengthening of fire intervals after the late 1800s has been noted in ponderosa pine-Douglas-fir habitat types throughout much of western North America (Arno 1976; Stokes and Dieterich 1980; Martin 1982). This trend has also occurred on the Boise National Forest (Figure 1). Three factors have been suggested as responsible for the lengthened fire interval: (1) fires started by native Americans decreased; (2) organized fire suppression began in 1905 in central Idaho, and in 1908 two fire lookouts were built in the Boise Basin (Smith 1983); and (3) widespread livestock grazing removed large portions of the grass and other light fuels (Steele et al. 1986).

Longer periods without fire within ponderosa pine-Douglas-fir habitat types has substantially altered the forest structure from presettlement conditions. Exclusion of understory fire since settlement has allowed understory fuels to accumulate and dense late successional tree species (Covington and Moore 1992) and brush to establish in formerly open stands (Cooper 1960; West 1969; Hall 1976; Gruell et al. 1982). These changes led to a shift from the low intensity surface fires to intense stand replacement fires.

The Boise National Forest is currently suffering from catastrophic insect, disease, and extensive stand replacement wildfires. Insect activity is widespread throughout central Idaho and particularly centered on the Boise National Forest. Of the eight major pests occurring on the Forest, six were at epidemic levels in 1992. Over half a million trees on the Boise National Forest have died since 1988 from bark beetle attacks (*Scolytus ventralis, Dryocetes confusus, Dendroctonus spp, Ips spp*), and trees on more than a quarter million acres have been defoliated by Douglas-fir tussock moths (*Orgyia pseudotsuga*).

The size and intensity of fires on the Boise National Forest have also increased dramatically since the 1950s (Figure 2). Prior to 1986, approximately 3,000 acres burned annually on the Forest. Since 1986, approximately 56,000 acres burned annually. Excluding the August 1992 Foothills fire (139,955 acres on the Boise National Forest), an average of 28,000 acres have burned on the forest annually–well above the historic average (Boise National Forest fire records). In the last 7 years, 414,767 acres have burned on the Boise National Forest (16% of the entire Forest). Of this, 346,618 acres (83%) were from fires larger than 5,000 acres (Boise National Forest fire records).

While six years of drought contributed substantially to the conditions described above, drought events are cyclic, and the Boise National Forest did not experience these same levels of mortality or fire effects during previous drought periods. Fire data (based on fire records) on the Boise National Forest indicate that, although large fires occurred historically (1926-1935 drought period), the frequency of large wildfires since 1986

FIGURE 2. Acres burned annually on the Boise National forest by time period.

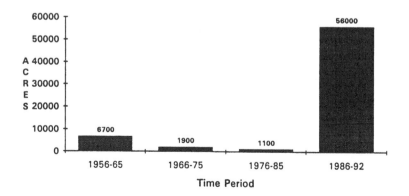

far exceeds the historic range of variability experienced over the last 70 years (Figure 3).

Changes in tree stand density and species composition within ponderosa pine-Douglas-fir habitat types have been suggested as the primary cause of the severe insect, disease and fire situation experienced during the last six years. If true, without changes in forest stand structure, the Boise National Forest can expect these events to recur during the next drought period.

Continued extensive stand replacement fires could substantially alter successional patterns on the Boise National Forest, favoring homogeneous landscapes. These changes could substantially affect wildlife species distribution, diversity, and viability as extensive areas are converted to a single successional stage. Figure 4 shows conceptually how an increase in fire size could influence pileated woodpecker distribution and dispersal by reducing the number of suitable habitat sites within the dispersal range of juveniles.

EFFECTS ON PLANT AND ANIMAL DIVERSITY AND LOCAL POPULATION VIABILITY

The 15,000 acre Logging Gulch Timber Sale on the Boise National Forest provided an opportunity to evaluate effects of forest management, fire exclusion, and future risks of extensive wildfire on wildlife species

FIGURE 3. Comparison of acres burned by fire size by time period since 1926. The average size of large fires includes all fires 300 acres or greater. Historically, large wildfires occurred on the Boise National Forest (shown by the 5,000 + acre line in between 1926-1935); however, the average fire size of large fires was substantially lower in the 1926-1935 drought period.

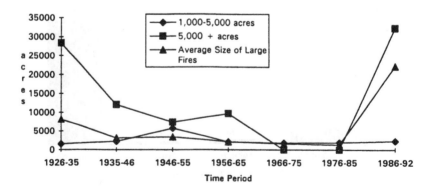

FIGURE 4. Conceptual comparison of the effects on pileated woodpecker distribution and dispersal from an increase in fire size.

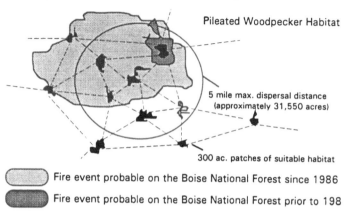

Pileated Woodpecker Habitat

5 mile max. dispersal distance
(approximately 31,550 acres)

300 ac. patches of suitable habitat

Fire event probable on the Boise National Forest since 1986

Fire event probable on the Boise National Forest prior to 198

distribution, dispersal and viability/persistence. An environmental analysis of several management options was conducted for the Logging Gulch area (Logging Gulch Timber Management Project Environmental Assessment). Six different alternatives (named A through F) were considered; three (Alternatives A, D, and F) are compared here in terms of their effect on piletated woodpecker (species associated with mature closed forest) and flammulated owl (species associated with mature open forest) distribution and likelihood of persistence. The three management strategies are described below:

Alternative A: Continue current management of the 15,000 acre project area. Since a large portion of the project areas is roadless, this decision basically results in a protection or preservation prescription with no treatment. Fire suppression would continue.

Alternative D: Apply a forest health prescription involving thinning 5,487 acres (60% of the forested area) of the project area, precommercial thinning 3,854 acres, and understory fuel reduction 5,487 acres.

Alternative F: Apply a forest health prescription involving thinning 2,885 acres (32% of the forested area) of the project area, precommercial thinning 3,364, and understory fuel reduction 2,885 acres.

The effects of forest management and risk of extensive wildfire on long term viability and likelihood of persistence for pileated woodpecker and flammulated owl were evaluated utilizing the following factors:

Demographics: Genetics, life history, distribution (density and extent), birth and death rates, sex ratios, population isolation.

Habitat: Patch size, quality, and distribution of habitat within the landscape.

Environmental: Vulnerability to catastrophic events (wildfire, flood, landslide, etc.).

Demographics, habitat, and environmental factors rarely function independently. Often, one of the factors may be dominant, making a population vulnerable to the effects of one or both of the other factors. The population viability and persistence decision tree was used on the Boise National Forest to evaluate the likelihood of persistence (Figure 5, Table 2).

The primary disturbance factor influencing wildlife habitat distribution across the Logging Gulch project area is fire. Stand replacement fire hazard ratings were developed by evaluating current stand structure and estimating changes immediately following harvest and 30 years after harvest. The hazard rating is based upon tree density, habitat type, aspect, slope, understory/overstory fuel levels, and stand structural stage (development of laddering fuels in the understory). Stands were given a rating of low, moderate, or high.

Analysis evaluated the number of acres, and location of high hazard conditions for each management strategy (Figure 6). Approximately 1,491 acres had a high fire hazard rating. Immediately after harvest, the acres with a high fire hazard rating were reduced to 951 under alternative D and 1,441 acres under alternative F. In 30 years, acres considered to be a high fire hazard are expected to increase to 6,733 (45% of project area, 75% of forested acres in Logging Gulch) under Alternative A (No Action). Alternate D and F were assumed to maintain the current conditions for 30 years (after 30 years, the number of acres expected to have a high fire hazard gradually increased).

It was assumed that, as stands with high fire hazard become connected, the risk of extensive stand replacement fire is increased. Because stands in Logging Gulch are currently at a low or moderate hazard rating, implementation of the action alternatives had little immediate effect on hazard ratings. With no action, the number of acres considered to be at high risk increased dramatically throughout the planning period.

FIGURE 5. Process used on the Boise National Forest to estimate the likelihood of persistence for selected wildlife species.

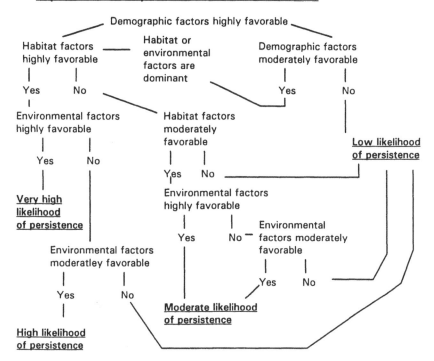

The risk to maintaining pileated woodpecker and flammulated owl habitat was analyzed by superimposing acres with the 30 year high fire hazard projection over the current and estimated future conditions. Maps 1, 2, and 3 display the current and projected distribution of suitable pileated woodpecker habitat and acres of high fire hazard. Maps 4, 5, and 6 display the current and projected distribution of suitable flammulated owl habitat and acres of high fire hazard.

Based upon an analysis of habitat maintained and the risk of stand replacement fire, a likelihood of persistence rating was made for each alternative. The rating was applied immediately after harvest and 30 years after harvest (Table 3). Because of the increased risk of extensive stand replacement fire in the upper basins of Logging Gulch, the likelihood of

TABLE 2. Summary of three levels of persistence for the pileated woodpecker and flammulated owl based on varying conditions of demographic, habitat, and environmental factors (modified from Remain and McIntyre 1993).

Characteristics of the population	Nature of Risk	Likelihood of Persistence		
		LOW	MODERATE	HIGH
Variability in survival and recruitment	Demographic	Pop. survival & recruitment responds sharply to normal environmental events. Year class failures common. Moderate changes to demographics cause elimination of recruitment or survival.	Continued existence of populations is moderately vulnerable to small changes in recruitment and/or survival. Periodic environmental disturbances cause moderate changes in recruitment or survival	Pop. are not vulnerable to disturbances that are within the normal range. Periodic environmental extremes do not result in substantial shifts in pop. distribution or local persistence.
Growth and Survival	Demographic	Populations do not have the resilience to respond to catastrophic events. Major long term shifts (greater than 100 years) in local species distribution are likely.	Populations have the resilience to recover from disturbances to demographics; however, response is expected to be slow.	Population has the resilience to recovery from catastrophic events within 5-10 years. Substantial changes in distribution unlikely.
Life History	Demographic	Habitat is not available for all life stages. Survival or reproduction is substantially reduced.	Habitat is available for all life stages; however, mortality and reproducing territories limit reproduction and recruitment.	Habitat is available for all life stages and mortality is within natural levels.
Size	Habitat	Habitat only meets min. patch size requirements and is isolated through portions of the planning area.	Patch size exceeds min. requirements. Habitat is not isolated; however, disturbances do lead to substantial changes in habitat distribution affecting species distribution and dispersal.	Patch size exceeds min. requirements. Minor disturbances do not affect distribution of suitable habitat. Species dispersal and distribution are not affected.
Quality	Habitat	Habitat only marginally meets requirements. Minor decline in habitat quality renders habitat unsuitable.	Habitat meets requirements for all life stages; however, key components are limiting and any further loss will result in unsuitable conditions. Habitat may improve in 10-20 years.	Habitat quality is high, meeting requirements for all life stages. A decline in some structural features does not render the habitat unsuitable.

	Low likelihood	Moderate likelihood	High likelihood
Distribution	Habitat disturbances are likely to eliminate the species over portions of its range. Habitat distribution only meets min. requirements.	Moderate disturbances are likely to cause a reduction in species distribution Habitat distribution exceeds min. requirements.	Habitat distribution is high through the area and natural disturbances do not effect the distribution of the species through the planning area.
Habitat			
Environmental — Natural disturbances (within range of historic variability)	Minor flucuations in local species distribution occur as a result of disturbances that are within the the range of historic variability.	No fluctuations in species distribution occur as a result of disturbances that are within the range of historic variability.	No fluctuations in species distribution occur as a result of disturbances that are within the range of historic variability.
Environmental — Catastrophic Events (wildfire, flood, drought)	Disturbances are likely to eliminate species from portions of its range. Continued distribution within the planning area is vulnerable to even limited disturbances in habitat.	Moderate disturbances within the natural range of variability are unlikely to eliminate species distribution within the planning area. However, continued distribution of the species is influenced by extreme events within the planning area.	Disturbances, including extreme events, are unlikely to eliminate the species from portions of its range. Continued distribution within the planning area is not vulnerable to disturbances anticipated..

Remarks:
The determination of low, moderate or high likelihood of persistence requires professional judgment. However, the factors outlined above should be used to help document the rationale for the determination. Generally, a species with a low likelihood of persistence meets only minimum requirements for each of the three factors identified. A species with a moderate likelihood of persistence may fully meet requirements for some of the factors yet only meets minimum requirements in one or two others and is vulnerable to changes in habitat, disturbance factors or demographics. A species with a high likelihood of persistence is one that has strong populations that are not isolated. Habitat size, distribution and quality is above minimum levels such that dispersal and recruitment to the population is maintained. As a result, the population can rebound from disturbances that are within the historic range of variability and persist given catastrophic events. The Boise National Forest intends to manage habitat to meet a high likelihood of species persistence over time. This means that the continued existence of a population is accomplished through assuring that suitable habitat is well distributed and able to withstand environmental factors such as wildfire.

FIGURE 6. Comparison of fire risk ratings (number of acres at high risk of stand replacement fire) immediately after harvest and 30 years after harvest.

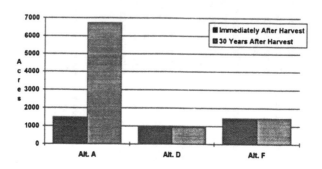

persistence for pileated woodpecker shifted from a high to a moderate rating in 30 years under the No Action Alternative (Alternative A). Analysis of flammulated owl persistence showed a shift from high to low in 30 years under the No Action alternative based upon the loss of habitat suitability due to increased tree density, closed forest canopy, and increased hazard of stand replacement fire.

Both action alternatives (D and F) resulted in a high persistence rating for the flammulated owl because of the ability to maintain more mature open stands of suitable patch size across the landscape and with a reduced risk of extensive stand replacement fire. Action alternatives could yield opportunities to increase species persistence ratings for the pileated woodpeckers through mitigative marking guides. For example, by reducing the number of trees thinned and leaving more large live trees and large snags, habitat suitability can be maintained while reducing the hazard of extensive stand replacement fire. As a result, a higher persistence rating can be achieved. The effects of Alternative A (no action) can only be mitigated through increased fire suppression.

Generally, Alternative A (no action) tended to decrease plant community diversity as stands progressed towards late successional stages and disturbances became more extensive. Action alternatives (D and F) had the potential to increase plant diversity by breaking up homogeneous areas. However, implementation of any single action alternative that treats large portions of the project area has the potential to reduce plant community diversity (Turner et. al. 1989) by changing large portions of the landscape to a single age class. In the Logging Gulch project area, selection of Alternative D would have reduced plant and animal diversity by creating more homogenous habitat than presently exists.

Map 1. Current pileated woodpecker habitat distribution in Logging Gulch and location of stands with high stand replacement fire hazard in 30 years under Alternative A.

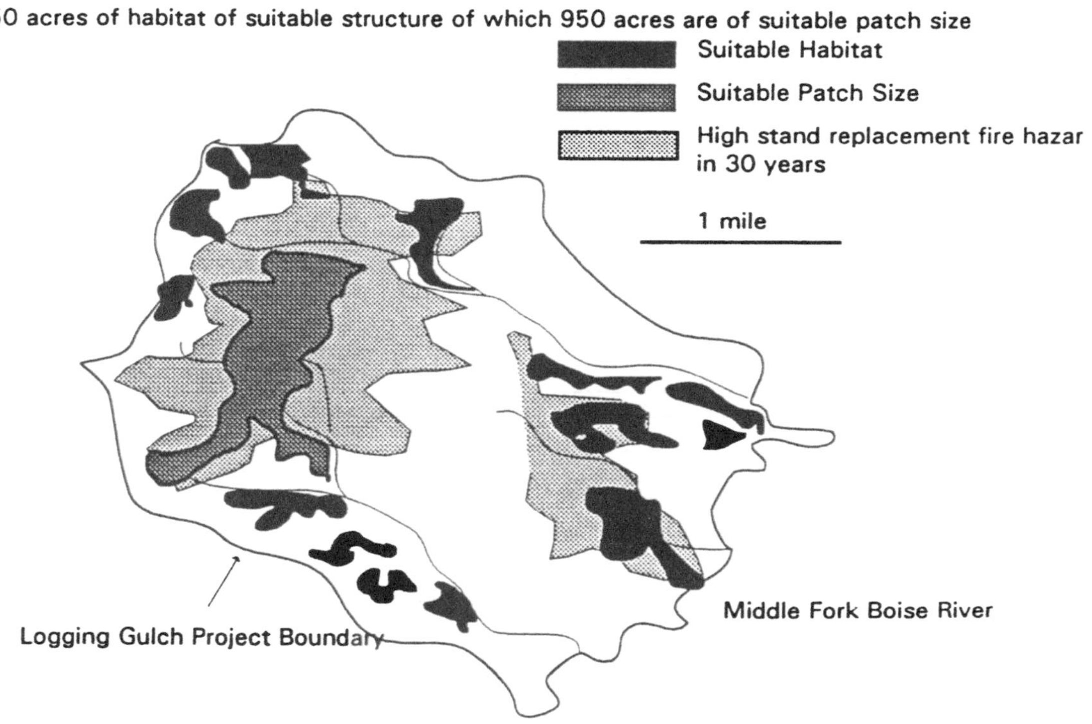

2,850 acres of habitat of suitable structure of which 950 acres are of suitable patch size

Suitable Habitat

Suitable Patch Size

High stand replacement fire hazard in 30 years

1 mile

Middle Fork Boise River

Logging Gulch Project Boundary

Map 2. Predicted pileated woodpecker habitat distribution in Logging Gulch and location of stands with high stand replacement fire hazard in 30 years under Alternative D.

Predicted 904 acres of suitable habitat of which 190 acres are of sufficenct patch size for the pileated woodpec

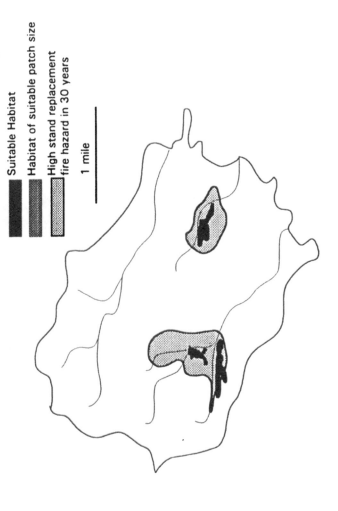

Suitable Habitat

Habitat of suitable patch size

High stand replacement
fire hazard in 30 years

1 mile

Map 3. Predicted pileated woodpecker habitat distribution in Logging Gulch and location of stands with high stand replacement fire hazard in 30 years under Alternative F.

Predicted 1,690 suitable acres of which 380 acres are of suitable patch size

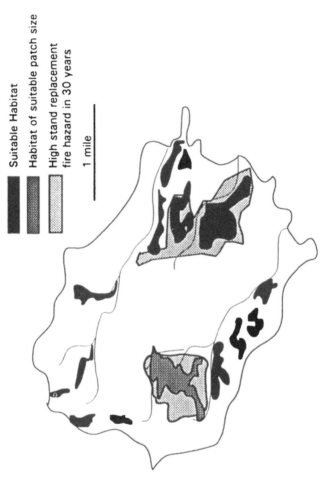

Suitable Habitat

Habitat of suitable patch size

High stand replacement fire hazard in 30 years

1 mile

Map 4. Current flammulated owl nesting habitat distribution in Logging Gulch and location of stands with high stand replacement fire hazard in 30 years under Alternative A.

Currently approximately 2,000 acres of suitable flammulated owl habitat exists which decreases to approximately 1,400 acres in 30 years.

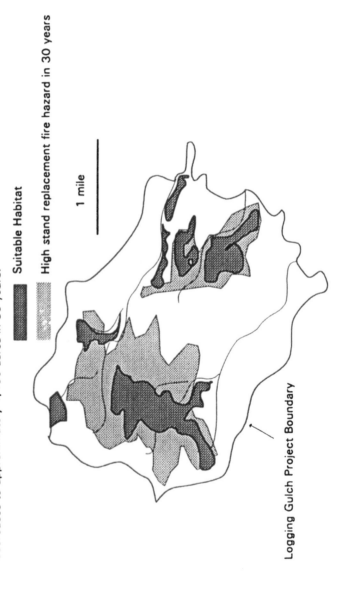

■ Suitable Habitat

▨ High stand replacement fire hazard in 30 years

1 mile

Logging Gulch Project Boundary

Map 5. Predicted flammulated owl nesting habitat distribution in Logging Gulch and location of stands with high stand replacement fire hazard in 30 years under Alternative D.

Predicted 1,042 acres of suitable flammulated owl nesting habitat maintained after harvest which invreases to 1,700 acres in 30 years.

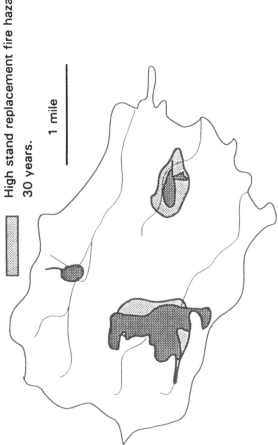

Suitable Habitat

High stand replacement fire hazard 30 years.

1 mile

Map 6. Predicted flammulated owl nesting habitat distribution in Logging Gulch and location of stands with high stand replacement fire hazard in 30 years under Alternative F.

Approximately 1,222 acres of suitable habitat which increases to 1,800 acres 30 years

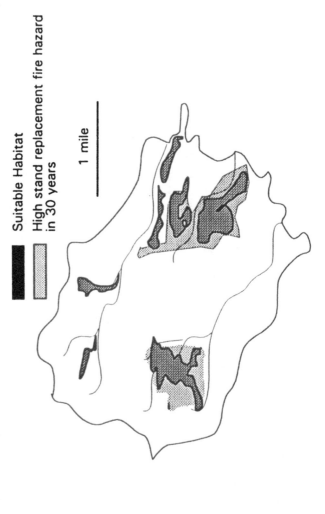

Suitable Habitat

High stand replacement fire hazard in 30 years

1 mile

TABLE 3: Summary of current likelihood of persistence over time for three alternatives considered for pileated woodpecker and flammulated owl in Logging Gulch.

Pileated woodpecker

Determination Alternative	Demographic	Habitat	Environmental	Overall Persistence 5 Years	30 Years
Alt. A (No Action)	High	High	Moderate	High	Moderate
Alt. F	High	High	High	High	High
Alt. D	High	Low	Moderate	Low	Low

Flammulated Owl

Determination Alternative	Demographic	Habitat	Environmental	Overall Persistence 5 Years	30 Years
Alt. A (No Action)	High	High	Moderate	High	Low
Alt. F	High	High	High	High	High
Alt. D	High	Moderate	Moderate	Mod.	High

RECOMMENDATIONS FOR THE FUTURE

Forest managers are custodians of the public trust on public lands. All management actions must be weighed not only in terms of the ability of managers to meet specific short- and long-term objectives; they must also be evaluated in terms of public concerns, whether those concerns have been mandated on the forest legislatively (as through the NFMA) or reflect changing social demands for products–commodity or non-commodity–from those lands. Forest management decisions must also be viewed in the broad context of space and time, beyond specific project boundaries. The traditional view of resource management has not always considered the effects of disturbance patterns at a landscape level. Increasingly, land managers will need to consider decisions in view of not only human effects to resources but also how decisions influence disturbance patterns and what changes in disturbance factors do to desired outputs and outcomes. These changes have consequences to wildlife species distribution, dispersal, and local population viability. The distribution of suitable habitat for species such as the flammulated owl, whiteheaded woodpecker, and foraging habitat for goshawk has declined as forest stand structure has shifted towards denser stands of Douglas-fir with more snags, closed canopy, and understory fuels. These shifts have temporarily increased the distribution of wildlife species associated with mature, closed canopy forests (pileated woodpecker and marten). However, the trend toward increased hazard of extensive stand replacement fire threatens the continued habitat distribution and dispersal which could affect local population viability.

Commodity products (such as wood fiber, minerals, and livestock forage) have traditionally been amenable to economic modelling as a decision-making tool, and thus have been featured prominently in decision analyses. Non-commodity products (such as recreation and plant and animal diversity) have been less amenable to economic analyses, and thus have had less direct application in formal decision analysis. However, these values also weigh heavily in decisions made in the public interest.

Public concern about, and participation in, forest management decisions regarding public lands is increasing. Many rural communities, and particularly those dependent on timber harvest, face an uncertain economic future. County governments and local Chambers of Commerce are very aware of the importance of timber harvest receipts and timber harvest jobs to local economies. Reductions in timber harvests and reductions in local job opportunities have a direct and immediate impact on local economies, while fluctuations in timber supply produce economic uncertainty about future conditions and may lead to either failure to invest in local economies or loss of existing businesses. Similarly, those who value non-commodity resources on public lands view the future of those resources as uncertain. These groups have championed more direct analyses of non-commodity values in the formal decision making process. Both groups have been increasingly willing to go to court to secure judicial decision that reduce uncertainty about the resources they value.

The "No Action" alternative has often been perceived as maintaining the status quo. We have demonstrated that the "No Action" alternative, even when evaluated in light of non-commodity resources such as wildlife, carries risks. Plant communities and even physical features of the landscape change through time in somewhat predictable ways. Thus, the "No Action" alternative cannot be evaluated without identification of the foreseeable risks and opportunities forgone. We have attempted to present an example of how the No Action and specific action alternatives may be analyzed to allow decision-makers more complete information on which to base decisions.

REFERENCES

Arno, S.F. 1976. The historical role of fire on the Bitterroot National Forest. USDA Forest Service Resource Pap. INT-187. 29p.

Boise National Forest. 1993. Logging Gulch timber management project environmental assessment. USDA Forest Service. Boise, ID.

Boise National Forest. 1993. Fire records: 1956-1992. USDA Forest Service. Boise, ID.

Cooper, C.F. 1961. Patterns in ponderosa pine forests. *Ecology* 42:493-499.

Covington, W. W. and M. M. Moore. 1992. Postsettlement changes in natural fire regimes: Implications for restoration of old-growth ponderosa pine forests. Paper presented at the old-growth forests in the Southwest and Rocky Mountain Regions, Portal, AZ.

Dieterich, J. M. 1980. Chimney spring forest fire history. USDA Forest Service Research Paper RM-220, 8p.

Franklin, J. F. and R. T. T. Forman. 1987. Creating landscape patterns by forest cutting: ecological consequences and principles. *Landscape Ecol.* 1:5-18.

Gruell, G.R., W. C. Schmidt, S. F. Arno and W. J. Reich. 1982. Seventy years of vegetative change in a managed ponderosa pine forest in western Montana-implications for resource management. USDA Forest Service Gen. Tech. Rep. INT-130. 47 p.

Hall, F. C. 1976. Fire and vegetation in the Blue Mountains-implications for land managers. Proc. Tall Timber Fire Ecology Conf. 15:155-170.

Krummel, J. R., R. H. Gardner, G. Sugihara, R. V. O'Neill and P. R. Coleman. 1987. Landscape patterns in a disturbed environment. *Oikos* 48: 321-324.

Martin, R. E. 1982. Fire history and its role in succession. In: p. 92-99. Forest succession and stand development research in the Northwest, Symposium proceedings, Oreg. State University. Corvallis, OR.

Pickett, S. T. A. and P. S. White, 1985. The Ecology of natural disturbance and patch dynamics. Academic Press. New York.

Remilard, M.M., G. K. Gruendling and D. J. Bogucki. 1987. Disturbance by beaver (Castor canadensis Kuhl) and increased landscape heterogeneity. In: pp. 103-122. Turner, M. G. (Ed.). Landscape Heterogeneity and disturbance. Springer. New York.

Rieman, B. E. and J. D. McIntyre. 1993. Demographic and habitat requirements for conservation of bull trout. General Technical Report INT-302. 38 p.

Smith, E. M. 1983. History of the Boise National Forest 1905-1976. Idaho State Historical Society. Boise, ID. 116 p.

Stokes, M. A. and J. H. Dieterich. 1980. Proceedings of the fire history workshop. Ft Collings, CO. USDA For. Serv. Gen. Tech. Rep. RM-81. 142 p.

Steele, R., S. F. Arno and K. Geier-Hayes. 1986. Wildfire patterns change in central Idaho's ponderosa pine-Douglas-fir Forest. *Western Journal of Applied Forestry* 1:16-18.

Turner, M.G., R.H. Gardner, V. H. Dale and R. V. O'Neill. 1989. Predicting the spread of disturbance across heterogeneous landscapes. *OIKOS* 55(1): 121-129.

Turner, M.G. and S. P. Bratton. 1987. Fire, grazing and the landscape heterogeneity of a Georgia barrier island. In: pp. 85-101. Turner, M.G. (Ed.). Landscape heterogeneity and disturbance. Springer. New York.

West, N.E. 1969. Successional changes in montane forests of the central Oregon cascades. *Amer. Midl/Nat.* 81:265-271.

Advance Regeneration in the Inland West: Considerations for Individual Tree and Forest Health

Dennis E. Ferguson

ABSTRACT. Advance conifer regeneration readily survives release from overstory competition in the Inland West, but foresters are concerned about the ability of released trees to attain normal growth rates. There are also concerns about forest health issues associated with managing advance regeneration. The best pre-release predictors of post-release growth response are pre-release vigor and crown ratio at release. The fastest growing trees before release grow the fastest following release. A general recommendation is that at least a 0.50 crown ratio at release is necessary for good response to release. A 0.50 crown ratio means that trees will have the necessary photosynthetic capability to respond to changes in stand conditions. The best time to release advance regeneration is after budset in the fall but before budburst in the spring. Release during this time period means needles that develop the first summer after release will be adapted to full sunlight. A conceptual framework of physiological response to release is presented.

INTRODUCTION

Forests of the Inland West have recently been the focus of discussions on forest health. Tree mortality caused by insects and diseases has been at

Dennis E. Ferguson is Research Silviculturist, USDA Forest Service, Intermountain Research Station, Moscow, ID, USA 83843.

[Haworth co-indexing entry note]: "Advance Regeneration in the Inland West: Considerations for Individual Tree and Forest Health." Ferguson, Dennis E. Co-published simultaneously in the *Journal of Sustainable Forestry* (The Haworth Press, Inc.) Vol. 2, No. 3/4, 1994, pp. 411-422; and: *Assessing Forest Ecosystem Health in the Inland West* (eds: R. Neil Sampson and David L. Adams) The Haworth Press, Inc., 1994, pp. 411-422. Multiple copies of this article/chapter may be purchased from The Haworth Document Delivery Center [1-800-3-HAWORTH; 9:00 a.m. - 5:00 p.m. (EST)].

unacceptably high levels, and large stand-replacing wildfires have been common. Overly dense stands of trees contribute to problems with insects, diseases, and wildfires. These dense, multi-layered stands often are the result of regeneration and growth of shade-tolerant species. Managing this regeneration has associated advantages and disadvantages. Better management decisions can be made by knowing how advance regeneration responds to release.

Advance regeneration are seedlings and saplings that become established prior to partial or total removal (or death) of the overstory. They are usually shade-tolerant species that become established under mature trees, although shade-intolerant species can be present in canopy gaps of even-age stands, under shelterwood harvests, and in uneven-age silvicultural systems. When the overstory is removed, advance regeneration is released to conditions of more sunlight, nutrients, and moisture while having less interference from larger trees. Seedlings can be considered advance regeneration in shelterwood removals, harvests in selection systems, and thinnings from above because they are released from competition of larger trees.

Reproduction by advance growth is the only way some species become established on some sites. For example, on a harsh subalpine fir (*Abies lasiocarpa*) habitat type, lodgepole pine (*Pinus contorta*) provides the necessary shade and thermal cover for subalpine fir to become established.

The occurrence of advance regeneration in forest ecosystems of the Inland West has been documented in several studies. Seidel (1979), in a northeastern Oregon study, reported 11 percent of 642 stocked milacre plots contained at least one advance regeneration-size tree in cutover stands. Seidel and Head (1983), in eastern Oregon, reported 32 percent of 811 stocked milacre plots contained at least one advance tree. For cutover forests in central Idaho, northern Idaho, and western Montana, advance regeneration was present on 40 to 46 percent of 5,244 stocked 1/300-acre plots (Ferguson and Stage 1982; Carlson et al. 1982; Ferguson et al. 1986).

There are advantages and disadvantages of advance regeneration in the context of forest health. Leaving advance regeneration keeps more of the "pieces in place" in forest ecosystems. Advance regeneration can quickly provide hiding and thermal cover for wildlife. Soil disturbance can be minimized because site preparation is not needed to secure regeneration.

A disadvantage of advance regeneration for forest health is that succession is usually accelerated to shade-tolerant species, with associated insect and disease problems such as tussock moth (*Orgyia pseudotsugata*), western spruce budworm (*Choristoneura occidentalis*), and various root and stem diseases. Shade-tolerant species often have slower growth rates than shade-intolerant species; therefore, leaving advance regeneration can be a dysgenic practice. According to Zobel and Talbert (1984), "One of the most serious types of dysgenic selection is the harvesting of desired species from mixed stands, leaving only the undesired species." Leaving advance regeneration can also be a dysgenic practice if genetically inferior trees are left on the site to grow and reproduce.

Another disadvantage of releasing advance regeneration relates to decreased use of fire in forest ecosystems and the shortening of the regeneration period. One beneficial process of fire is that it raises soil pH for a few years, allowing many species of trees, shrubs, forbs, and grasses to become established. Without fire, and with continued site occupancy by most conifers, soil pH gradually decreases. If conifers occupy a site for hundreds of years, the resulting low soil pH can cause forest health problems such as aluminum toxicity and nutrient imbalances (Binkley 1987).

The same effect of lowering soil pH over time can happen if the usual successional period for shrubs, forbs, and grasses is eliminated or shortened following a regeneration harvest. Advance regeneration can lead to rapid site occupancy, thus eliminating the number, size, and duration of shrubs, forbs, and grasses. The higher proportion of time that the site is occupied by conifers would lead to lower soil pH. A study by Nelson (in preparation) in northern Idaho shows the beneficial role of shrubs in raising soil pH. Rainfall having a pH of 5.2 decreased to 4.8 after being intercepted by the conifer canopy, but was raised to 5.1 after being intercepted by shrubs.

RESPONSE TO RELEASE

Response of advance regeneration to release has been reviewed by McCaughey and Ferguson (1988). This continuing collection of information has application to the general topic of keeping trees and forests healthy. The review pointed out some generalities about released trees.

Few trees die from the sudden exposure caused by overstory removal (when death was not caused by logging damage). For example, studies have found 2 percent mortality for red fir (*Abies magnifica*) in 8 years since release (Oliver 1986), 0.3 percent for grand fir (*Abies grandis*) and red fir in 5 years (Seidel 1977), and 7 percent for grand fir (Seidel 1980) in 5 years.

Tree size at release also is a factor in survival. Gordon (1973) found that released red fir and white fir (*Abies concolor*) that were less than 1 foot tall had a much higher probability of death from the sudden exposure of overstory removal. Tesch et al. (1990) reported the same trend for Douglas-fir

(*Pseudotsuga menziesii*). Because few trees die, a full array of responses to release is possible, from barely surviving to vigorous growth. Each tree has to be assessed individually to determine its probable response to release.

Diameter growth response usually occurs in the first or second year after release. This is followed in 2 to 5 years by increased growth in height. Height growth response is not evident until the second year after release for most conifers because terminal buds are pre-determined. Terminal bud expansion the first year after release does not reflect site conditions following harvest because terminal bud primordia were laid down the previous fall (Kozlowski 1964). Gordon (1973) felt that, after release, stems buttress rapidly as a response to counteract bending forces; therefore, initial diameter growth after release may not be an early indicator of response.

When diameter or height growth response is plotted over time, the result is a sigmoid response curve. Growth before release is suppressed, but has yearly fluctuations. Following release, trees respond with increased growth until they reach a post-release growth rate, which also has yearly fluctuations. Maranto (1993) has shown that Douglas-fir takes longer to reach a post-release growth rate on cold/wet sites than on warm/moist sites. This site effect may be true for other species as well (McCaughey and Schmidt 1982).

Some degree of physiological shock is typical for released regeneration, which results in a few years of slower growth immediately after release. Physiological shock is even common in thinnings (Harrington and Reukema 1983). The degree of physiological shock will, in large part, determine the number of years required to reach the post-release growth rate.

Studies suggest that released advance regeneration attains growth rates comparable to site potential for the species and site conditions. Oliver (1986) reported that 41 percent of released red fir reached or exceeded the expected periodic annual increment for the sites. Released western hemlock (*Tsuga heterophylla*) showed a slight growth lag, then grew like trees of a similar size that had not been suppressed (Oliver 1976).

PREDICTING RESPONSE TO RELEASE

Two primary tree characteristics can be used to predict the probable response to release: pre-release vigor and crown ratio at release. The best and easiest to measure indicator of pre-release vigor is 5-year height increment prior to release. Faster growing trees prior to release will grow faster following release (Ferguson and Adams 1980; Helms and Standiford 1985; Maranto 1993). Helms and Standiford (1985) also found the pattern of pre-release growth was important. Trees having constant or increasing growth for the 5-years prior to release responded better in height growth following release than did trees having a declining growth pattern. Another important benefit of measuring 5-year pre-release vigor is that it negates the need to quantify the amount of overstory competition the tree is receiving. This finding was documented for grand fir in northern Idaho (Ferguson and Adams 1980) and for Douglas-fir in central Idaho (Maranto 1993).

For some species it is difficult or impossible to measure the previous 5 years of growth accurately. In these cases, response to release should be better for the faster growing trees and can be quantified by measuring 1, 2, or 3 years of height growth prior to release, or by measuring diameter increment (Graham 1982; Carlson and Schmidt 1989).

The other useful tree characteristic for predicting response to release is crown ratio at release. Seidel (1980, 1985) found at least a 0.50 crown ratio was needed for true fir and mountain hemlock (*Tsuga mertensiana*) to respond well to release. Helms and Standiford (1985) found that released Douglas-fir regeneration with less than 0.45 crown ratio did not grow well following release. Oliver (1986) reported that red fir with at least a 0.40 crown ratio grew more rapidly after release. Thus, a minimum of 0.50 crown ratio should be adequate for releasing regeneration-size trees.

Other variables are useful in predicting response to release. Cool, moist sites are generally better for release because adequate moisture and/or shade are available to allow trees to respond well to overstory removal. Sites that are naturally cool and moist are northerly aspects and higher elevation habitat types. Sites can be silviculturally manipulated to remain cool by retaining partial overstories and leaving snags, cull trees, shrubs, and competing advance regeneration on the site for 2 to 5 years to provide shade while released regeneration adjusts to new site conditions.

On sites where moisture is limiting, seral species respond better to release than climax species. For example, Douglas-fir responds better to release on a grand fir habitat type than on a Douglas-fir habitat type (Maranto 1993) and grand fir responds better to release on western redcedar (*Thuja plicata*) and western hemlock habitat types where it is not the climax species (Ferguson and Adams 1980). When moisture is not limiting, climax species such as western redcedar and subalpine fir can respond well to release (Herring 1977; Graham 1982).

Tree age at release is not as important as the variables mentioned above for predicting response to release. When tree age at release is significant in regression equations that predict response to release, older trees respond less to release than younger trees. Successful response to release depends more on crown ratio, pre-release vigor, and site conditions than tree age.

PHYSIOLOGY OF RELEASE

A conceptual framework that explains the physiology of release is discussed below. Ideas for this section grew out of the work of Tucker and Emmingham (1977) in their study of western hemlock. Details from other published research are included to complete the story for conifers in the Inland West.

Crown Geometry

Advance regeneration is able to survive in the understory because of sunflecks that move across the forest floor (Emmingham and Waring 1973). When sunlight strikes the foliage, seedlings are able to begin photosynthesis quickly. Once the sunfleck passes, photosynthesis ceases.

Suppressed trees do not receive enough sunlight to support full crowns. Maximum photosynthesis is attained by capturing as much sunlight as possible when sunflecks shine on the foliage. Needles of advance conifers are most often displayed in single planes with little mutual shading of needles, as noted by Keller and Tregunna (1976) for western hemlock. This arrangement of foliage means that nearly every needle will be exposed to full sunlight after release.

Sun and Shade Needles

There are important morphological differences between needles that develop in the shade and those that develop in sunlight. These differences are important for trees responding to release. Shade needles have fewer, but larger, stomata than sun needles (Tucker and Emmingham 1977). After release, stomata in shade needles cannot completely close to control evapotranspiration, so needles abscise prematurely because of moisture stress (Keller and Tregunna 1976; Tucker and Emmingham 1977). There are more stomata on sun needles, and they are smaller. The smaller stomata close to control evapotranspiration.

Another difference between sun and shade needles is the amount of palisade mesophyll that protects the chlorophyll in the needle. Shade needles of western hemlock have a single row of palisade mesophyll while sun needles have a double row (Tucker and Emmingham 1977). Maranto (1993) found more palisade mesophyll on sun needles than on shade needles of Douglas-fir. The extra palisade mesophyll protects the chlorophyll that lies deeper within the needles. Without this extra protection, shade needles on released trees can die from sun scald.

Sun and shade needles can also differ in the amount of cuticular waxes on the outside of the needles. Maranto (1993) found that shade needles of Douglas-fir did not have as much wax as did sun needles. The additional wax reduces evapotranspiration.

Shade needles that are dropped prematurely or are killed by sun scald reduce the photosynthetic area (crown) that is available for the tree to adjust to release. Signs of physiological stress the first year after release are premature needle drop, sun scald, and smaller-than-normal current year needles. Current-year needles may be shorter in length because of moisture stress (Tucker and Emmingham 1977). Short current-year needles affect height growth the second year after release because photosynthates for terminal bud expansion come from 1-year old foliage (Kozlowski and Winget 1964).

Two important attributes of the development of sun needles are subject to silvicultural control. First, the protective double row of palisade mesophyll in needles is formed under shelterwoods (Tucker and Emmingham 1977). Thus, partial opening of the canopy should provide enough sunlight for the development of additional palisade mesophyll on needles that develop under the shelterwood.

The second attribute is that sun versus shade needles are not pre-determined when the bud is set in the fall (Tucker et al. 1987). Needles develop into sun or shade needles depending on the amount of sunlight received while they are expanding. Trees released after budset in the fall but before budburst in the spring would have current-year needles that are adapted to sunlight. Needles on trees that are released after budburst would have only shade needles.

In summary, released trees must make rapid adjustments to respond well to release. Keller and Tregunna (1976) noted tissue damage to western hemlock the first day after release. Released trees must be able to maintain a favorable water balance and not suffer extensive sun scald damage in order to retain enough crown to adjust to release. The degree of physiological shock determines whether the tree lives or dies, and the number of years surviving trees need to reach post-release growth.

The first summer following release is important because of physiological shock. A cool, moist summer would result in less physiological shock than a hot, dry summer. While we cannot know the weather the first summer after release, we can provide partial shade for released trees by removing the overstory in 2 or more cuttings, and by leaving snags, cull trees, shrubs, and other regeneration-size trees on the site for a few years. Releasing trees after buds are set in the fall but before budburst in the spring will mean that the current-year foliage will be sun needles.

There will be at least a 1-year delay in height growth response to release because most conifers in the Inland West have pre-determined buds. Height growth the first summer after release reflects conditions when the bud was developed the previous fall. Height growth the second year after release will reflect the new growing conditions for the tree and the degree of physiological shock that occurred the first year after release.

LONG TERM CONSIDERATIONS

Increases in Shade-Tolerant Species

Shade-tolerant advance regeneration survives harvests in the Inland West and is able to respond well to release. There are other trends shifting species compositions towards shade-tolerant species. A second trend is past harvesting practices that removed the more valuable shade-intolerant species, while leaving shade-tolerant species to stock the site and regenerate in high numbers. This trend was recognized even in the 1920s in the northern Rocky Mountains (Koch 1923; Neff 1928), which led to recommendations for killing unmerchantable shade-tolerant overstory trees (Foiles 1950).

A third trend is control of wildfires. Human intervention has reduced the number and size of wildfires. Fire exclusion has allowed shade-tolerant species to regenerate and grow in the understory, whereas previously, low intensity ground fires periodically removed these trees.

A fourth trend is introduction of non-native pests such as white pine blister rust (*Cronartium ribicola*). Blister rust has drastically decreased the proportion of western white pine (*Pinus monticola*) in the Rocky Mountains (Bingham 1983), and is currently decimating whitebark pine (*Pinus albicaulis*) at higher elevations (Keane and Arno 1993). Both of these species are relatively intolerant of shade, and more shade-tolerant species are growing in their place.

A trend now beginning is ecosystem management. Under the Forest Service's new ecosystem management policy, clearcutting will be replaced, where possible, by partial cuts. Partial cutting will tend to favor shade-tolerant species over shade-intolerant species.

These long-term trends will shift species compositions to shade-tolerant species. This shift can be either good or bad, depending on the definition of forest health and land management objectives. What is clear is that land managers will have to counteract these trends if they wish to encourage shade-intolerant regeneration in forest ecosystems of the Inland West.

Insect and Disease Considerations

Insects and diseases are major factors to consider when releasing advance regeneration. Insect problems such as western spruce budworm and Douglas-fir tussock moth are associated with shifts in species composition to shade-tolerant species. If regeneration is defoliated prior to release, reduced crown area makes it that much more difficult for trees to respond well to release.

There are several decay fungi that can infect trees through logging injuries. Readers are referred to McCaughey and Ferguson (1988) for discussions of decay fungi that infect shade-tolerant conifers of the Inland West.

Dwarf mistletoe (*Arceuthobium* sp.) can infect advance regeneration before release, thus perpetuating mistletoe into the next generation of trees. The longer the exposure to dwarf mistletoe from overstory trees, the higher the probability of infection in the understory. Advance regeneration that is infected on the trunk often are topkilled (Scharpf 1982). Dwarf mistletoe reduces growth of advance regeneration and limits the ability of the tree to respond well to release.

The life cycle of Indian paint fungus (*Echinodontium tinctorium*) makes this heart rot an important concern when managing advance regeneration. Indian paint fungus infects true firs and hemlocks in the Inland West. Infection occurs on stubs of shade-killed branchlets of suppressed trees (Etheridge and Craig 1976; Filip et al. 1983). A minimum of 40 years of suppressed growth is necessary to produce suitable infection courts. After the stub heals over, the fungus becomes dormant for up to 50 years. Infections are reactivated by mechanical injury such as logging scars, broken tops, broken branches, frost cracks, or even insect attacks.

Thus, Indian paint fungus is present in suppressed trees but does not become a problem until many years after release. If decay-free trees are a management goal, true fir and hemlocks that have been suppressed for more than 40 years are poor choices because of the high probability of dormant Indian paint fungus. Filip et al. (1983) provide guidelines for managing advance grand and white fir regeneration for Indian paint fungus and other decay fungi.

CONCLUSIONS

The occurrence of advance regeneration is a normal part of secondary succession. There are advantages and disadvantages of releasing advance regeneration that must be weighed when deciding how to achieve desired future conditions in Inland West forests. The purpose of this paper was to present a summary of information for managing advance regeneration so that better decisions can be made.

If the decision is made to release advance regeneration, there are several key points to remember. Faster growing trees prior to release will grow better following release. At least a 0.50 crown ratio is recommended for good response to release, and it is important to minimize physiological shock that would cause premature needle loss or sun scald.

Releasing advance regeneration can be a dysgenic practice in two ways. The first is when genetically inferior trees are left on the site to grow and reproduce. Second is a change in species composition. Shade-tolerant species often do not grow as well as seral shade-intolerant species. This shift in species composition can dramatically decrease yields. While released trees may grow as well as subsequent trees of the same species and size, they may not grow as well as other species on the same site. Counteracting this is the uncertainty of establishing subsequent regeneration.

A trend in human management of Inland West forests has been a shift to shade-tolerant species. This trend will likely continue into the future under ecosystem management concepts because of more partial cuts. Forest managers need to be aware of these long-term trends so that positive steps can be taken to encourage establishment of adequate numbers of shade-intolerant species to maintain the desired proportion of species in Inland West forests.

REFERENCES

Bingham, R.T. 1983. Blister rust resistant western white pine for the Inland Empire: the story of the first 25 years of the research and development program. USDA For. Serv. Gen. Tech. Rep. INT-146. 45 p.

Binkley, D. 1987. Forest soils and acid deposition–an overview and synthesis. In: pp. 222-236. Proceedings of the forest-atmosphere interaction workshop. U.S. Dept. of Energy. Lake Placid, NY.

Carlson, C.E., L.J. Theroux and W.W. McCaughey. 1982. The influence of silvicultural practices on the susceptibility and vulnerability of northern Rocky Mountain forests to western spruce budworm: comprehensive progress report. USDA For. Serv., Intermountain Research Station. Missoula, MT. 55p.

Carlson, C.E. and W.C. Schmidt. 1989. Influence of overstory removal and western spruce budworm defoliation on growth of advance conifer regeneration in Montana. USDA For. Serv. Res. Pap. INT-409. 13 p.

Emmingham, W.H. and R.H. Waring. 1973. Conifer growth under different light environments in the Siskiyou Mountains of southwestern Oregon. *Northwest Sci.* 47:88-99.

Etheridge, D.E. and H.M. Craig. 1976. Factors influencing infection and initiation of decay by the Indian paint fungus (*Echinodontium tinctorium*) in western hemlock. Can. J. For. Res. 6:299-318.

Ferguson, D.E. and D.A. Adams. 1980. Response of advance grand fir regeneration to overstory removal in northern Idaho. *Forest Sci.* 26:537-545.

Ferguson, D.E. and A.R. Stage. 1982. The effects of spruce budworm on regeneration success in Idaho's forest ecosystems. Progress report. USDA For. Serv., Intermountain Research Station. Moscow, ID. 25 p.

Ferguson, D.E., A.R. Stage and R.J. Boyd. 1986. Predicting regeneration in the grand fir-cedar-hemlock ecosystem of the northern Rocky Mountains. *Forest Sci.* Mono. 26. 41 p.

Filip, G.M., P.E. Aho and M.R. Wiitala. 1983. Indian paint fungus: a method for recognizing and reducing hazard in advanced grand and white fir regeneration in eastern Oregon and Washington. USDA For. Serv., Pacific Northwest Region. 24 p.

Foiles, M.W. 1950. Recommendations for poisoning western hemlock. Res. Note 77. USDA For. Serv., Northern Rocky Mountain Forest and Range Exp. Sta. 1 p.

Graham, R.T. 1982. Influence of tree and site factors on western redcedar's response to release: a modeling analysis. USDA For. Serv. Res. Pap. INT-296. 19 p.

Gordon, D.T. 1973. Released advance reproduction of white and red fir . . . growth, damage, mortality. USDA For. Serv. Res. Pap. PSW-95. 12 p.

Harrington, C.A. and D.L. Reukema. 1983. Initial shock and long-term stand development following thinning in a Douglas-fir plantation. *Forest Sci.* 29:33-46.

Helms, J.A. and R.B. Standiford. 1985. Predicting release of advance reproduction of mixed conifer species in California following overstory removal. *Forest Sci.* 31:3-15.

Herring, L.J. 1977. Studies of advance subalpine fir in the Kamloops forest district. Province of British Columbia, Ministry of Forests, Res. Note 80. 22 p.

Keane, R.E. and S.F. Arno. 1993. Rapid decline of whitebark pine in western Montana: evidence from 20-year remeasurements. *West. J. Appl. For.* 8:44-47.

Keller, R.A. and E.B. Tregunna. 1976. Effects of exposure on water relations and photosynthesis of western hemlock in habitat forms. *Can. J. For. Res.* 6:40-48.

Koch. E. 1923. The inferior species in the white pine type in Montana and Idaho. *J. Forestry* 21:588-599.

Kozlowski, T.T. 1964. Shoot growth in woody plants. *Bot. Rev.* 30:335-392.

Kozlowski, T.T. and C.H. Winget. 1964. Contributions of various plant parts to growth of pine shoots. Univ. Wis. Forest Res. Note 113. 44 p.

Maranto, J.C. 1993. Response of Douglas-fir advance regeneration to overstory removal in central Idaho. Univ. of Idaho. M.S. thesis. 60 p.

McCaughey, W.W. and W.C. Schmidt. 1982. Understory tree release following harvest cutting in spruce-fir forests of the Intermountain West. USDA For. Serv. Res. Pap. INT-285. 19 p.

McCaughey, W.W. and D.E. Ferguson. 1988. Response of advance regeneration to release in the Inland Mountain West: a summary. In: pp. 255-266. Schmidt, W.C. (Comp.). Proceedings–future forests of the Mountain West: a stand culture symposium. USDA For. Serv. Gen. Tech. Rep. INT-243.

Neff, P. 1928. The inferior-species problem in the northern Rocky Mountains. *J. Forestry* 26:591-599.

Nelson, J.A. (in preparation). The role of understory shrub species in nutrient cycling and site fertility in a *Pseudotsuga menziesii/Physocarpus malvaceus* habitat type in central Idaho. Univ. of Idaho. M.S. thesis.

Oliver, C.D. 1976. Growth response of suppressed hemlocks after release. In: pp. 266-272. Atkinson, W.A. and R.J. Zasoski (Eds.). Western hemlock management. College For. Resources, Inst. For. Prod. Contribution 34. Univ. Wash.

Oliver, W.W. 1986. Growth of California red fir advance regeneration after overstory removal and thinning. USDA For. Serv. Res. Pap. PSW-180. 6 p.

Scharpf, R.F. 1982. Problems of dwarf mistletoe in advance regeneration of true firs. In: pp. 33-36. proceedings of the 29th Western International Forest Disease Work Conference. Vernon, B.C., Canada.

Seidel, K.W. 1977. Suppressed grand fir and Shasta red fir respond well to release. USDA For. Serv. Res. Note PNW-288. 7p.

Seidel, K.W. 1979. Regeneration in mixed conifer clearcuts in the Cascade Range and the Blue Mountains of eastern Oregon. USDA For. Serv. Re's. Pap. PNW-248. 24p.

Seidel, K.W. 1980. Diameter and height growth of suppressed grand fir saplings after overstory removal. USDA For. Serv. Res. Pap. PNW-275. 9 p.

Seidel, K.W. 1985. Growth response of suppressed true fir and mountain hemlock after release. USDA For. Serv. Res. Pap. PNW-344. 22p.

Seidel, K.W. and S.C. Head. 1983. Regeneration in mixed conifer partial cuttings in the Blue Mountains of Oregon and Washington. USDA For. Serv. Res. Pap. PNW-310. 14 p.

Tesch, S.D., M.S. Crawford, K. Baker-Katz and J.W. Mann. 1990. Recovery of Douglas-fir seedlings from logging damage in southwestern Oregon: preliminary evidence. *Northwest Sci.* 64:131-139.

Tucker, G.F. and W.H. Emmingham. 1977. Morphological changes in leaves of residual western hemlock after clear and shelterwood cutting. *Forest Sci.* 23:195-203.

Tucker, G.F., T.M. Hinckley, J.W. Leverenz and Shi-Mei Jiang. 1987. Adjustments of foliar morphology in the acclimation of understory Pacific Silver Fir following clearcutting. *Forest Ecology and Management.* 21:249-268.

Zobel, B. and J. Talbert. 1984. Applied forest tree improvement. John Wiley and Sons. New York. 505 p.

Beyond Even- vs. Uneven-Aged Management: Toward a Cohort-Based Silviculture

Gregory H. Aplet

ABSTRACT. The effect of forest management has been to override the dynamics produced by natural disturbances with a dynamic of human design. Under even-aged management, the forest is held in an early successional state, providing some habitat for species of open field and edge but little for species of the forest interior. Alternatively, uneven-aged management may maintain forest interior conditions but does not provide the large-scale heterogeneity created by pre-settlement disturbance processes. The traditional choice between even- and uneven-aged management approaches condemns the landscape to one of two mutually exclusive dynamics, causing the loss of natural landscape heterogeneity and the biodiversity it supports. The conservation of biodiversity will require that silviculture move beyond the artificial dichotomy of even- versus uneven-aged management to accommodate the full range of patterns, intensities, and frequencies of the natural disturbance regime. Traditional emphases on "regenerating the stand" and "optimizing growing stock" must be replaced by an emphasis on establishing the *appropriate-sized cohort* in the context of landscape-scale objectives.

INTRODUCTION

An important conclusion to emerge from recent discussions of forest health, including the Eastside Forest Ecosystem Health Assessment (Ever-

Gregory H. Aplet is Forest Ecologist, The Bolle Center for Forest Ecosystem Management, The Wilderness Society, 900 Seventeenth Street NW, Washington, DC 20006.

[Haworth co-indexing entry note]: "Beyond Even- vs. Uneven-Aged Management: Toward a Cohort-Based Silviculture." Aplet, Gregory H. Co-published simultaneously in the *Journal of Sustainable Forestry* (The Haworth Press, Inc.) Vol. 2, No. 3/4, 1994, pp. 423-433; and: *Assessing Forest Ecosystem Health in the Inland West* (eds: R. Neil Sampson and David L. Adams) The Haworth Press, Inc., 1994, pp. 423-433. Multiple copies of this article/chapter may be purchased from The Haworth Document Delivery Center [1-800-3-HAWORTH; 9:00 a.m. - 5:00 p.m. (EST)].

ett 1993) and the Society of American Foresters's Task Force Report on Sustaining Long-term Forest Health and Productivity (SAF 1993a), is that the maintenance of forest health cannot be achieved except in the broader context of ecosystem management. These reports have led to the conclusion that the primary objective of ecosystem management is the conservation of biological diversity. The societal goal of sustaining forest health implies sustaining ecosystems, and sustaining ecosystems requires maintaining biodiversity.

Recent attempts to sustain some of the more critically imperiled elements of biodiversity, such as the northern spotted owl and red-cockaded woodpecker, have led to the conclusion that maintaining biodiversity means maintaining habitat. Maintaining habitat for the vast array of elements of biodiversity will require maintaining habitat diversity. Historically, habitat diversity was maintained through the heterogeneous distribution, across space and time, of the primary elements affecting ecosystem structure: climate, soils, species availability, and disturbances such as fire, wind, and flood. Recently, the historical model which viewed ecosystems either existing in a persistent steady state or returning to one, has been replaced by the contemporary model in which ecosystem structure and function is strongly influenced by occassional disturbances. The transition from the "equilibrium" to the "non-equilibrium" paradigm in ecology has had a profound effect on the way ecologists perceive the structure and function of ecosystems.

THE CONTEMPORARY MODEL OF FOREST DYNAMICS

Over the last few decades, a general model of forest dynamics has emerged (Bormann and Likens 1979; Oliver 1981; Peet and Christensen 1987) and has recently been documented by Oliver and Larson (1990). Under this model, forested landscapes are understood to be composed of a patchwork of individual stands, each of which is distinguished by its physical environment and disturbance history. Each stand moves through a sequence of stages following the catastrophic disturbance (e.g., crown fire, windstorm, insect epidemic) that helped determine the stand's boundary. The *stand initiation stage* occurs immediately following the major disturbance event. During this stage, a new population of trees establishes on the open site. This can take as short a time as a single year, or it may last many decades, but it results in the establishment of a *single cohort* of trees with essentially even-aged structure.

Recruitment into the cohort halts when the "growing space"—the intangible, multi-dimensional complex of factors that sustain plant life—is fully occupied. At this point, the single cohort is established, and the stand enters the *stem exclusion stage.* Further establishment is prevented until growing space becomes available again, usually through the death of over-story trees. Eventually, this mortality opens the canopy enough to allow a second cohort of small trees to establish and grow, and the stand enters the third stage of stand development, the *understory reinitiation stage.* With the passage of more time, the original cohort breaks up, understory trees grow to reach the canopy, a multi-layered canopy develops, and the stand enters the *old growth stage.* An old growth stand can be composed either of trees of many different ages or a few cohorts established following low-intensity disturbance events such as surface fires or insect outbreaks. Some disturbances may leave only a few live trees, with the next cohort forming a two-stage structure. A stand may exist in the old-growth or multi-cohort stage for many generations before the next stand-replacing disturbance.

These dynamics produce a landscape in various stages of stand devel-opment, distributed according to the interactions between the four major factors affecting forest ecosystems: Species, soils, local climates, and dis-turbance histories. The variety of stand types thus produced provides the habitat diversity that sustains biodiversity and a healthy forest ecosystem.

THE EFFECT OF TIMBER MANAGEMENT

Of the four major factors affecting forest ecosystems, humans have had their greatest influence on the structure and function of forests through their effect on disturbance. The ecological effect of forest management has been to alter the pattern, intensity, and frequency of disturbance to create a forest of desired composition, structure, and function. Traditionally, the objective of forest management has been timber production, and silvicul-ture has employed either even- or uneven-aged management to achieve that objective. Under even-aged management (e.g., clearcutting, seed tree, shelterwood, coppice), the objective is to establish an even-aged stand (single cohort) on a site where the canopy trees have been removed, grow trees until stand-level productivity drops to an unacceptable level, and again regenerate the site. Under uneven-aged management, the objective is to maintain an optimum level of "growing stock" in a stand that will yield harvests of consistent volume and composition over time without remov-ing the canopy.

These systems can effectively produce timber, but their widespread use can, and in some places does, have undesirable effects on the landscape-level heterogeneity that sustains biodiversity. The effect of even-aged management, applied broadly across the landscape, is to restrict habitat diversity to the first two stages of stand development. Understory reinitiation and old growth stages generally occur well after stand-level productivity has peaked and are not typically included in even-aged regimes. Likewise, uneven-aged management, by maintaining stands in a complex, multi-tiered structure, commits the landscape to homogeneous "quasi-old growth" conditions. In practice, perfectly even-aged or uneven-aged structures are difficult to maintain, so some variety is inevitable.

To some extent, landscape-level heterogeneity (i.e., habitat diversity) can be achieved intentionally by allocating some land to each system, maintaining young stands through even-aged management, and achieving some aspects of late-successional forest through uneven-aged management. However, this allocation is still at odds with the natural forces that shape ecosystems. The lesson of the non-equilibrium model is that forests change. Assigning parts of the forest to permanent structural stages opposes the forces of change, ultimately to the detriment of the entire ecosystem. For example, the exclusion of fire to maintain unnaturally dense forests in the Inland West now threatens the sustainability of historically fire-adapted ecosystems. Managing ecosystems to accommodate change will require a silvicultural perspective that goes beyond the artificial dichotomy of even- versus uneven-aged management.

COHORT MANAGEMENT

The four stage classification described by Oliver and Larson (1990) provides a point of departure from the traditional dichotomy, described above, toward methods that incorporate the full process of stand development into a silvicultural system. Rather than characterizing stands as either even-aged or uneven-aged based on structure, this model describes stands based on the processes occurring within them. Under this model, a stand is an areal unit, not simply with distinct physical characteristics, but with a distinct dynamic history. Further, stand development involves a time gradient from even-aged to uneven-aged dynamics within a single stand. The two traditionally disparate visions of forest structure and dynamics are united by a gradient of disturbance intensity from catastrophe to gaps, each disturbance resulting in the establishment of a characteristic cohort of new individuals.

Viewing stand development as a process involving a gradient of disturbance intensity which results in the establishment of cohorts of various sizes puts forest management in a different light. The objective of management no longer needs to be "regenerating the stand" or "optimizing levels of growing stock," but *establishing the appropriate-sized cohort* for a particular stage of stand development.

This "cohort management" requires viewing silvicultural prescriptions as points on a disturbance gradient (Figure 1). For example, a traditional, even-aged management clearcut represents one extreme of the gradient in which an intense disturbance results in the complete removal of the previously existing stand and the establishment of a single cohort. At the other extreme is a thinning operation, in which trees are removed, but no cohort is established; the objective is not to establish a new cohort, but to reallocate growing space to the existing cohort. Between these extremes is a full range of options from "new forestry" cuts (i.e., "shelterwood with reserves" *sensu* SAF 1993b), in which a small number of trees from the residual cohort is left, to single-tree selection cuts in which the established cohort is small relative to the number of residuals. From left to right across Figure 1, disturbance size, intensity, and regeneration decrease, while the number of residual trees increases. Thinking of the stand in terms of cohorts builds a bridge across the gap between even- and uneven-aged management.

Cohort management provides the framework in which to apply silvicultural treatments that are consistent with the natural dynamics of the forest. Once silvicultural treatments are understood as points on a disturbance continuum, they can easily be employed to emulate the natural disturbance processes that maintain habitat diversity and/or to achieve more traditional

FIGURE 1. Traditional silvicultural treatments reflect a range of disturbance sizes and intensities.

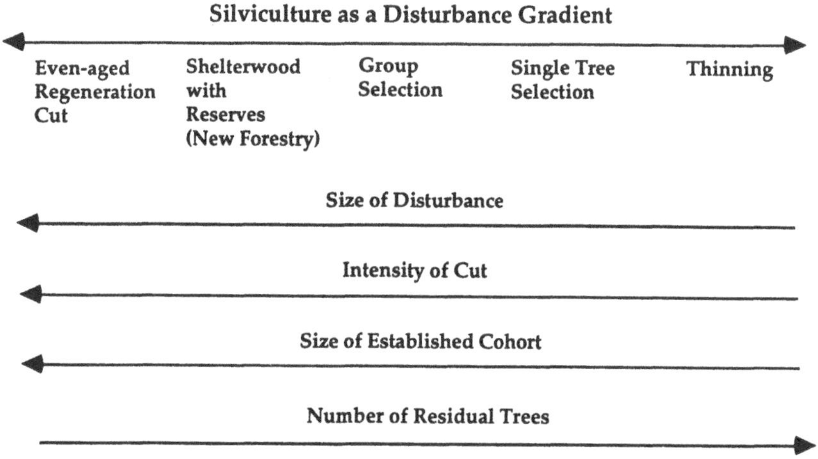

Silviculture as a Disturbance Gradient

Even-aged Regeneration Cut	Shelterwood with Reserves (New Forestry)	Group Selection	Single Tree Selection	Thinning

Size of Disturbance

Intensity of Cut

Size of Established Cohort

Number of Residual Trees

management objectives. Many natural disturbances already have management analogs in the existing lexicon of forestry (Figure 2). For example, catastrophic disturbances, such as stand replacing fires, correspond to large regeneration cuts; widespread windthrow and some insect epidemics correspond to intermediate harvests; and group selection may approximate bark beetle mortality. These events occur with characteristic frequencies and patterns that can be applied to silvicultural treatments. The adoption of cohort management does not necessarily require new methods, but, rather, a new perception of the role of silviculture in the forest. Cohort management requires the perspective of working *with* nature rather than battling *against* it.

RECONCILIATION

Cohort management provides the means to reconcile forest management with natural stand dynamics to maintain habitat diversity across the landscape. It uses time-tested and well understood methods to emulate the processes with which species evolved and communities developed. The innovation comes in finding new ways to use old methods that are consis-

FIGURE 2. Cohort management utilizes traditional silvicultural treatments in a system that sustains natural disturbance and stand dynamics.

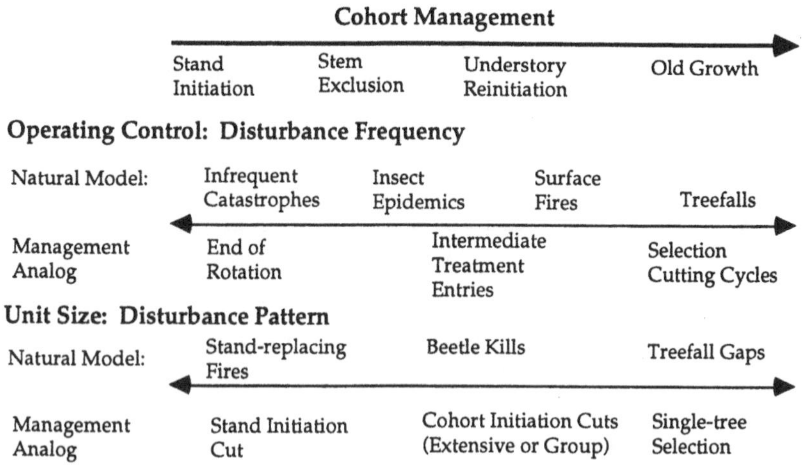

Cohort Management →

	Stand Initiation	Stem Exclusion	Understory Reinitiation	Old Growth

Operating Control: Disturbance Frequency

Natural Model:	Infrequent Catastrophes	Insect Epidemics	Surface Fires	Treefalls
Management Analog	End of Rotation	Intermediate Treatment Entries		Selection Cutting Cycles

Unit Size: Disturbance Pattern

Natural Model:	Stand-replacing Fires	Beetle Kills	Treefall Gaps
Management Analog	Stand Initiation Cut	Cohort Initiation Cuts (Extensive or Group)	Single-tree Selection

tent with ecosystem processes. For instance, many single-species communities such as lodgepole pine, aspen, and jack pine, rarely progress beyond the stem exclusion stage before succumbing to insects, disease, and/or fire. These forests respond well to catastrophic disturbances and may depend on them for continued dominance (Elliott-Fisk 1988; Peet 1988). In these systems, clearcutting (with provision of a woody legacy) may be the most appropriate regeneration method. This method is consistent with the widely acknowledged germination and establishment requirements for many species of an open canopy and mineral seedbed.

In other systems, such as the Engelmann spruce-subalpine fir forests of the central Rockies, forest development following catastrophic fire is more complex. Regeneration can be slow in these subalpine forests; the stand initiation stage can last over a century in this type (Aplet et al. 1988), resulting in considerable concern about the widespread use of clearcutting (Carey 1984). The complex mortality patterns that accompany stand development may provide opportunities for silviculturists to implement new approaches. For instance, spruce-fir stand dynamics are characterized by a period of mortality in the shorter-lived fir in the third or fourth century of development (Aplet et al. 1988). Rather than clearcutting stands of this age, natural mortality can be emulated by harvesting enough fir to open the canopy, release suppressed fir saplings, and establish a new cohort of spruce seedlings. Younger stands can be lightly thinned without disrupting processes, and older stands can be managed on a selection basis.

Other forests, such as the hardwood forests of the Appalachians, also show evidence of catastrophic disturbance, but structural development is dominated by an old-growth dynamic long before the loss of the original cohort (Runkle 1985). Disturbance is highly variable and may range from single treefalls to small windthrows to defoliation events, fires, landslides, and tornadoes. The diverse species composition ensures the survival of some individuals, and even in the event of major disruptions such as chestnut blight, recovery is rapid. Records of early travelers (e.g., Bartram 1791) suggest that gap-phase dynamics dominated the presettlement landscape, explaining the positive response of these forests to single tree and small group selection (Lamson and Smith 1991). Much of the Appalachian old growth forest was selectively logged in the past, and while some stands have lost certain species, many others have recovered well (Runkle 1982). As in many eastern forests, stand replacing disturbances were rare historically; maintaining natural disturbance regimes will keep management activities concentrated on the right side of Figure 2. Once an abundance of late successional stands is restored, occasional intensive harvesting can be used to maintain landscape heterogeneity.

These examples are offered to illustrate the potential for working with the disturbance regime to effect extraction without disrupting processes. Adoption of this approach may require some reduction in harvest level, as the landscape will not be maintained in highly productive, early successional conditions; however, the variety of stand conditions can be expected both to maintain habitat diversity and to produce a variety of high-quality products (Oliver 1992).

RESEARCH NEEDS

Determination of the level of disturbance needed to establish the appropriate-sized cohort is a complex process requiring considerable knowledge of both stand and landscape processes. Development of site-specific prescriptions requires indepth understanding of the usual course of stand development in each community and the condition of the stand in question. Studies of type-specific natural stand dynamics are not uncommon (see Oliver and Larson 1990), but more research is needed based on the four-stage model. Prior to recommending treatment, the silviculturist must consider the managed stand in light of its place in the developmental sequence.

Though proper management will require an understanding of stand processes, stand-level recommendations cannot be made in isolation. Determination of the proper silvicultural prescription will depend on the desired future condition not just of the stand, but of the landscape. The silviculturist must consider the distribution of stand stages on the landscape prior to recommending treatment. Much of the forested landscape of North America is now in stands less than 100 years old, and restoration of a diverse landscape will require growing stands into later stages of development. The distribution of stand stages in the presettlement landscape mosaic represents a tangible model of a healthy, functioning ecosystem and will be a useful guide in setting desired future conditions.

Cohort management also depends on an acute understanding of natural (presettlement) disturbance regimes. Disturbance size, frequency, and intensity, and the variability of these parameters should be understood prior to scheduling silvicultural treatments. These details are very poorly understood for most forest systems; however, considerable progress has been made recently in the Pacific Northwest (Morrison and Swanson 1990; Everett 1993; Agee 1993), the Sierra Nevada (Kilgore and Taylor 1979; Bonnickson and Stone 1982), and the Lake States (Frelich and Lorimer 1991), among others. Similar research efforts should be undertaken in all natural to seminatural forest ecosystems.

Finally, research is needed in the area of treatment. Target stand structures and landscape mosaics will remain only aspirations unless silvicultural objectives can be achieved on the ground. Much is already known about the response of stands to various treatments, but more research is needed into the state of post-disturbance environments and silvicultural means of emulating those environments. Attention should be given to the effects of intermediate harvests of various intensities on subsequent stand dynamics. Managers will need to move beyond the narrow concerns of adequate stocking of commercial species in even-aged systems and sustained harvests of less-and-less valuable species in uneven-aged systems (Lamson and Smith 1991). The future lies in a sophisticated understanding of the effects of disturbance on cohort establishment.

CONCLUSION

Recently, both the USDA Forest Service and the Bureau of Land Management endorsed the practice of ecosystem management on federal lands (Unger 1993; Baca 1993). Ecosystem Management in the Forest Service seeks to "take care of the land by protecting or restoring the integrity of its soils, air, waters, biological diversity, and ecological processes" (Robertson 1992). Cohort management offers a way to restore some of these ecological processes to the ecosystem.

But forest stands do not function in isolation. They are parts of an integrated whole–a landscape mosaic reflecting a continuum of disturbance sizes and intensities. Traditional forest management has ignored the critical role that landscape pattern plays in ecosystem function. Ecosystem management will be achieved only when foresters manage stands as components of a larger whole (e.g., a district, forest, or county). Much of the landscape may not prove suitable for the production of fiber; but where production is appropriate, cohort management provides a framework for managing stands as dynamic units. Managing those units as part of a dynamic landscape represents an even greater challenge.

REFERENCES

Agee, J. K. 1993. Fire ecology of Pacific Northwest forests. Island Press. Covelo, CA.

Aplet, G. H., R. D. Laven and F. W. Smith. 1988. Patterns of community dynamics in Colorado Engelmann spruce-subalpine fir forests. *Ecology* 69:312-319.

Baca, J. 1993. Testimony before a hearing on the definition and implementation of ecosystem management, United States Senate Subcommittee on Agriculture Research, Conservation, Forestry, and General Legislation, November 9, 1993.

Bartram, W. 1791. Travels through North and South Carolina, east and west Florida, the Cherokee country, the extensive territories of the Muscogulges, or Creek Confederacy, and the country of the Chactaws. Unabridged 1955 Dover reprint. New York.

Bonnickson, T. M. and E. C. Stone. 1982. Reconstruction of a presettlement giant sequoia-mixed conifer forest community using the aggregation approach. *Ecology* 63:1134-1148.

Bormann, F. H. and G. E. Likens. 1981. Pattern and process in a forested ecosystem. Springer-Verlag. New York.

Carey, H. H. 1984. Regeneration failure and silvicultural mythology: a case study from the spruce-fir type in New Mexico. Forest Trust Research Report 1.

Elliott-Fisk, D. L. 1988. The boreal forest. In: pp. 33-62. Barbour, M. G. and W. D. Billings (Eds.). North American Terrestrial Vegetation. Cambridge University Press. New York.

Everett, R. (Comp.). 1993. Eastside Forest Ecosystem Health Assessment. USDA Forest Service.

Frelich, L. E. and C. G. Lorimer. 1991. Natural disturbance regimes in hemlock-hardwood forests of the Upper Great Lakes region. *Ecological Monographs* 61:145-164.

Kilgore, B. M. and D. Taylor. 1979. Fire history of a sequoia-mixed conifer forest. *Ecology* 60:129-142.

Lamson, N. I. and H. C. Smith. 1991. Stand development and yields of Appalachian hardwood stands managed with single-tree selection for at least 30 years. USDA Forest Service Research Paper NE-655. 6p.

Morrison, P. H. and F. J. Swanson. 1990. Fire history and pattern in a Cascade Range landscape. USDA Forest Service General Technical Report PNW-GTR-254. 77p.

Oliver, C. D. 1981. Forest development in North America following major disturbances. *Forest Ecology and Management* 3:153-168.

Oliver, C. D. 1992. Achieving and maintaining biodiversity and economic productivity. *Journal of Forestry* 90:20-25.

Oliver, C. D. and B. C. Larson. 1990. Forest stand dynamics. McGraw-Hill. New York.

Peet, R. K. 1988. Forests of the Rocky Mountains. In: pp. 63-101. Barbour, M. G. and W. D. Billings (Eds.). North American Terrestrial Vegetation. Cambridge University Press. New York.

Peet, R. K. and N. L. Christensen. 1987. Competition and tree death. *Bioscience* 37:586-595.

Robertson, F. D. 1992. Letter from F. Dale Robertson to regional foresters and station directors. June 4, 1992.

Runkle, J. R. 1982. Patterns of disturbance in some old-growth mesic forests of eastern North America. *Ecology* 63:1533-1546.

Runkle, J. R. 1985. Disturbance regimes in temperate forests. In: pp. 17-33.

Pickett, S.T.A. and P. S. White (Eds.). The ecology of natural disturbance and patch dynamics. Academic Press. San Diego.

SAF. 1993a. Task force report on sustaining long-term forest health and productivity. Society of American Foresters. Bethesda, MD.

SAF. 1993b. Silviculture terminology. SAF Silviculture Working Group Newsletter, October 1993. Society of American Foresters. Bethesda, MD.

Unger, D. 1993. Testimony before a hearing on the definition and implementation of ecosystem management, United States Senate Subcommittee on Agriculture Research, Conservation, Forestry, and General Legislation, November 9, 1993.

Index

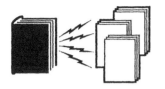

Haworth
DOCUMENT DELIVERY
SERVICE
and Local Photocopying Royalty Payment Form

This new service provides (a) a single-article order form for any article from a Haworth journal and (b) a convenient royalty payment form for local photocopying (not applicable to photocopies intended for resale).

- *Time Saving:* No running around from library to library to find a specific article.
- *Cost Effective:* All costs are kept down to a minimum.
- *Fast Delivery:* Choose from several options, including same-day FAX.
- *No Copyright Hassles:* You will be supplied by the original publisher.
- *Easy Payment:* Choose from several easy payment methods.

Open Accounts Welcome for . . .
- Library Interlibrary Loan Departments
- Library Network/Consortia Wishing to Provide Single-Article Services
- Indexing/Abstracting Services with Single Article Provision Services
- Document Provision Brokers and Freelance Information Service Providers

---

MAIL or *FAX* THIS ENTIRE ORDER FORM TO:

Attn: **Marianne Arnold**
Haworth Document Delivery Service
The Haworth Press, Inc.
10 Alice Street
Binghamton, NY 13904-1580

or FAX: (607) 722-1424
or CALL: 1-800-3-HAWORTH
(1-800-342-9678; 9am-5pm EST)

PLEASE SEND ME PHOTOCOPIES OF THE FOLLOWING SINGLE ARTICLES:
1) Journal Title: _____
    Vol/Issue/Year:_____Starting & Ending Pages:_____
Article Title:_____

2) Journal Title: _____
    Vol/Issue/Year:_____Starting & Ending Pages:_____
Article Title:_____

3) Journal Title: _____
    Vol/Issue/Year:_____Starting & Ending Pages:_____
Article Title:_____

4) Journal Title: _____
    Vol/Issue/Year:_____Starting & Ending Pages:_____
Article Title:_____

(See other side for Costs and Payment Information)

COSTS: Please figure your cost to order quality copies of an article.

1. Set-up charge per article: $8.00
   ($8.00 × number of separate articles) _____

2. Photocopying charge for each article:

   1-10 pages: $1.00 _____

   11-19 pages: $3.00 _____

   20-29 pages: $5.00 _____

   30+ pages: $2.00/10 pages _____

3. Flexicover (optional): $2.00/article _____

4. Postage & Handling: US: $1.00 for the first article/

   $.50 each additional article _____

   Federal Express: $25.00 _____

   Outside US: $2.00 for first article/

   $.50 each additional article _____

5. Same-day FAX service: $.35 per page _____

6. Local Photocopying Royalty Payment: should you wish to
   copy the article yourself. Not intended for photocopies made
   for resale. $1.50 per article per copy
   (i.e. 10 articles x $1.50 each = $15.00) _____

   GRAND TOTAL: _____

METHOD OF PAYMENT: (please check one)

❏ Check enclosed   ❏ Please ship and bill. PO # _____
(sorry we can ship and bill to bookstores only! All others must pre-pay)

❏ Charge to my credit card: ❏ Visa; ❏ MasterCard; ❏ American Express;

Account Number:_____ Expiration date:_____

Signature: X_____ Name: _____

Institution: _____ Address: _____

City: _____ State:_____ Zip:_____

Phone Number: _____ FAX Number: _____

MAIL or *FAX* THIS ENTIRE ORDER FORM TO:

Attn: **Marianne Arnold**
Haworth Document Delivery Service
The Haworth Press, Inc.
10 Alice Street
Binghamton, NY 13904-1580

or FAX: (607) 722-1424
or CALL: 1-800-3-HAWORTH
(1-800-342-9678; 9am-5pm EST)